A guidebook to lipoprotein technique

LABORATORY TECHNIQUES IN BIOCHEMISTRY AND MOLECULAR BIOLOGY

Volume 14

Edited by

R.H. BURDON - *Department of Biochemistry, University of Glasgow*

P.H. van KNIPPENBERG - *Department of Biochemistry, University of Leiden*

ELSEVIER
AMSTERDAM · NEW YORK · OXFORD

A GUIDEBOOK TO LIPOPROTEIN TECHNIQUE

G.L. Mills *and* Patricia A. Lane

S.A. Courtauld Institute of Biochemistry, Middlesex Hospital Medical School, London (Great Britain)

and

P.K. Weech

Institut de Recherches, Clinique de Montréal, Montréal (Canada)

1984

ELSEVIER
AMSTERDAM · NEW YORK · OXFORD

ISBN 0-444-80574-5 (pocket edition)
0-444-55490-4 (Paperback)
series: 0-7204-4200-1

Published by:

ELSEVIER SCIENCE PUBLISHERS B.V.
PO BOX 211
1000 AE AMSTERDAM
THE NETHERLANDS

Sole distributors for the USA and Canada:

ELSEVIER SCIENCE PUBLISHING COMPANY, INC.
52 VANDERBILT AVENUE
NEW YORK, NY 10017
USA

Library of Congress Cataloging in Publication Data

Mills, G.L.
A guidebook to lipoprotein technique
(Laboratory techniques in biochemistry and molecular biology; v. 14)
Bibliography: p.
Includes index.
1. Blood lipoproteins — Laboratory manuals.
I. Lane, Patricia A. II. Weech, P.K. III. Title. IV. Series.
QP519.L2 vol. 14 574.19'2'028s [612'.12] 84-6057
[QP99.3.L52]
ISBN 0-444-55490-4
ISBN 0-444-80574-5 (pbk.)

Printed in The Netherlands

Preface

Charles Greville, writing in 1839 of a book by Lady Blessington, sourly remarked that 'while it is very difficult to write good books, it is not easy to compose even bad ones'. In writing this guide to lipoprotein technique, we have come to appreciate the truth of this observation, not only because writing is a difficult art, but because the plasma lipoproteins are a peculiarly intractable subject. To begin with, there is no such thing as 'a lipoprotein'. Like 'wine', lipoproteins abound in an almost infinite variety, many of them differing in only the most subtle ways. The realisation of this complexity and the need to isolate the 'purest' possible lipoproteins for detailed studies of their chemistry has, since about 1945, led to the development of many methods for their isolation, identification and estimation. The newcomer to the field is therefore faced with a choice that he may find difficult to resolve unaided. It is primarily to such a novice that this guidebook is directed; we shall have comparatively little to say to the experienced worker in a large, well-established lipoprotein laboratory.

Unfortunately, the most powerful of the available techniques are often those which the most sophisticated and expensive equipment. They are therefore not always available to small teams or to isolated investigators. Moreover, they are often time-consuming and not well suited to routine clinical analysis. Consequently, many simple methods have been developed which will give the rough-and-ready results that often suffice for para-scientific experiments. In this book, we have tried to present a range of techniques that can be used for work at these different levels of sophistication. At the same time, we have tried to present even the most advanced techniques in their simplest and most widely accessible forms. But it must be clearly understood that very simple methods cannot be used for work of the highest calibre; for example, it has to be accepted that it is very difficult to do significant work on lipoproteins without a preparative ultracentrifuge.

In the end, it is the user who must select the technique most appropriate for his needs and, to this end, we have devoted a part of our text to a discussion of the relative merits and limitations of the methods available. The method having been chosen with the aid of this digest, the experimental details can then be obtained from the relevant section of the book. In these sections, we have tried to present every significant detail. But significance in this sense is subjective and some may think that our instructions are sometimes self-evident. Our excuse is that, in a world in which a post-doctoral researcher has been known to try running a single tube in the preparative ultracentrifuge, it is unwise to assume too high a level of expertise. Moreover, although techniques are sometimes described in careful detail by their originators, later users tend to degrade them by omitting steps, or by changing conditions without checking that it is valid to do so.

In composing this guide, we have inevitably drawn on the work of many scientists, as the list of references will show. However, we are glad to acknowledge the special influence that has been exerted by Dr. F.T. Lindgren and by Dr. P. Alaupovic, and their respective colleagues, both on the development of our subject and on our treatment of it. Many other colleagues have contributed by passing on practical tricks of expertise that are often vital to the success of an experiment, but are not so often made public. In this context, we are particularly indebted to Dr. Sonia Goldstein and to Dr. M.J. Chapman for their encouragement and for help with those sections that deal with their special interests. Finally, it is a pleasure to recognise the willing and expert help of the librarians and photographers of the Middlesex Hospital.

Glossary of lipoprotein nomenclature as used in this manual

For details of the different systems refer to Section 1.1.

(1) The density nomenclature will generally be used, either in specific terms, e.g. 'the fraction of density 1.02–1.04 g/ml', or in the orthodox generalised forms HDL, LDL, VLDL.

(2) Where it is more appropriate to use the electrophoretic nomenclature, the terms α-, β-, and pre-β-lipoprotein will be used.

(3) Fractions that are isolated by methods that do not define the product will be referred to the most relevant term of the electrophoretic nomenclature in quotation marks, as follows: 'α-lipoprotein', etc.

(4) Lipoproteins that are defined by their protein moieties will be named according to a modification of the Alaupovic system i.e. lipoprotein A, lipoprotein B, etc. (LP-A, LP-B, etc.).

Note that LP-A, LP-B, LP-C etc. each contain only the mixture of proteins that is designated A, B or C, etc. respectively. Lipoproteins formed from the individual proteins that compose these mixtures will be referred to as: lipoprotein A-I, lipoprotein A-II, lipoprotein C-I, etc. (LP-A-I, LP-A-II, etc.). In principle, this system can be extended to particles that bear more than one apo-lipoprotein, e.g. LP-B, C or LP-A, C, E, etc.

Likewise, it is possible to construct names for such complex particles as may be formed by the association of those that carry only one protein apiece, e.g. LP-B:LP-C, or LP-B:LP-C:LP-E, etc. or by the association of particles that carry more than one protein, e.g. LP-B, C:LP-C, E, etc.

Glossary of apo-lipoprotein nomenclature as used in this manual

For details refer to Section 1.2.

The prefix 'apo-' will be added to any term that includes the word lipoprotein or its abbreviation, e.g.

apo-high-density lipoprotein (apo-HDL), etc.
apo-lipoprotein A (apo-LP-A), etc.
apo-lipoprotein A-I apo-LP-A-I), etc.
apo-α-lipoprotein, etc.

Some of these protein moieties will be heterogeneous, for example apo-HDL, apo-LDL, apo-LP-A. Others will consist of a single protein, for example, apo-LP-A-II. Some of the apo-lipoproteins, of which A-I and E are the best known examples, are composed of a mixture of several closely similar proteins that are usually referred to as 'polymorphs'. It is arguable that this term is not properly used to describe a group of proteins whose common feature is certain immuno-reactive properties i.e. they are at least partially iso-immunogenic. Some authors have used the term 'isoform' instead of polymorph, but this is also not without its objections. In the absence of an entirely satisfactory, agreed nomenclature for these substances we shall generally use the more common term 'polymorph'.

Contents

Preface . . . . *v*

Glossary of lipoprotein nomenclature as used in this manual . *vii*

Glossary of apo-lipoprotein nomenclature as used in this manual *viii*

Chapter 1. Brief introduction to the plasma lipoproteins . . . *1*

1.1. Nomenclature of the lipoproteins 6
1.1.1. Electrophoretic mobility 7
1.1.2. Hydrated density 7
1.1.3. Flotation rate 10
1.1.4. Particle size 11
1.1.5. 11
1.1.6. Lipoproteins named specifically 13
1.2. Nomenclature of apo-lipoproteins 13

Chapter 2. The isolation and purification of plasma lipoproteins 18

2.1. Isolation by ultracentrifugation 20
2.1.1. Serial ultracentrifugation 25
2.1.1.1. Preparation of specific diluents 26
2.1.1.2. Adjustment of density with solid salts 36
2.1.1.3. Measurement and final adjustment of density 38
2.1.1.4. Choice of rotor and tubes 43
2.1.1.5. Loading the tubes 47
2.1.1.6. Loading the rotor 49

2.1.1.7. Unloading the rotor 50
2.1.1.8. Specimen procedure for serial centrifugation 55
2.1.1.9. Subfractionation of chylomicrons and VLDL 60
2.1.1.10. Serial centrifugation at different temperatures 64
2.1.2. Density gradient centrifugation 64
2.1.2.1. Choice of rotor and tubes 69
2.1.2.2. Making the gradient 70
2.1.2.3. Loading and unloading the rotor 73
2.1.2.4. Unloading the gradient 74
2.1.2.5. The profile of the gradient 75
2.1.2.6. Rate-zonal ultracentrifugation 77
2.2. Isolation by precipitation 78
2.2.1. Choice of method 85
2.2.2. Re-solution of precipitated lipoproteins 87
2.2.3. Precipitation with polyanions 87
2.2.3.1. Precipitation of β- and pre-β-lipoproteins with heparin 88
2.2.3.2. Precipitation of α- and β-lipoproteins with sodium phosphotungstate 90
2.2.3.3. Precipitation of α- and β-lipoproteins with dextran sulphate . . 91
2.2.4. Procedure for low-temperature ethanol precipitation 93
2.3. Isolation by chromatography 94
2.3.1. Gel filtration chromatography 95
2.3.2. Affinity chromatography 98
2.3.2.1. Immunoadsorbents 98
2.3.2.2. Immobilised heparin 101
2.3.2.3. Immobilised lectin 103
2.3.3. Adsorption chromatography 105
2.4. Isolation by preparative electrophoresis 106
2.5. The isolation of Lp(a) and LP-X 109
2.5.1. Lp(a) 110
2.5.2. LP-X 112
2.6. The preservation of isolated lipoproteins 114

Chapter 3. The characterisation of plasma lipoproteins: introduction to principles and methods 117

3.1. The purity of lipoprotein preparations 117
3.2. Physical and chemical characterisation 118
3.2.1. Physical methods 119
3.2.2. Chemical characterisation 122
3.3. Heterogeneity and polydispersity 125
3.3.1. Contamination with lipoproteins that lie outside the nominal bounds of the preparation 128

3.3.2. Heterogeneity within the nominal bounds of the preparation 129
3.3.2.1. Electron microscopy 129
3.3.2.2. Ultracentrifugation 130
3.3.2.3. Gel filtration chromatography 133
3.3.2.4. Adsorption chromatography 134
3.3.2.5. Electrophoresis 135

Chapter 4. Physical characterisation of lipoproteins 139

4.1. The analytical ultracentrifuge 139
4.1.1. Estimation of polydispersity in the analytical ultracentrifuge 141
4.1.2. Sedimentation rate, particle weight and particle diameter 145
4.1.2.1. Measurement of flotation (sedimentation) rate 148
4.1.2.2. Measurement of particle weight and diameter 149
4.2. Partial specific volume (hydrated density) 153
4.3. The analysis of polydispersity in the preparative ultracentrifuge 156
4.3.1. Equilibrium centrifugation 160
4.4. Gel filtration 163
4.4.1. The estimation of polydispersity by gel filtration chromatography . 165
4.5. Electron microscopy 166

Chapter 5. Electrophoretic analysis of intact lipoproteins . . . 171

5.1. Preparation of samples 173
5.2. Electrophoresis on paper 174
5.3. Electrophoresis on cellulose acetate membranes 178
5.4. Electrophoresis in agarose gel 180
5.5. Electrophoresis in polyacrylamide gel 185
5.5.1. Preparation of acrylamide gels 187
5.5.2. Staining procedures 192
5.5.3. Acrylamide gels for lipoprotein analysis 194
5.5.3.1. Gel of uniform pore size 194
5.5.3.2. Discontinuous separating gels 196
5.5.3.3. Polyacrylamide gradients gels 198
5.6. Immunoelectrophoresis 202
5.6.1. Electrophoresis of lipoprotein-X (LP-X) 203
5.7. Two-dimensional immuno-electrophoresis 210
5.8. Isoelectric focussing 219

Chapter 6. The chemical characterisation of lipoproteins . . . 221

6.1. Estimation of total lipoprotein 224
6.1.1. Estimation of lipoprotein by refractometry 226
6.2. Separation of the lipid and protein moieties 228
6.3. Analysis of the lipid moiety 228
6.3.1. Chromatographic separation of lipids 229
6.3.1.1. Chromatography on columns of silicic acid 230
6.3.1.2. Thin-layer chromatography 231
6.3.1.3. Gas chromatography 235
6.3.2. Estimation of lipids 236
6.3.2.1. Cholesterol 237
6.3.2.2. Glycerides 238
6.3.2.3. Phosphorus 240
6.3.2.4. Fatty acid analysis 241
6.3.2.5. Determination of unusual components 242
6.4. Analysis of the protein moiety 242
6.4.1. Quantitative estimation of total apo-lipoproteins 246
6.4.1.1. Total protein 247
6.4.1.2. Chemical estimation of apo-lipoprotein B 249
6.4.2. Electrophoretic analysis of apo-lipoproteins 252
6.4.2.1. Techniques for staining gels 253
6.4.2.2. Recording the electrophoretic patterns 258
6.4.2.3. The polyacrylamide gel–urea system 259
6.4.2.4. The polyacrylamide gel–sodium dodecyl sulphate (SDS) system . 266
6.4.2.5. Estimation of mobility and molecular weight by SDS-gel electrophoresis 270
6.4.2.6. Isoelectric focussing (IEF) in polyacrylamide gel 273
6.4.3. The identification of apo-lipoproteins by electrophoretic analysis . . 276
6.5. Qualitative analysis of apo-lipoproteins by immunodiffusion 288
6.6. Identification and estimation of carbohydrates 295
6.7. Quantitative analysis of lipoproteins by micro-scale thin-layer chromatography 300

Chapter 7. The estimation of lipoproteins 308

7.1. Estimation of lipoproteins by refractometry 312
7.1.1. Analytical ultracentrifugation 315
7.2. Nephelometry 351
7.3. The estimation of lipoproteins by chemical methods 357
7.3.1. Precipitation methods for 'HDL'–cholesterol 358
7.3.2. The estimation of LDL– and VLDL–cholesterol 363
7.3.3. Derived estimates of lipoprotein concentration 366

7.3.4. Quantification of electrophoretic analyses 367
7.3.4.1. Elution of lipids 367
7.3.4.2. Staining and photometric estimation 368
7.3.4.3. The lipoproteins are precipitated in the agarose gel with polyanions 376
7.4. Estimation of LP–X 378
7.5. Immunoassay of lipoproteins 380

Chapter 8. Isolation and assay of apo-lipoproteins 384

8.1. Isolation of apo-lipoprotein A-1 394
8.1.1. Chromatographic procedure 394
8.1.2. Isoelectric focussing 399
8.2. Isolation of apo-lipoprotein A-II 403
8.3. Isolation of apo-lipoprotein A-IV 404
8.4. Isolation of apo-lipoprotein B 405
8.4.1. Chromatographic procedure 406
8.4.2. Non-oxidative procedure 408
8.5. Isolation of the C apo-lipoproteins 410
8.5.1. Chromatographic procedure 412
8.5.2. Isoelectric focussing 414
8.6. Isolation of apo-lipoprotein D 416
8.7. Isolation of apo-lipoprotein E 418
8.8. Isolation of apo-lipoprotein F 421
8.9. The estimation of apo-lipoproteins 421
8.9.1. Apo-lipoprotein assay by electrophoresis 435
8.9.2. Radial immuno-diffusion 438
8.9.3. Electro-immunoassay 440
8.9.4. Radio-immunoassay 445

Appendix 1. The collection and preservation of blood plasma 449

Appendix 2. Methods of concentrating lipoprotein solutions . 460

Appendix 3. The extraction of lipid from lipoproteins 463

Appendix 4. The preparation of stains 472

Appendix 5. The photography of electrophoresis and immunodiffusion gels 479

Appendix 6. Radio-labelling 482

List of suppliers mentioned in the text 488

References 491

Subject index 508

CHAPTER 1

Brief introduction to the plasma lipoproteins

One of the characteristics of vertebrate animals is a highly developed system for the production of stable dispersions of fat in water. This ability plays an important part in the transport of fat from the gut to the bloodstream and in the intravascular transport of energy. It is also used by mammals in the production of milk. In each case, the dispersion of the fat is brought about by surrounding small globules with a hydrophilic shell that consists mainly of phospholipid and protein. The particles that perform the transport function in blood have come to be known as lipoproteins, and are present in essentially the same form in all vertebrates. It appears that the main difference between these plasma lipoproteins and the particles produced by the mammary gland is their smaller average size. In this respect the plasma lipoproteins have much in common with the microemulsions of Schulman (see Bowcott and Schulman, 1943; Prince, 1977), whereas the particles from milk are more akin to the conventional emulsions that are optically opaque.

The way in which plasma lipoproteins are synthesised is unknown. At first sight, the production of the interface between the particle and the surrounding medium would require considerable energy but, in many cases, appropriate mixtures of lipids and emulsifiers can be induced to form microemulsions spontaneously in the laboratory, and this may be a clue to the mechanism of lipoprotein synthesis by living cells. Whatever their genesis may be, the general structure of the plasma lipoproteins seems clear. They consist of a core of hydrophobic material that is surrounded by a hydrophilic envelope. The core is mainly composed of triglyceride and cholesteryl esters, although in lipoproteins from some of the more primitive vertebrates it may also contain hydrocarbons and glyceryl ethers. The envelope is composed of phospholipid and a mixture of various specific proteins (apo-lipoproteins), together with some unesterified cholesterol. As with any emulsion, these substances are bound by non-covalent forces and their proportions are variable, at least within fairly wide limits. Moreover, the lipoproteins produced by biosynthesis, like the emulsion particles made in the laboratory, are not uniform in size.

Unfortunately for the biochemist, this heterogeneity at the site of synthesis is not the only source of variation in the size and composition of the plasma lipoproteins. These substances are unusual, if not unique, in that they undergo a considerable degree of metabolic alteration while they are circulating in the blood. Although some of this modification may take place while the particles are passing through the liver, it is now generally thought to be mainly brought about within the extra-hepatic vascular bed. This is a process that results in a progressive decrease in the size of the lipoprotein particle at a rate that is itself related to the size of the particle. Thus, a steady state distribution of lipoproteins is established in the plasma, in which the particles are of similar chemical structure but differ in size and composition. These differences in composition are, in turn, the cause of variation in such constitutive properties as density and electric charge. The extent of this variation in the major properties of the lipoproteins is summarised in Fig. 1.1 over the whole range that is normally found in plasma. Note that the abscissae are not necessarily on a linear scale and that, because the electrophoretic mobility of the lipoproteins is dependent on the experimental system used, it is expressed as a fraction of the mobility of albumin. It is these differences in the properties of the lipoproteins that are used to separate them into different fractions. But it will be evident from the figure that, because the distribution of the particles is essentially continuous, the differentiation of these fractions will be quite arbitrary.

It is important to note that there is no fixed relationship between the size of a lipoprotein and its composition, and that in consequence there is no fixed correlation between size and the density or relative

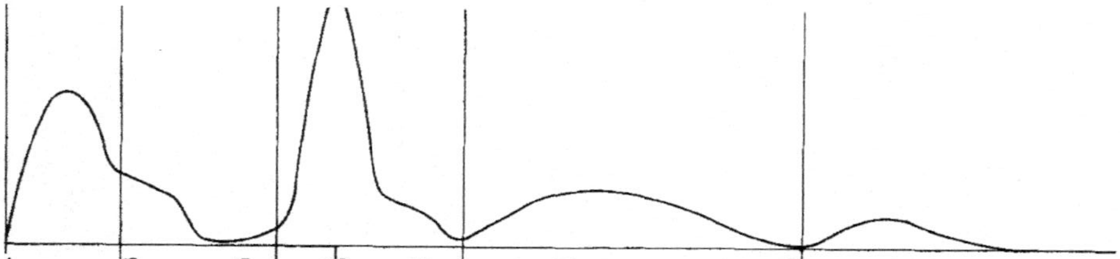

Particle diam. (nm)	4	10, 15	20	25	35	50, 100
Particle wt. (× 10^6)	0.2	0.4	2, 3		4.5	100, 1000
Hydrated density (g/ml)	1.15	1.09	1.063, 1.035		1.0, 0.96	0.95, 0.93
Flotation rate at: solvent density 1.063			0, 12		20, 100	400
solvent density 1.20		3.5, 9	20, 50		70	
Mean chemical composition (%)						
CE	12	16	38	23	12	5
UC	3	6	10	8	7	2
TG	4	4	9	32	55	84
PL	26	30	22	21	18	7
PR	55	44	21	16	8	2
Apo-lipoprotein content (% total PR)						
A	94.6	82.4	1.4		Tr	1
B	-	3.7	88.2		54	40
C	2.8	8.9	6.7		38	47
D	2.2	2.9	Tr		Tr	Tr
E	0.4	2.1	3.7		8	12
Electrophoretic mobility (%) relative to albumin	85 - 95		35 - 55		55 - 70	0 - 5

Fig. 1.1. The major physical and chemical properties of human plasma lipoproteins. The upper diagram shows a characteristic quantitative distribution of the different particles. Abbreviations: CE, cholesteryl ester; UC, unesterified cholesterol; TG, triglyceride; PL, phospholipid; PR, protein.

weight of the particles. Thus, it is theoretically possible for a lipoprotein preparation that is homogeneous with respect to particle size to be heterogeneous with respect to composition and density. It follows therefore, that the implication in Fig. 1.1 that these properties are distributed in essentially the same way is not necessarily strictly true. Unfortunately, the physics and chemistry (as opposed to the biochemistry) of the lipoproteins are neglected subjects and the extent to which the distributions actually differ is not often known. However, a practical example of heterogeneity in a superficially homogeneous preparation is illustrated in Fig. 1.2, which shows the analysis by two-dimensional immunoelectrophoresis of a β-lipoprotein fraction. Although this gave one band on simple electrophoresis in agarose gel, it is clearly immunochemically heterogeneous.

It is easily seen that the dispersive potential of such a complex system is immense, and it is conceivable that no two lipoprotein particles are exactly alike. Such heterogeneous mixtures are not only difficult to characterise, but pose grievous problems for the terminologist, and it was to provide a rational description of just such systems that Gibbons (1963, 1972) developed his concept of inhomogeneity.

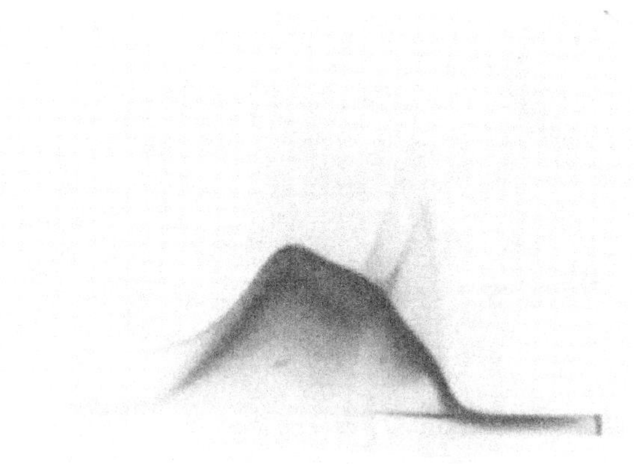

Fig. 1.2. An illustration of heterogeneity within a lipoprotein fraction, disclosed by the method of two-dimensional immunoelectrophoresis (Section 5.7).

This regards a population of particles as heterogeneous if one of its parameters has a polymodal distribution, or has a distribution curve with more than two points of inflection. According to this definition, the whole plasma lipoprotein profile taken together (Fig. 1.1) is heterogeneous with respect to size, density and electrophoretic mobility. Likewise, the HDL fraction must also be regarded as heterogeneous because its distribution curve has three inflections. This is often, though not invariably true of the LDL as well.

By contrast, a system in which the distribution of a parameter is unimodal is termed 'polydisperse' (with respect to that parameter). This is usually true of the size distribution of the VLDL and sometimes of the LDL also. However, it is important to note in this context that even a homogenous substance will give a unimodal distribution when analysed by methods in which the effects of more than one physical phenomenon are superimposed, as for example in ultracentrifugation or gel filtration chromatography, where the sharpness of the boundaries is degraded by diffusion. A sophisticated examination of the data may then be needed to determine whether the semblance of homogeneity is real.

The lipoprotein distribution that is shown in Fig. 1.1 is a generalised example of the steady state that is established in the plasma as a resultant of many metabolic reactions. In a real group of human beings however, it would be rare for any two subjects to have exactly the same distribution. This raises practical difficulties for the student of lipoproteins, since preparations that do not originate from the same person will be different, even though they may have been isolated in the same way. This is exemplified in Fig. 1.3, which compares the ultracentrifugally determined distribution of the LDL fractions obtained from several apparently normal individuals. It is evident that every such preparation should be characterised by the measurement of such properties as will allow it to be identified and will establish the extent of its divergence from others that are nominally identical. This is particularly important if reproducibility between laboratories is to be maintained, but imposes an unwelcome addition to the labour of the experimenter. It is therefore important, (a) to

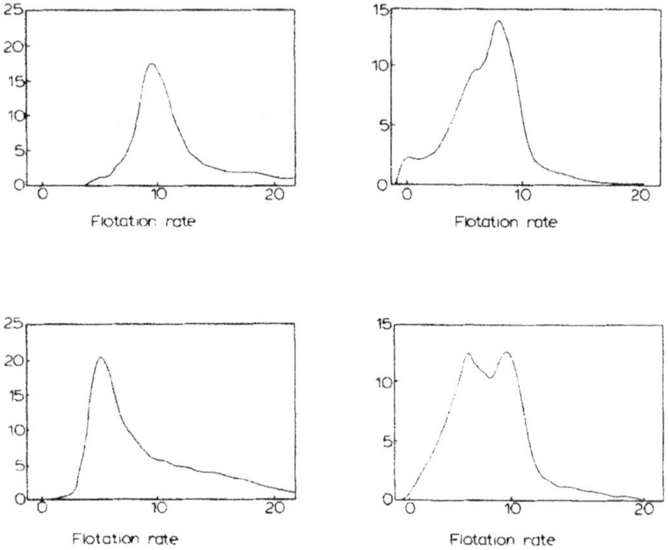

Fig. 1.3. The varying distribution of LDL in different subjects, illustrated by analytical ultracentrifugation of four different human sera.

define very precisely the way in which the lipoproteins are isolated, and (b) to decide on a strategy of characterisation that will identify the preparation with an appropriate degree of certainty without being unrealistically burdensome. It is one of the purposes of this manual to describe a selection of reliable techniques for the isolation and characterisation of lipoproteins, with a sufficient discussion of their merits to allow the most appropriate to be chosen.

However, before proceeding to the main part of the text, it will be necessary to consider the nomenclature of the plasma lipoproteins and of their various protein components.

1.1. Nomenclature of the lipoproteins

The polydispersity of the lipoproteins makes their classification very difficult and there is at present no workable systematic nomenclature.

However, there are several operational systems that are based on methods of fractionating lipoprotein mixtures. These all attempt to classify the sum total of the plasma lipoproteins into groups by setting arbitrary boundaries to distinguish one from another. The commonly used systems may be summarised as follows.

1.1.1. Electrophoretic mobility

Like other proteins, the lipoproteins are electrically charged, and will migrate in an electric field. In this way they can be resolved into four main classes by zone electrophoresis in a supporting medium, namely;

Chylomicrons: these are the particles that are too large to migrate into the pores of the stabilising medium.

α-Lipoproteins: these were so named because they have the mobility of α-globulins. More precisely, their mobility is that of the α_1-globulins and they are therefore sometimes referred to as α_1-lipoproteins.

Pre-β-lipoproteins: so-called because their mobility is a little greater than that of the β-lipoproteins, they are also sometimes called α_2-lipoproteins because their mobility is close to that of the α_2-globulins.

β-Lipoproteins: these migrate at the same rate as the β-globulins.

In addition to these, two minor classes have been described under the names *pre-albumin* and *γ-lipoprotein*.

It is important to note that all these classes are polydisperse with respect to all known properties.

1.1.2. Hydrated density

Lipoproteins are substances of relatively low density and will float if they are ultracentrifuged in a solvent of greater density. By carrying out successive centrifugations in which the density of the solvent is increased by a known amount each time, it is possible to isolate lipoproteins fractions that are defined as having a density that is less than one value but greater than a second. This commonly used method of isolating purified lipoproteins is the basis of what has come to be the most widely used system of lipoprotein nomenclature. Five standard classes have been defined in this way by arbitrarily chosen limits of density, namely:

Chylomicrons: particles of density less than 0.94 g/ml.

Very-low-density lipoproteins (VLDL): lipoproteins of density greater than 0.94 g/ml but less than 1.006 g/ml.

Low-density lipoproteins (LDL): particles of density between 1.006 and 1.063 g/ml. In some laboratories this class has been subdivided. Thus, the fraction of density 1.006 g/ml to about 1.02 g/ml has been called *Intermediate-density lipoprotein* (IDL) by some, and LDL_1 by others. The complementary fraction of density 1.02 to 1.063 g/ml is then called LDL or LDL_2 respectively. On the whole, independent and irrational extensions to the basic system of nomenclature, such as these, tend to confuse and are to be discouraged on that account.

High-density lipoproteins (HDL): lipoproteins of density between 1.063 and 1.20 g/ml. This class is heterogeneous, and can be separated into two sub-classes of density 1.063–1.125 g/ml (HDL_2) and 1.125–1.20 g/ml (HDL_3). The term HDL_1 has also been applied to a small class of lipoproteins that has a mean density close to 1.06 g/ml. It consequently overlaps both the LDL and HDL classes, and it is debateable to which it should be assigned.

Very-high-density lipoproteins (VHDL): lipoproteins of density greater than 1.20 g/ml but less than about 1.25 g/ml.

This system is a very flexible one since almost any lipoprotein fraction that is prepared by sequential ultracentrifugation can be defined by the two densities at which this process is carried out. For example, if the lower limiting density is 1.024 g/ml and the upper limiting density is 1.045 g/ml, the lipoproteins of density between these two values can be referred to as the 1.024–1.045 fraction. However, it must be remembered that although it is the hydrated density of the lipoprotein particle that enables us to give it a name in this system, in practice it is the density of the solvent that is measured. During ultracentrifugation, there is a re-distribution of salt towards the bottom of the tube that causes the density of the solvent at the top to fall slightly. Thus, when a lipoprotein fraction is defined in this way, it is the solvent density *after* ultracentrifugation that should be used.

It must also be noted that the density of a lipoprotein particle is a function of temperature, the value of which should be quoted when a lipoprotein is referred to this system of nomenclature. For technical reasons, the Donner laboratory used a standard temperature of 26 °C for the analytical ultracentrifugation of lipoproteins, and the densities at which the preparative centrifugations were carried out were also defined at this temperature (DeLalla and Gofman, 1954; Ewing et al., 1965). However, Hatch and Lees (1968) apparently re-defined the standard temperature to 20 °C and lipoproteins isolated according to their protocols are not identical with those obtained by the 'Donner' system. The general practice in this context is obscure, since the temperature at which the density of the plasma is adjusted is rarely reported, but it seems likely that an ill-defined 'room temperature' is often used, i.e. somewhere near 20 °C. If this assumption is correct, the adoption of 20 °C as the standard at which lipoprotein densities should be reported may be both practical and realistic. Moreover, it is the conventional standard temperature at which the densities of other substances and the results of analytical ultracentrifugation are both reported. Until a standard temperature for the definition of lipoprotein densities is officially decided, each experimenter must determine for himself what conditions he will use. In this manual however, we shall adopt a temperature of 20 °C. Unfortunately, samples prepared according to this definition may give a slightly different profile in the analytical ultracentrifuge, at low flotation rates, as compared with samples defined at 26 °C. However, with the improved temperature control that is available on modern ultracentrifuges, there seems to be no reason why consistency should not be achieved by running the lipoprotein analyses at 20 °C.

Note that all the fractions to which we have been referring are polydisperse with respect to all known properties.

1.1.3. Flotation rate

In this system, the lipoproteins are identified by the rate at which they float in the analytical ultracentrifuge (in Svedberg units). Two standard conditions of centrifugation are needed, as follows:

(a) lipoproteins of density less than 1.063 g/ml (conventionally at 26 °C, see Section 1.1.2) are centrifuged at 26 °C in NaCl solution of density 1.063 g/ml;

(b) the high-density lipoproteins are analysed in a mixture of NaCl and NaBr of density 1.20 g/ml, also at 26 °C.

Traditionally, four main classes of lipoproteins have been defined in this way by arbitrarily fixing the limits of flotation rate as follows:

S_f *> 400:* these large particles, of flotation rate greater than 400 Svedbergs in NaCl solution of density 1.063 g/ml, are approximately equivalent to the 'chylomicrons' of other classifications.

S_f *20–400:* the low-density lipoproteins that have a flotation rate of between 20 and 400 Svedbergs are closely equivalent to VLDL.

S_f *0–20:* the second main class of low-density lipoproteins has a flotation rate between zero and 20 Svedbergs at solvent density 1.063 g/ml, and is virtually identical with the LDL. This class is often sub-divided into the S_f 0–12 and S_f 12–20 fractions.

$F_{1.20}$ *0–9:* these lipoproteins are nearly equivalent to the HDL, and float in a solvent of density 1.20 g/ml with velocities between zero and 9 Svedbergs. The class is usually sub-divided into the $F_{1.20}$ 0–3.5 (approx. HDL_3) and $F_{1.20}$ 3.5–9 (approx. HDL_2).

This classification is also very versatile since almost any fraction can be defined by its limiting flotation rates. In a similar system that was used in some laboratories in the early 1950s, the analytical centrifugation of the whole lipoprotein profile (low- plus high-density fractions) was carried out in KBr solution of density 1.21 g/ml. However, this procedure has some technical drawbacks and is now of historical interest only (Lewis et al., 1952).

Note again that all the conventional fractions defined above are polydisperse with respect to all known properties.

1.1.4. Particle size

A simple classification based on size has been developed by Stone et al. (1970, 1971), in which the diameters of the lipoproteins are estimated by a combination of their ability to scatter light under defined conditions and to pass through membrane filters of specified pore size. The term *large particle* is used for lipoproteins of diameter greater than 100 nm, which effectively correspond to the chylomicrons. Lipoproteins of diameter 20–100 nm are referred to as *medium* particles, and those smaller than 20 nm as *small* particles. The *medium* class is roughly equivalent to the S_f 20–400 class, or VLDL, while the *small* particles include both the LDL and HDL. Unfortunately, this simple scheme is limited by the availability of filters of differing pore size and by the unsuitability of filters as a means of resolving the *small* particles. Moreover, the intensity of scattered light is dependent on both the size and the concentration of the lipoprotein particles. In its present state of development therefore, this system of nomenclature is of value only in a clinical context.

1.1.5.

It cannot be too strongly emphasised that, although the terms used in these different systems of nomenclature are often roughly equivalent, they are not strictly interchangeable. For example, HDL and α-lipoprotein are not the same thing, and the term HDL should only be applied to material that has been isolated by ultracentrifugation.

The operational classification also makes for difficulty when lipoproteins are isolated by methods other than centrifugation or electrophoresis, e.g. precipitation or gel filtration. Since these methods do not provide the basis for a nomenclature of their own, it is tempting to classify the preparations according to one of the other existing systems. But any correspondence between the preparations and the chosen system of nomenclature will be at best, imprecise and at worst, unknown. It is therefore unacceptable to say, as in the following extract, that 'Serum lipoproteins were isolated and fractionated by gel chromatography into three major fractions: VLDL, LDL and HDL', because the density classes are exactly defined and the chromatographic fractions are not. The correct procedure in these circumstances is to analyse each fraction by a reference method and to define it accordingly. However, when describing these *non-reference* methods in this manual, we shall apply the generalised terms *α-lipoprotein* and *β-lipoprotein* to fractions that are known to approximate to the electrophoretic classes.

In an attempt to overcome the defects associated with an operational nomenclature based on physical principles, Alaupovic (1968, 1972, 1974) has proposed that the most rational way of classifying lipoproteins is by reference to their apo-lipoprotein content. In principle, this contention has much to commend it. It is based on the observation (Lee and Alaupovic, 1970, 1974; Kostner and Alaupovic, 1972; McConathy and Alaupovic, 1976) that it is possible, by exhaustive immunochemical adsorption, to isolate classes of lipoprotein that bear only one apo-lipoprotein, i.e. that are homogeneous with respect to protein, although polydisperse with respect to their other properties. If they are intact and native, each of these classes of lipoprotein may be legitimately named after their apo-lipoprotein, for example, Lipoprotein B or Lipoprotein A-I. (See Section 1.2 for the nomenclature of apolipoproteins.) These classes have been called *families* by Alaupovic and differ significantly from ordinary lipoprotein preparations in being homogeneous with respect to one chemical property. However, many of the lipoprotein particles that are normally present in the plasma carry more than one apo-lipoprotein. Alaupovic has suggested that these particles are *association complexes* that are formed from the particles that compose the *families,* but this concept has many vague and ill-defined features. In a more specific proposal, Osborne and Brewer (1977) suggest that lipoproteins which carry a single apo-lipoprotein should be termed *primary* lipoproteins. They then postulate that these can reversibly associate, according to the Mass Action law, to form *secondary* lipoproteins. Unfortunately, although this provides a plausible explanation for the existence of some of the lipoproteins that have more than one apo-lipoprotein, it does not solve the problem of naming them. As it stands, the *lipoprotein family* concept provides a good way of classifying some lipoproteins that are often quantitatively minor and are rarely isolated. In practice, it does nothing to solve the problem of classifying the antigenically heterogeneous preparations with which most experimenters are forced to work. Future developments in lipoprotein technology may change this however.

1.1.6. Lipoproteins named specifically

Lp (a) This lipoprotein, which has a hydrated density of about 1.085 g/ml and a relative particle weight of about 5×10^6, was originally thought to be a polymorphic form of lipoprotein B. It is now thought to contain apolipoprotein B and at least one additional specific protein. It migrates as a pre-β-lipoprotein on electrophoresis and, because of its high density, has been termed 'sinking pre-β-lipoprotein'. It is a normal component of plasma.

Lipoprotein X (LP-X) This abnormal lipoprotein is present in the plasma of patients with obstructive jaundice. It has the ultracentrifugal properties of an LDL, but migrates towards the cathode on electrophoresis at pH 8 in agar. It contains apo-lipoprotein C, apolipoprotein D and albumin.

1.2. Nomenclature of apo-lipoproteins

The protein components of the lipoprotein particles are known as apo-lipoproteins. The prefix *apo-* can be legitimately applied to any name that explicitly contains the term *lipoprotein* or its abbreviation e.g. high-density lipoprotein or its contraction HDL. Unfortunately, there is no other generic name that is more succinct than *apo-lipoprotein.* Note that the term *apo-protein* should be regarded as jargon; useful enough in laboratory conversation, but to be avoided in writing. Moreover, it is incorrect to use *apo-* as a prefix to the name of an apo-lipoprotein as, for example, in *apo-B*. The correct abbreviation should take the form *apo-LP-B*, which means *the apo-lipoprotein designated as 'B'*, and not that the protein is necessarily isolated from lipoprotein B. Where there can be no chance of confusion, it may be acceptable to refer to the apo-lipoproteins by their names only, as in *the proteins A, B and C*.

A considerable number of proteins can be isolated from lipoprotein particles, some of which are loosely attached by non-specific bonds and are of no structural or functional significance. Criteria are therefore needed by which these proteins can be distinguished from true apo-lipoproteins. This distinction can be difficult, since some apo-lipoproteins are also loosely attached and can exchange between particles, whereas albumin, which is probably not an apo-lipoprotein, can bind very tightly under some circumstances. It seems to be tacitly accepted that an apo-lipoprotein should be present on the lipoprotein particle for a purpose, such as the stabilising of the particle. However, it is not yet clear to what extent the presence of protein is really necessary for stability, although there is evidence that it cannot be entirely dispensed with. Probably the most valuable criterion at present available is one proposed by Alaupovic, namely that an apo-lipoprotein must be capable of acting as the only protein component of a lipoprotein particle i.e. that it can form a lipoprotein family (Section 1.1.5). If this condition is satisfied, there is strong evidence that the protein can stabilise the structure of the particle. However, this is technically a very difficult test to apply. Moreover, it is of uncertain value if the protein in question can easily exchange between different lipoprotein particles and is consequently of doubtful structural significance. A way out of this dilemma is suggested by the fact that some loosely bound apo-lipoproteins are now known to be regulators of enzymes that are concerned with lipoprotein metabolism. If this metabolic role is accepted as implying a functional requirement for the presence of the protein on the particle, it can be invoked as an additional test for an apo-lipoprotein. It is not an essential criterion however, and it is clearly much easier to detect the presence of a protein on the lipoprotein particle than to discover its metabolic role.

Despite this problem of definition, the existence of at least six apo-lipoproteins is now accepted, and a system of nomenclature is therefore necessary. The earlier literature on the subject is confused by the existence of two such systems. In the first, each protein was identified, at the time of its discovery, by a letter of the alphabet. In the second, they were identified by their C-terminal aminoacids. The fact that some of these C-terminals were at first wrongly identified did nothing to promote the use of the latter system and, as it is also rather cumbersome, it appears to have become extinct. However, the alternative is not without problems of its own that arise from the fact that many apo-lipoproteins are now known to be heterogeneous. For example, apo-lipoprotein A was eventually found to consist of two separate and quite different proteins that were designated A-I and A-II. At first sight, this would appear to present no difficulty since each can be regarded as an apo-lipoprotein. However, Alaupovic disagreed with this view on the grounds that it was an essential part of the definition of an apo-lipoprotein that it should be able to form a lipoprotein 'family', i.e. that it can be the sole protein present in a class of lipoproteins. At first, it was thought that proteins such as A-I and A-II did not fit this definition and that they required an alternative name, for which the clumsy term *constitutive polypeptides* was coined. On this view, the term *apo-lipoprotein* is a noun of assembly that represents a mixture of two or more different proteins whose only common feature is the fact that they often occur together of the same lipoprotein particle. But, since each of these proteins is presumably capable of forming a bond of some kind with the surface of the particle, the grounds on which they are denied the status of apo-lipoproteins appear to be arguable. Moreover, some of these so-called polypeptides have since been shown to be capable of forming *families*. Many authors have therefore applied the term apo-lipoprotein to proteins such a A-I and A-II, and we shall also adopt this usage.

The apo-lipoproteins A-I and C-III have also been shown to exist

TABLE 1.1
The principal apo-lipoproteins

Apo-lipoprotein		Polymorphic forms	Main source	Mol. wt. $\times 10^{-3}$	C-terminal aminoacid	Function
A	A-I	five	HDL	28.5	Gln	LCAT activation
	A-II	-	HDL	17.5	Gln	?
B		-	LDL	?	Ser	Binding to specific receptors
C	C-I	-	*Chylomicrons*	7.0	Ser	LCAT activation
	C-II	-	and VLDL	8.5	Glu	LPL activation
	C-III	three		8.5	Ala	LPL inactivation
D	(A-III)	-	HDL_3	22.0	?	LCAT activation?
E		four?	VLDL and HDL_1	36.5	?	Receptor binding?
F		-	HDL	ca. 30?	?	?

LCAT, lecithin:cholesterol acyl transferase.
LPL, lipoprotein lipase.

in polymorphic forms, which are designated by adding an Arabic numeral as in C-III-0, C-III-1 etc. Apo-lipoprotein E has likewise been separated into several forms by isoelectric focussing but it is not yet clear whether these are polymorphs or are different proteins. If is therefore uncertain whether they should be distinguished by Arabic or Roman numerals.

At the time of writing, there are nine proteins that have been established as apo-lipoproteins according to our definition of the term, and these are listed in Table 1.1. Note however, that apo-lipoprotein D is also known to some workers as A-III. In addition to these, several other proteins have been proposed as candidates. For example, there are the proteins tentatively designated as apo-lipoproteins G and A-IV, as well as β_2-glycoprotein-1 (which is thought by some to be apo-lipoprotein H), *glycine-rich peptide, proline-rich peptide* and *threonine-poor peptide.* But it has yet to be convincingly shown that any of these are genuine components of the lipoprotein particle and are not present as the fortuitous result of non-specific adsorption. The status of the protein moiety of Lp(a) is also uncertain; originally it was thought to be a polymorph of apo-lipoprotein B, but there is now some evidence that it is a mixture of B with at least one other specific protein. Whether the latter are strictly to be regarded as apo-lipoproteins is undecided.

In this manual we shall use the terms and conventions of the alphabetic system of nomenclature as they are summarised in the glossary at the beginning of this chapter.

CHAPTER 2

The isolation and purification of plasma lipoproteins

There is probably little practical difference between plasma and serum as a source of isolated lipoproteins if small losses of some particles by entrainment in the clot can be tolerated. However, the use of plasma has the advantage that the blood can be chilled and the cells removed immediately after collection, thus minimising the enzymic degradation that may occur while the clot is retracting. Even under the best circumstances, the isolation of a particular class of lipoproteins may take several days, during which they will be open to degradation in three ways:

(1) by enzymes that may be either native to the plasma, or of microbiological origin

(2) as a result of oxidation

(3) since the components of the lipoprotein complex are bound together only by non-covalent forces, it is possible for some disproportionation to occur during the preparative manipulations.

As explained in Appendix 1, the first two degradative effects can be minimised by storing the plasma between 0 and 4 °C and by adding the following preservatives:

Ellman's reagent	0.6 mg/ml
Phenylmethyl sulphonyl fluoride	0.35 mg/ml
Thimerosal	0.08 mg/ml
Sodium azide	0.13 mg/ml
EDTA	0.37 mg/ml

(All these reagents can be obtained from laboratory suppliers such as BDH Chemicals, Eastman Organic Chemicals, E. Merck and Sigma Chemical Co.)

A prompt start to the isolation will remove the lipoproteins from the main source of degradative enzymes, but it must be remembered that some of these may remain adsorbed to the isolated particles. Oxidation is probably the most important hazard once isolation has begun, and EDTA must be added to all the solutions used. Even this precaution will not completely prevent oxidative effects and the most scrupulous workers will make up their reagents in de-aerated water and store the lipoprotein solutions under nitrogen. Sterility must also be maintained and all solutions should therefore contain sodium azide and thimerosal.

It is during the isolation of the lipoproteins that their intrinsic instability becomes important. The possibility that HDL particles may be eroded during ultracentrifugation was discussed by Scanu and Hughes in 1960, and was later confirmed by Levy and Fredrickson (1965). Later evidence suggests that loss of protein may occur under these circumstances through the perturbation of the equilibrium between apo-lipoprotein that is bound to the particles and the low concentration of free protein in solution (Pownall et al., 1978). This equilibrium probably involves only the smaller apo-lipoproteins. At present there seems to be little that can be done to prevent this disproportionation of the lipoproteins.

Broadly speaking, there are three general ways of isolating lipoproteins in practicable quantities:

(1) Ultracentrifugation
(2) Precipitation
(3) Chromatography

When used individually these methods yield 'lipoproteins' in which the dispersion of one property has been limited. By using several methods in sequence, the overall dispersion can be further reduced but, because these extra stages are laborious, they are rarely undertaken.

2.1. Isolation by ultracentrifugation

This technique separates the lipoproteins according to differences in their hydrated density. The most widely used procedure is the serial centrifugation at different solvent densities that originated in the Donner laboratory and has been described in exquisite detail by Lindgren (1975), although it is to Havel et al. (1955) that the popularity of the method is probably due. Although the technique is not well suited to the fractionation of very large quantities of plasma, it offers the best combination of capacity and resolution that is available. Centrifugation on density gradients may give a greater resolution but is probably more suitable for the analysis of small quantities of lipoprotein than for large-scale preparative work.

The centrifugal methods have great flexibility insofar as it is possible to isolate any fraction that can be defined by two densities that are greater than that of the solvent density of the plasma i.e. 1.006 g/ml. However, it is evident that definition of the lipoprotein fraction will be impaired if adjustment of the solvent density is inaccurate. It is therefore significant that there is usually a degree of uncertainty about the solvent density of the plasma. In mammals this is assumed to be 1.006 g/ml, but there seems to be little information on its variation from man to man or, more particularly, from genus to genus. The same value has been reported for some species of lower vertebrates though there may be some doubt whether it can be generally applied to reptiles or amphibians. In fish however, the solvent density of plasma is known to be relatively high (1.015–1.025 g/ml), though the value may vary among species. Fortunately, it can quite easily be determined, either by measuring the density of an ultrafiltrate of plasma, or by the simple equilibrium dialysis procedure described by Jensen and Smith (1976). A precise adjustment of the density of plasma to any desired value can, of course, be made by dialysis but this has the disadvantage that it effectively increases the time for which the plasma is stored, and will probably lead to unwanted changes in the volume of the sample.

A more important error in the final density of the adjusted plasma can arise from the fact that about 6% of its volume is occupied by the proteins. Strictly therefore, it is the partial solvent volume (0.94 times the actual volume) that should be used when calculating the amount of salt to be added to raise the density. Even then, there will be a residual error due to the variation from subject to subject, but this is usually small enough to be ignored. So far as can be judged from published papers, this correction for the partial volume of the protein is rarely applied and the reported densities underestimate the actual values. Since it is easy to correct for at least the major part of this error, it seems desirable that this should be more widely done. However, it may then be necessary to have two parallel adjustment procedures in use, one for plasma and another for the washing or fractionation of isolated lipoproteins.

In some experiments, it may be possible to reduce these uncertainties in density by first isolating lipoproteins of a higher density than is required and then sub-fractionating them.

Even when the initial adjustment of the density can be carried out precisely, it must be remembered that the re-distribution of salt during the centrifugation brings about a decrease in the density of the solvent at the top of the tube. The magnitude of this effect can be gauged from the fact that, during an 18 hour centrifugation at 40 000 rev/min in the Spinco 40.3 rotor, the density of a NaCl solution will fall from 1.065 to 1.062 g/ml at the top of the tube. Where the conditions of centrifugation are different, or the solvent contains a different mixture of salts, the extent of the re-distribution must be specially measured. This can easily be done by determining the refractive index of the solution at different points down the tube after the centrifugation (Section 2.1.2.5). The precision densitometer could also be used (Section 2.1.1.3). The refractometric method has been routinely used by Lindgren (Lindgren et al., 1964; Lindgren, 1975) to estimate the solvent density of the lipoprotein fractions obtained by the centrifugation of plasma. This enables a correction to be made for manipulative errors or variations in the solvent density of plasma (Section 2.1.1.3).

It is not always realised that a lipoprotein particle and its solvent expand at different rates when their temperature is raised. As a result, the solvent density will only be equal to the limiting density of the lipoprotein fraction if the ultracentrifugation is done at the same temperature as that at which the density adjustment was made (say 20 °C). If this is not the case, the differential expansion of the two components of the system can introduce an important error in the definition of the lipoprotein fraction that is isolated. For example, in the experiments of Chapman et al. (1978), serum was adjusted first to a density of 1.024 g/ml at 20 °C and then centrifuged at 5 °C. After removal of the lipoproteins, the serum was re-adjusted to 1.045 g/ml at 20 °C and centrifuged at 5 °C. Subsequent measurements (at 26 °C) of the hydrated density of the resulting lipoprotein fraction gave a mean value of 1.028 g/ml rather than one close to 1.035 g/ml as might have been expected. A similar effect can be seen in the experiments reported by Lee and Alaupovic (1974).

The decrease in density of LDL on heating through the range 5–25 °C is approximately -7×10^{-4} g/ml/°C (Toro-Goyco, 1958; Mills, 1977). Over the same range of temperature, the corresponding value for a sodium chloride solution of density (at 20 °C) between 1.00 and 1.08 g/ml lies between the values -1.65×10^{-4} g/ml/°C and -3.93×10^{-4} g/ml/°C, and can be obtained to a good approximation from the equation $d\rho/dt = (26.85 - 28.5\rho)10^{-4}$ where ρ is the density of the sodium chloride solution at 20 °C and t is temperature.

The practical consequences of these differences in expansivity are summarised in Table 2.1 from which it can be seen that an LDL which has a density of 1.03 g/ml at 20 °C has a density of 1.0257 g/ml at 26 °C. Because the lipoprotein expands more rapidly than the solvent, it is necessary, when the ultracentrifugation is to be performed at a temperature below 20 °C, to adjust the solvent to a density (at 20 °C) that is greater than 1.03 g/ml. Thus, if the centrifugation is to be carried out at 5 °C, this adjusted value should be 1.0364 g/ml.

Unfortunately, we do not at present have the information with which to compensate accurately for this differential expansivity in all centrifugations. In the first place, the LDL are the only lipoproteins for which any attempt to determine the thermal expansivity has been made. At present therefore, a correction can only be applied to VLDL and HDL if it is assumed that their coefficients of expansivity are the same as that for LDL. From the evidence available, it appears that the latter may be effectively the mean of the values for the components of the lipoprotein. Moreover, since these values are not very different, the variation in expansivity across the range of lipoproteins from VLDL to HDL may prove to be quite small, and that the value for LDL may be acceptable as a first approximation for them all. However, this has yet to be demonstrated.

Secondly, the correlation between temperature and density is known only for a restricted range of sodium chloride solutions. For solutions of other substances that are used for the preparative centrifugation of lipoproteins, such as sodium bromide, mixtures of bromide with chloride, or sucrose, the required information is generally lacking. It is worth noting however, that the expansivity of these solutions is greatest when the salt concentration is high. The differential effect of temperature therefore tends to be least when HDL are centrifuged, and may be less than that which results from the re-distribution of the salt. Nonetheless, in work of the highest calibre, the expansivity of the solvent to be used should either be measured, or the differential effect eliminated by centrifuging at a temperature close to 20 °C.

Temperature is also one of the factors that determine the minimum duration of the centrifugation, through its effect on the viscosity of the solvent. For example, a centrifugation of 24 hours at 18 °C will need to be extended to at least 36 hours at 4 °C. It is, of course, important to ensure that the duration of the experiment is adequate. If it is too short, an unduly large number of particles will remain in the infranatant solution, from which they will ultimately be recovered as contaminants of a fraction of higher density.

To summarise these observations, it may be said that experiments are often reported to have been on lipoprotein preparations that are not quite what they purport to be. Ignoring the partial volume of the plasma proteins will result in the adjusted solvent density being slightly too high, whereas the redistribution of salt and the expansivity effect both reduce it. The magnitude of the resultant error will depend on the circumstances of the experiment, but may be exemplified as follows: assume that a sample of plasma is to be adjusted to a solvent density of 1.063 g/ml at 26 °C. Errors arising from pipetting and similar manipulations should fall within the range ±0.0005 g/ml. If the partial volume of the plasma proteins is ignored, the actual density would be approximately 1.064 g/ml. Redistribution of salt would then reduce this to about 1.062 g/ml and, if the centrifugation was done at 5 °C, the final limiting particle density would be about 1.055 g/ml at 26 °C. It is tempting to dismiss this error as being of little consequence when the product is a heterogeneous mixture which varies from one blood sample to another. But, as Fig. 2.1 shows, it can make a considerable difference to the nature of the fraction that is isolated and make it difficult, or perhaps impossible, for others to reproduce the experiment exactly. An accurate knowledge of the limiting densities of the lipoprotein fraction will be particularly important if these are the only characteristics used to define it.

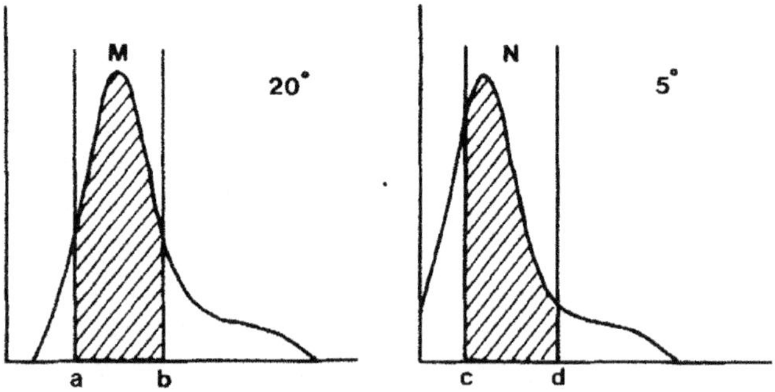

Fig. 2.1. This figure illustrates the error that can be made in defining a lipoprotein fraction if the difference between the thermal expansion of lipoprotein and its solvent is ignored during serial ultracentrifugation. The hatched area M represents the fraction of LDL that it is proposed to isolate, defined by the limiting densities (a) and (b), as measured at 20 °C. If the centrifugations are then actually carried out at 5 °C, the lipoprotein fraction that is recovered will have the distribution represented by the hatched area N.

2.1.1. Serial ultracentrifugation

The principle of this method is as follows. The solvent density of plasma (or other solution of lipoproteins) is adjusted to the lower limit of the fraction that it is proposed to isolate and it is then centrifuged at about 100 000 *g* for several hours. The lipoproteins of density less than that to which the solution was adjusted float to the top of the tube, while other substances sediment towards the bottom. The floating lipoproteins are removed. The residue in the tube is stirred and its solvent density adjusted to the upper limit of the required lipoprotein fraction. The centrifugation is repeated and the wanted lipoproteins are recovered from the top of the tube. By readjusting the solvent density of the infranatant solution to successively higher values, a sequence of lipoprotein fractions can be isolated. However, no matter how carefully the experiment is performed, these fractions always overlap because particles with a density close to that of the solvent migrate only very slowly into (or out of) the upper part of the tube. The usual solution to this problem is to wash the lipoproteins by repeated centrifugation at the upper density limit. This also tends to remove adsorbed plasma proteins, but has the disadvantage that the prolonged manipulation may lead to degradation of the lipoproteins.

This technique has the draw-back that, like equilibrium centrifugation on gradients, it cannot be used to sub-fractionate chylomicrons or VLDL, which are less dense than physiological salt solution. For this purpose it is necessary to use a procedure in which the solution is centrifuged for successively longer periods of time at defined rotor speeds. This technique fractionates the lipoproteins, not by density, but by the rate at which they float up the centrifuge tube, which is a function of both density and size. This procedure is described in detail in Section 2.1.1.9.

Before embarking on the preparation, it may be necessary to adjust the solvent density of the plasma, or of a solution of isolated lipoproteins, to some specified value. The best way of doing this is by equilibrium dialysis. Unfortunately, this not only requires large volumes of salt solution of accurately known density, but also takes at least 24 hours, during which the lipoproteins may be subject to degradation (Appendix 1 and Section 2.5). It is therefore important to use de-aerated water to prepare the salt solutions and to conduct the dialysis in a stoppered flask that is completely filled. Antioxidants can also be added to the solution (Appendix 1 and Section 2.5). Despite the objections, this is the procedure to use for work of the highest precision. For the routine analysis or preparation of lipoproteins however, dialysis is un-attractively time-consuming and the solvent density is usually manipulated by mixing the lipoprotein solution either with an appropriate amount of a concentrated salt solution of known density, or with a weighed quantity of solid salt.

2.1.1.1. Preparation of specific diluents All the diluents that are used to adjust the density of plasma, or of other lipoprotein solutions, must contain both EDTA and a bacteriostatic agent. It may also be advisable to add antioxidants (Section 2.5). The most commonly used bacteriostat is sodium azide, although sodium ethyl mercurithiosalicylate (Thimerosal) may be used instead. Many workers prefer to use both of these and may also add an antibiotic like chloramphenicol (50 μg/ml).

The composition of these diluents is therefore of the form:

NaCl or NaBr (analytical grade) to give the require density, plus
Disodium ethylenediamine tetra-acetic acid (EDTA): 0.372 g (1.0 millimolal)
Sodium azide: 0.13 g (2.0 millimolal)
Thimerosal: 0.081 g (0.2 millimolal)
Water: 1.0 kg
Sodium hydrogen carbonate: sufficient to adjust pH to 7.0–7.5.

It is customary to use solutions of NaCl for densities below about 1.19 g/ml and to add NaBr to raise the density above this value. However, the use of mixed solvents of this kind makes the estimation of density by refractometry (Section 2.1.2.5) more difficult because the refractive index cannot be directly related to the total salt concentration as it can if only one solute is present. Moreover, if the isolated lipoproteins are to be quantitatively estimated by refractometry (Section 7.1), it must be remembered that their refractive increment is known only for a limited number of simple salt solutions. In addition, it is more tedious to calculate the amount of a mixture of salts that is needed to obtain a solution of specified density than when a single salt is used. It may therefore often be more convenient to use NaBr for the whole range of solvent densities that is needed. Even so, since plasma is the source of the lipoproteins, it is inevitable that NaCl will be present and, to minimise these difficulties, Lindgren (1975) recommends that a large volume of NaCl solution of density 1.0063 g/ml should be prepared and that all other solutions should be made from this by the addition of NaBr. Whatever procedure is adopted, the effect on the density of adding EDTA and NaN_3 may be ignored, since it is within the limits of experimental error.

To calculate the salt concentration of the diluent required to adjust a lipoprotein solution to a specified density

This can be most easily explained by an example.

Suppose that we require to isolate lipoproteins of density less than 1.063 g/ml from plasma after its solvent density has been adjusted by the addition of a sodium chloride solution. What should the density of the diluent be?

First, decide the relative proportions of the plasma and the diluent. It is customary to use either 1 or 2 volumes of salt solution to 1 of plasma, but other ratios may be used if these are more convenient. In this example the ratio will be 2:1.

Now decide at what temperature the ultracentrifugation is to be carried out. If this is 10 °C, Table 2.1 shows that it is necessary to make the final solvent density of the diluted plasma 1.0665 g/ml at 20 °C in order to isolate lipoproteins of limiting density 1.063 g/ml at 20 °C.

If we ignore, for the moment, the redistribution of salt during centrifugation, the question is now: what is the density of the NaCl solution that will raise the solvent density of plasma to 1.0665 g/ml at 20 °C when added in the proportion 2:1 by volume?

TABLE 2.1

The first column shows the limiting density of the lipoprotein that it is desired to float in the ultracentrifuge, defined at a standard temperature of 20 °C. The second column gives the density of the same lipoprotein at 26 °C. The subsequent columns give the density of the solvent (measured at 20 °C) needed to float this lipoprotein when the centrifuge is run at the temperatures quoted at the head of the columns. Note that no allowance is made here for the redistribution of salt during the centrifugation.

		Density (g/ml) of solvent at 20 °C when centrifugation is to be at the following temperature		
d_{20}	d_{26}	15 °C	10 °C	5 °C
1.0050	1.0007	1.0076	1.0101	1.0125
1.0063	1.0020	1.0089	1.0114	1.0137
1.0100	1.0057	1.0125	1.0150	1.0173
1.0150	1.0107	1.0175	1.0198	1.0221
1.0200	1.0157	1.0224	1.0247	1.0269
1.0250	1.0207	1.0273	1.0295	1.0316
1.0300	1.0257	1.0322	1.0344	1.0364
1.0350	1.0307	1.0372	1.0392	1.0412
1.0400	1.0357	1.0421	1.0441	1.0460
1.0450	1.0407	1.0470	1.0490	1.0508
1.0500	1.0457	1.0520	1.0538	1.0556
1.0550	1.0507	1.0569	1.0587	1.0604
1.0600	1.0557	1.0618	1.0635	1.0651
1.0630	1.0587	1.0648	1.0664	1.0680
1.0650	1.0607	1.0667	1.0684	1.0699
1.0675	1.0632	1.0692	1.0708	1.0723
1.0700	1.0657	1.0717	1.0732	1.0747
1.0800	1.0757	1.0815	1.0830	1.0843
1.0900	1.0857	1.0914	1.0927	1.0939
1.1000	1.0957	1.1012	1.1024	1.1034
1.1100	1.1057	1.1111	1.1121	1.1130
1.1200	1.1157	1.1210	1.1218	1.1226

Note: In this example we shall follow Lindgren in assuming that 6% of the volume of plasma is occupied by the proteins i.e. the solvent volume is 0.94 of the measured volume.

The most commonly used solution to this problem is an approximation that is obtained by means of the equation

$$DV = d_1v_1 + d_2v_2 \quad (1)$$

where D, d_1 and d_2 are the densities of the final mixture, the plasma and the diluent respectively. V, v_1 and v_2 are the corresponding volumes. If the data relevant to the present case are put into this equation, we get

$$1.0665\ (2 + 0.94) = 1.0063\ (0.94) + 2d_2$$

whence $d_2 = 1.0948$ g/ml at 20 °C. The concentration of this solution can then be obtained by interpolation from Table 2.2, where the values are quoted in molal units (gram-moles/kg water) to avoid the need to account for the change in volume that is involved if the solution is made up at a different temperature.

The foregoing procedure may suffice for much preparative work but it assumes that the relation between the concentration and density of a salt solution is linear i.e. that the final volume of the mixture is equal to the sum of the volumes of the constituent solutions. However, this is not the case (Baxter and Wallace, 1916) and, for the best comparative work or the most exacting physico-chemical studies, a correction should be made for this non-linearity. The following method is a modification of the iterative procedure described by Lindgren (1975), in which convergence is slightly more rapid. It makes use of the equations that express the conservation of the masses of salt and water when the two solutions are mixed, namely:

$$S = s_1 + s_2 \quad (2)$$

and

$$W = w_1 + w_2 \quad (3)$$

where S, s_1 and s_2 are the weights of salt in the final mixture, the plasma and the diluent respectively, while W, w_1 and w_2 are the corresponding weights of water. It is also necessary to know the weight concentration of salt (C_s) and of water (C_w) in mg/ml of solution, and the ratio C_s/C_w. These parameters are listed for NaCl and NaBr in Tables 2.2 and 2.3, from which intermediate values can be obtained with sufficient accuracy by linear interpolation.

As an example of this calculation we shall take the problem 'what is the density of the NaCl solution that will raise the solvent density of plasma to 1.125 g/ml when added in the proportion 2:1?'

TABLE 2.2

The properties of sodium chloride solutions, the concentration of which is expressed in molal units i.e. gmoles/kg of water. Intermediate values can be obtained with sufficient accuracy by linear interpolation. Abbreviations are as follows:
d, the density of the NaCl solution (g/ml).
C_s, the concentration of salt in mg/ml of solution.
C_w, the concentration of water in mg/ml of solution.
n_{20}, the refractive index of the salt solution for Na_D line, at 20 °C.
n_{26}, the refractive index for Na_D line at 26 °C.
η_{20}, viscosity at 20 °C in centipoises.
η_{26}, viscosity at 26 °C in centipoises.
Mol. wt., 58.44
Note that the values quoted in this table may differ slightly from those given by Lindgren (1975). This is due to the different method of interpolation that has been used in the preparation of the table.

d_{20}	Molality	C_s	C_w	C_s/C_w	n_{20}	n_{26}	η_{20}	η_{26}
1.0050	0.1643	9.56	995.44	0.0096	1.33461	1.33403	1.0181	0.8837
1.0063	0.1962	11.42	994.89	0.0115	1.33493	1.33435	1.0206	0.8863
1.0100	0.2877	16.71	993.30	0.0168	1.33584	1.33525	1.0280	0.8936
1.0150	0.4126	23.91	991.10	0.0241	1.33707	1.33645	1.0387	0.9038
1.0200	0.5389	31.15	988.85	0.0315	1.33829	1.33766	1.0500	0.9141
1.0250	0.6666	38.44	986.56	0.0390	1.33951	1.33885	1.0620	0.9247
1.0300	0.7957	45.77	984.22	0.0465	1.34073	1.34005	1.0745	0.9356
1.0350	0.9263	53.16	981.84	0.0541	1.34194	1.34124	1.0876	0.9468
1.0400	1.0583	60.58	979.41	0.0619	1.34315	1.34242	1.1013	0.9583
1.0450	1.1918	68.05	976.95	0.0697	1.34436	1.34361	1.1155	0.9703
1.0500	1.3268	75.56	974.43	0.0775	1.34556	1.34478	1.1302	0.9827
1.0550	1.4632	83.11	971.88	0.0855	1.34675	1.34596	1.1454	0.9956
1.0600	1.6012	90.71	969.29	0.0936	1.34794	1.34713	1.1612	1.0090
1.0630	1.6847	95.28	967.71	0.0985	1.34866	1.34784	1.1709	1.0173
1.0650	1.7406	98.34	966.65	0.1017	1.34913	1.34830	1.1775	1.0229
1.0673	1.8053	101.86	965.43	0.1055	1.34968	1.34884	1.1852	1.0295
1.0700	1.8816	106.01	963.98	0.1100	1.35032	1.34947	1.1944	1.0374
1.0750	2.0241	113.72	961.28	0.1183	1.35150	1.35064	1.2118	1.0525
1.0800	2.1682	121.46	958.53	0.1267	1.35268	1.35181	1.2298	1.0683
1.0850	2.3138	129.24	955.75	0.1352	1.35385	1.35297	1.2485	1.0847
1.0900	2.4610	137.06	952.93	0.1438	1.35502	1.35413	1.2679	1.1018
1.0950	2.6097	144.91	950.08	0.1525	1.35619	1.35529	1.2880	1.1196
1.1000	2.7600	152.79	947.20	0.1613	1.35736	1.35645	1.3088	1.1382
1.1050	2.9119	160.71	944.29	0.1702	1.35852	1.35761	1.3305	1.1576

TABLE 2.2 *(continued)*

d_{20}	Molality	C_s	C_w	C_s/C_w	n_{20}	n_{26}	η_{20}	η_{26}
1.1100	3.0654	168.65	941.34	0.1792	1.35968	1.35877	1.3531	1.1778
1.1150	3.2206	176.63	938.37	0.1882	1.36083	1.35993	1.3766	1.1988
1.1200	3.3773	184.64	935.36	0.1974	1.36198	1.36109	1.4013	1.2207
1.1250	3.5357	192.67	932.33	0.2067	1.36313	1.36224	1.4270	1.2435
1.1300	3.6958	200.73	929.27	0.2160	1.36428	1.36340	1.4539	1.2672
1.1350	3.8575	208.82	926.18	0.2255	1.36542	1.36455	1.4822	1.2918
1.1400	4.0208	216.94	923.07	0.2350	1.36656	1.36570	1.5119	1.3174
1.1450	4.1859	225.07	919.94	0.2447	1.36770	1.36685	1.5431	1.3440
1.1500	4.3526	233.24	916.78	0.2544	1.36883	1.36799	1.5759	1.3715
1.1550	4.5211	241.42	913.59	0.2643	1.36997	1.36913	1.6105	1.4001
1.1600	4.6912	249.63	910.39	0.2742	1.37109	1.37027	1.6469	1.4297
1.1650	4.8631	257.86	907.16	0.2842	1.37222	1.37140	1.6854	1.4604
1.1700	5.0367	266.10	903.92	0.2944	1.37335	1.37253	1.7260	1.4922
1.1750	5.2121	274.37	900.65	0.3046	1.37447	1.37364	1.7688	1.5250
1.1800	5.3892	282.66	897.37	0.3150	1.37559	1.37475	1.8141	1.5590
1.1850	5.5681	290.96	894.07	0.3254	1.37671	1.37585	1.8620	1.5941
1.1900	5.7487	299.28	890.76	0.3360	1.37782	1.37693	1.9127	1.6303
1.1950	5.9312	307.61	887.42	0.3466	1.37893	1.37800	1.9663	1.6677

(A) To obtain a first approximation to the result, it is first necessary to determine from Table 2.2 the value of C_s and C_w for plasma and for the final, adjusted mixture. Thus, plasma of solvent density 1.0063 g/ml contains 11.42 mg/ml of NaCl and 994.89 mg/ml of water, and a NaCl solution of density 1.125 g/ml contains 192.67 mg/ml of NaCl and 932.33 mg/ml of water. The total weight of NaCl in the final mixture is therefore given by eqn. (2):

$$192.67\ (2 + 0.94) = 11.42\ (0.94) + 2C_s$$

i.e. the amount of salt in the required diluent is approximately 277.86 mg/ml. Likewise, from eqn. (3) the amount of water in the diluent is found to be about 902.93 mg/ml. An inspection of Table 2.2 will show that these two values do not correspond to a single salt solution. However, when the ratio of the two is taken, the error in the numerator almost compensates for that in the denominator and the resulting

TABLE 2.3

The properties of sodium bromide solutions. Abbreviations and units as in Table 2.2. Mol. wt. 102.91

d_{20}	Molality	C_s	C_w	C_s/C_w	n_{20}	n_{26}	η_{20}	η_{26}
1.0050	0.0847	8.67	996.31	0.0087	1.33428	1.33361	1.0058	0.8742
1.0063	0.1012	10.37	995.91	0.0104	1.33451	1.33383	1.0066	0.8749
1.0100	0.1486	15.21	994.78	0.0153	1.33517	1.33448	1.0091	0.8771
1.0150	0.2128	21.75	993.24	0.0219	1.33605	1.33534	1.0125	0.8801
1.0200	0.2772	28.30	991.69	0.0285	1.33694	1.33621	1.0160	0.8832
1.0250	0.3420	34.86	990.14	0.0352	1.33782	1.33708	1.0194	0.8865
1.0300	0.4071	41.43	988.57	0.0419	1.33871	1.33795	1.0230	0.8899
1.0350	0.4725	48.01	986.99	0.0486	1.33959	1.33881	1.0265	0.8934
1.0400	0.5382	54.60	985.41	0.0554	1.34047	1.33968	1.0301	0.8971
1.0450	0.6042	61.19	983.81	0.0622	1.34135	1.34054	1.0338	0.9009
1.0500	0.6706	67.80	982.21	0.0690	1.34223	1.34141	1.0375	0.9048
1.0550	0.7372	74.41	980.59	0.0759	1.34311	1.34227	1.0413	0.9089
1.0600	0.8041	81.03	978.97	0.0828	1.34399	1.34314	1.0452	0.9130
1.0630	0.8444	85.01	978.00	0.0869	1.34451	1.34366	1.0476	0.9156
1.0650	0.8714	87.67	977.34	0.0897	1.34487	1.34400	1.0492	0.9174
1.0673	0.9024	90.72	976.59	0.0929	1.34527	1.34440	1.0511	0.9194
1.0700	0.9390	94.31	975.70	0.0967	1.34574	1.34487	1.0533	0.9218
1.0750	1.0069	100.95	974.06	0.1036	1.34662	1.34573	1.0575	0.9264
1.0800	1.0751	107.61	972.40	0.1107	1.34749	1.34659	1.0618	0.9312
1.0850	1.1436	114.27	970.74	0.1177	1.34836	1.34745	1.0662	0.9360
1.0900	1.2125	120.94	969.06	0.1248	1.34924	1.34831	1.0707	0.9410
1.0950	1.2817	127.62	967.38	0.1319	1.35011	1.34918	1.0753	0.9462
1.1000	1.3513	134.31	965.69	0.1391	1.35098	1.35004	1.0801	0.9514
1.1050	1.4212	141.01	964.00	0.1463	1.35185	1.35090	1.0850	0.9569
1.1100	1.4914	147.71	962.29	0.1535	1.35271	1.35175	1.0901	0.9624
1.1150	1.5619	154.42	960.58	0.1608	1.35358	1.35261	1.0953	0.9681
1.1200	1.6328	161.14	958.86	0.1681	1.35445	1.35347	1.1007	0.9739
1.1250	1.7040	167.86	957.14	0.1754	1.35531	1.35433	1.1063	0.9799
1.1300	1.7756	174.59	955.40	0.1827	1.35618	1.35518	1.1120	0.9860
1.1350	1.8476	181.33	953.66	0.1901	1.35704	1.35604	1.1179	0.9923
1.1400	1.9198	188.08	951.91	0.1976	1.35790	1.35690	1.1240	0.9987
1.1450	1.9925	194.83	950.16	0.2051	1.35876	1.35775	1.1302	1.0053

TABLE 2.3 *(continued)*

d_{20}	Molality	C_s	C_w	C_s/C_w	n_{20}	n_{26}	η_{20}	η_{26}
1.1500	2.0655	201.59	948.39	0.2126	1.35962	1.35860	1.1367	1.0120
1.1550	2.1388	208.36	946.62	0.2201	1.36048	1.35946	1.1433	1.0188
1.1600	2.2125	215.13	944.85	0.2277	1.36134	1.36031	1.1502	1.0258
1.1650	2.2866	221.91	943.07	0.2353	1.36220	1.36116	1.1572	1.0330
1.1700	2.3610	228.70	941.28	0.2430	1.36305	1.36201	1.1645	1.0403
1.1750	2.4358	235.49	939.48	0.2507	1.36391	1.36286	1.1720	1.0477
1.1800	2.5110	242.29	937.68	0.2584	1.36476	1.36371	1.1797	1.0553
1.1850	2.5865	249.10	935.87	0.2662	1.36561	1.36456	1.1877	1.0631
1.1900	2.6624	255.91	934.06	0.2740	1.36646	1.36541	1.1958	1.0710
1.1950	2.7387	262.73	932.24	0.2818	1.36731	1.36626	1.2042	1.0791
1.2000	2.8153	269.55	930.41	0.2897	1.36816	1.36710	1.2129	1.0874
1.2500	3.6031	338.10	911.86	0.3708	1.37660	1.37552	1.3137	1.1793
1.3000	4.4311	407.13	892.83	0.4560	1.38491	1.38385	1.4433	1.2899
1.3500	5.3014	476.56	873.44	0.5456	1.39311	1.39209	1.6050	1.4216
1.4000	6.2164	546.29	853.80	0.6398	1.40118	1.40023	1.8014	1.5777
1.4500	7.1782	616.23	834.01	0.7389	1.40912	1.40828	2.0344	1.7618
1.5000	8.1891	686.28	814.18	0.8429	1.41693	1.41627	2.3049	1.9780

value of C_s/C_w (= 0.3077) is a close approximation to that required. Interpolation of Table 2.2 shows that the NaCl solution with this C_s/C_w ratio has a density of 1.1765 g/ml.

To refine this first estimate, it is necessary to repeat the sequence of calculations as follows:

(B) Determine from Table 2.2 the real values of C_s and C_w in a solution of density 1.1765 g/ml:

$$C_s = \frac{(0.3077-0.3046)\ (282.66-274.37)}{(0.3150-0.3046)} + 274.37 = 276.84 \text{ mg/ml}$$

and

$$C_w = \frac{(0.3077-0.3046)\ (897.37-900.65)}{(0.3150-0.3046)} + 900.65 = 899.67 \text{ mg/ml}$$

(C) From eqns. (2) and (3) the total weights of salt and water in the final mixture can now be calculated i.e.

$$2(276.84 + 0.94(11.42) = 564.41 \text{ mg NaCl}$$

and $$2(899.67) + 0.94(994.89) = 2734.54 \text{ mg water}$$

whence C_s/C_w in the final solution is 0.2064. Interpolation of Table 2.2 shows that this mixture will have a density of 1.1248 g/ml, which is almost within the acceptable tolerance limits. The approximate value of 1.1765 g/ml is therefore fairly close to the actual density of the required diluent. Note that, if the amount by which the density of the original solution is to be raised is smaller than in this example, the first approximation will often be within limits. For example, the first estimate of the density of the diluent needed to raise the density of plasma to 1.0665 g/ml is 1.0934 g/ml and this requires no further refinement. In the present case, the calculated value of the final density is not within the tolerance of ±0.0002 g/ml and a refined estimate for the density of the diluent must be calculated as follows:

(D) First, estimate the real volume of the mixture that results when the diluent is added to the plasma, by putting into eqn. (1) the density of the mixture that was calculated in (C):

$$V = \frac{2(1.1765) + 0.94(1.0063)}{1.1248} = 2.9329 \text{ ml}$$

Using this value of V, refine the estimates of C_s and C_w for the diluent by repeating the procedures of paragraph (A) i.e.

$$2.9329(192.67) = 0.94(11.42) + 2C_s$$

and $$2.9329(932.33) = 0.94(994.89) + 2C_w$$

whence $C_s = 277.17$ mg/ml, $C_w = 899.62$ mg/ml, $C_s/C_w = 0.3081$ and, by interpolation of Table 2.2, the revised estimate of the density of the diluent is 1.1767 g/ml. Now repeat calculations (B) and (C), which yield a density for the final mixture of 1.125 g/ml. The required diluent therefore has a density of 1.1767 g/ml and its concentration is 5.2723 molal NaCl.

If the density calculated for the adjusted plasma is still not within acceptable limits at this stage, the procedure in paragraph (D) must be repeated. Although the whole procedure is tedious to execute by hand, it is not one that must be frequently performed. Moreover, it can easily be carried out on a programmable calculator.

Solutions of sodium bromide are the most commonly used alternative to chloride for the centrifugation of lipoproteins in high density solvents, and relevant data for this salt are summarised in Table 2.3. Other substances such as potassium bromide and sucrose are also sometimes used and analogous tables of data for these materials can be found in references of which the Handbook of Chemistry and Physics (CRC Press) is an example. However, it must be remembered that eqns. (2) and (3) cannot be directly applied to mixtures of different salts. For example, if NaBr is used to raise the density of plasma to 1.0665 g/ml, the calculation must be made entirely in terms of NaCl or NaBr. Since all lipoprotein preparations effectively originate from plasma, Lindgren (1975) recommends that sodium bromide solutions should be made up by the addition of the solid salt to 0.196 molal NaCl solution. There is much to be said for this practice, since it simplifies not only the preparation of the solutions but also the refractometric estimation of solvent density or lipoprotein concentration. In calculating the amount of NaBr that is needed to prepare a mixture of specified density, the least objectionable course is to replace (in concept) the 0.196 molal NaCl by the bromide solution of the same density i.e. 0.101 molal NaBr (Table 2.3). This NaBr solution would contain 10.37 mg/ml of salt and 995.91 mg/ml of water. If the final density of the plasma is to be 1.0665 g/ml, the eqns. (2) and (3) then become:

$$2.94(89.66) = 0.94(10.37) + 2C_s$$

and

$$2.94(976.85) = 0.94(995.91) + 2C_w$$

whence, by following the procedure in paragraphs (A), (B), (C) etc., the density of the required diluent can be calculated to be 1.094 g/ml.

The next step is to determine what weight of NaBr must be added to 0.196 molal NaCl in order to make a solution of density 1.094 g/ml. A sodium bromide solution of this density is 1.2679 molal, i.e. it contains 130.48 g of salt/kg of water. Deduct from this the contribution of 0.101 molal NaBr namely 10.41 g/kg of water (which is the density-equivalent of 0.196 molal NaCl), to obtain the weight of NaBr required i.e. 120.07 g/kg of water. This weight should therefore be dissolved in the weight of 0.196 molal NaCl that contains 1 kg of water, namely 1011.45 g. Alternatively, since 1 litre of 0.196 molal NaCl contains 993.08 g of water at 20 °C, this volume of the NaCl solution can be added to 119.24 g of NaBr (i.e. 120.07 × 0.99308). It is worth noting however, that the foregoing calculation can also be done as described in Section 2.1.1.2.

No matter how carefully a specific diluent is prepared, it cannot be assumed that its density is within the acceptable limits of ±0.0002 g/ml. It should therefore be checked and, if necessary, adjusted by one of the methods described in Section 2.1.1.3.

2.1.1.2. Adjustment of density with solid salts If it is important to avoid undue dilution of the samples, their solvent density can be increased by the addition of a weighed amount of salt. This method is relatively tedious, and it has the disadvantage that it may be difficult to check that the desired density has actually been achieved. Note that, if the density of plasma is to be adjusted in this way, the following solution should first be added to the extent of 1% of the plasma by volume: 0.15 M NaCl containing 0.1 M EDTA and 0.2 M NaN_3.

It is possible to use the data given in Tables 2.2 or 2.3 to calculate the amount of salt needed to bring about a specified adjustment of density (Lindgren, 1975). However, the method of Radding and Steinberg (1960) is slightly simpler and is the one we shall describe here. It makes use of the fact that the weight of salt *(M)* to be added to a volume V of solution, to change its density from d_1 to d_2 (at a stated temperature) is given by the expression:

$$M = \frac{v\ (d_2 - d_1)}{1 - \bar{v}\, d_2}$$

where $\bar{v}$ is the partial specific volume of the salt at the relevant temperature and concentration. The values of $\bar{v}$ for NaCl, NaBr and KBr at temperatures between 5 and 25 °C can be obtained by interpolation of the Tables 2.4, 2.5 and 2.6, which are derived from the data of Baxter and Wallace (1916). As an example, we shall calculate the weight of NaBr required to raise the density of 10 ml of plasma (d_1 = 1.0063 g/ml) to 1.0655 g/ml al 20 °C.

TABLE 2.4

The partial specific volume ($\bar{v}$) of sodium chloride in aqueous solutions of different density (d) and temperature. Note that the value of d is the density of the solution at the temperature of the experiment, which is quoted as a subscript to $\bar{v}$ in the appropriate column. The values given are obtained by interpolation of data from Baxter and Wallace (1916).

d	$\bar{v}_5$	$\bar{v}_{10}$	$\bar{v}_{15}$	$\bar{v}_{20}$	$\bar{v}_{25}$
1.0100	0.2552	0.2697	0.2826	0.2939	0.3038
1.0250	0.2669	0.2804	0.2925	0.3032	0.3124
1.0500	0.2835	0.2956	0.3063	0.3159	0.3241
1.0630	0.2909	0.3022	0.3123	0.3213	0.3290
1.0650	0.2920	0.3032	0.3132	0.3220	0.3297
1.0750	0.2971	0.3077	0.3173	0.3257	0.3330
1.1000	0.3084	0.3178	0.3262	0.3337	0.3402
1.1250	0.3181	0.3264	0.3339	0.3406	0.3464
1.1500	0.3269	0.3344	0.3412	0.3473	0.3527
1.1900	0.3409	0.3477	0.3539	0.3596	0.3648

TABLE 2.5

The partial specific volume ($\bar{v}$) of sodium bromide in aqueous solutions of different density (d) and temperature. Note that the value of d is the density of the solution at the temperature of the experiment, which is quoted as a subscript to $\bar{v}$ in the appropriate column. The values given are obtained by interpolation of data from Baxter and Wallace (1916).

d	$\bar{v}_5$	$\bar{v}_{10}$	$\bar{v}_{15}$	$\bar{v}_{20}$	$\bar{v}_{25}$
1.0100	0.2074	0.2157	0.2235	0.2308	0.2376
1.0250	0.2110	0.2189	0.2263	0.2332	0.2396
1.0500	0.2165	0.2238	0.2306	0.2369	0.2427
1.0630	0.2192	0.2262	0.2327	0.2386	0.2442
1.0650	0.2195	0.2264	0.2329	0.2388	0.2443
1.0750	0.2214	0.2281	0.2344	0.2401	0.2454
1.1000	0.2257	0.2320	0.2377	0.2430	0.2477
1.1250	0.2297	0.2355	0.2407	0.2455	0.2497
1.1500	0.2332	0.2385	0.2434	0.2477	0.2516
1.2000	0.2392	0.2438	0.2478	0.2514	0.2544
1.2100	0.2403	0.2447	0.2487	0.2521	0.2551
1.2500	0.2443	0.2482	0.2516	0.2545	0.2569

TABLE 2.6

The partial specific volume ($\bar{v}$) of potassium bromide in aqueous solutions of different density (*d*) and temperature. Note that the value of *d* is the density of the solution at the temperature of the experiment, which is quoted as a subscript to $\bar{v}$ in the appropriate column. The values given are obtained by interpolation of data from Baxter and Wallace (1916).

d	$\bar{v}_5$	$\bar{v}_{10}$	$\bar{v}_{15}$	$\bar{v}_{20}$	$\bar{v}_{25}$
1.0100	0.2688	0.2753	0.2811	0.2863	0.2909
1.0250	0.2712	0.2776	0.2834	0.2885	0.2930
1.0500	0.2751	0.2813	0.2869	0.2919	0.2962
1.0630	0.2769	0.2831	0.2886	0.2934	0.2977
1.0650	0.2773	0.2834	0.2889	0.2938	0.2980
1.0750	0.2787	0.2847	0.2901	0.2949	0.2991
1.1000	0.2820	0.2878	0.2930	0.2976	0.3016
1.1250	0.2851	0.2906	0.2956	0.2999	0.3038
1.1500	0.2880	0.2932	0.2979	0.3020	0.3057
1.2000	0.2933	0.2979	0.3020	0.3058	0.3091
1.2100	0.2944	0.2988	0.3028	0.3064	0.3097
1.2500	0.2980	0.3020	0.3056	0.3089	0.3119

The partial volume of the solvent in the plasma after the addition of 1% of EDTA solution will be 9.5 ml. The required weight will therefore be:

$$M = \frac{9.5\ (1.0665 - 1.0063)}{1 - 0.2390(1.0665)} = 0.7675\ \text{g}$$

In practice, the required solvent density will depend on the temperature of the centrifugation and should be calculated as described in Section 2.1.1.1.

2.1.1.3. Measurement and final adjustment of density *(a) Pycnometry.* Either the 'density bottle' or the capillary pycnometer may be used. The former is somewhat more convenient to use but usually requires a larger volume of solution than the capillary type. It is therefore well suited for determining the density of bulk salt solutions, as described in detail by Schachman (1957). The capillary pycnometer has a volume of 2–5 ml, and will probably be the better choice if the density of a lipoprotein solution is to be measured e.g. for the determination of partial specific volume. Lindgren (1975) has given a concise description of the use of this pycnometer, on which the method given below is based.

Both types of pycnometer need to be kept thoroughly clean by frequent treatment with chromic acid mixture. Each time this is done, the apparatus must be exhaustively washed with distilled water and well dried before use. Drying can be done by washing several times with acetone and then blowing a stream of nitrogen through the vessel for several minutes. It must be re-weighed after each cleaning. Before use, the volume of the vessel must be determined at the temperature of use (e.g. 20 °C) by weighing it when full of distilled water in the usual way. When filling the vessel, it is important to ensure that no air bubbles are entrapped and that it is not heated by incautious handling. A density bottle should be filled at a temperature below that at which the measurement is to be made. Then, as thermal equilibrium is established, the excess liquid will expand out of the hole in the stopper and can be carefully wiped away. If the bottle is equilibrated in a water bath, it will be necessary to dry the outside without expressing further liquid either by squeezing the bottle or by heating it with the hand.

The same precautions must be observed when handling the capillary pycnometer, which is filled by gentle, controlled suction to a point just above the datum line. After thermal equilibrium has been reached, the meniscus can be set at the line by withdrawing liquid from the vessel by touching the tip with a filter paper. It is an advantage to observe this coincidence with a lens. In practice, it may be difficult to establish this coincidence exactly and Lindgren (1975) prefers to measure the distance of the meniscus from the datum line and to correct for the included volume of the capillary. Although this necessitates the prior calibration of the capillary, the technique may, in the long run, be less tedious than the exact adjustment of the meniscus to the datum line. The procedure is as follows: Fill the pycnometer with distilled water to a point above the datum, equilibrate at the operating temperature and measure the distance of the meniscus from the datum with a magnifying scale or cathetometer. Weigh the apparatus and then withdraw a little fluid from the tip. Re-equilibrate, measure the new position of the meniscus and weigh the pycnometer again. Repeat this process at least once more. Since the weight and density of the water that filled a known length of the capillary are now known, the volume of the pycnometer at the datum line and the volume per centimetre of the capillary can both be determined. The calibrated instrument can then be used to determine the density of a salt solution as in the following example:

Volume of the pycnometer to the datum line at 20 °C = 10.8105 ml.
Volume per cm of capillary = 0.005 ml.
Weight of empty pycnometer = 19.9713 g.
Weight when filled with salt solution to a point 3.1 mm above datum = 31.4861 g.
Volume of salt solution in pycnometer = 10.8105 + (0.31 × 0.005) = 10.8121 ml.
Whence the density of the salt solution = 1.065 g/ml.

All the weighings needed for these determinations should be done on a clean and properly maintained analytical balance, and should be made to an accuracy of ±0.1 mg. Densities measured with a 10 ml pycnometer should then be accurate to within ±0.00003 g/ml. However, this neglects the buoyant effect of the air which will diminish the true weight of the contents of a 10 ml pycnometer by about 10 mg, assuming that brass weights are used. Although this is partially offset by the corresponding error in the calibration of the pycnometer, the apparent density of a salt solution can be high by up to 0.0005 g/ml (Lindgren, 1975). This is smaller than some of the errors mentioned in Section 2.1 and can be neglected for most preparative purposes. For the best analytical work however, it may be desirable to correct all weighings for atmospheric buoyancy by means of the relation

$$W = w + w\rho \left(\frac{1}{d_1} - \frac{1}{d_2}\right)$$

where W is the corrected weight, w the measured weight, ρ is the density of air at the temperature, pressure and humidity of the measurement, and d_1 and d_2 are the density of the object being weighed and of the weights respectively. In many cases, it may suffice to use the approximate correction to the measured density of the salt solution that is given by the relation

$$\Delta d = \rho \, (1 - \frac{d_s}{d_w})$$

where d_s and d_w are the densites of the salt solution and water respectively (Bauer and Lewin, 1960). The density of dry air at 20 °C and 760 mm pressure is 0.001 g/ml. Values under other conditions can be found in many compilations of physical data e.g. the CRC Handbook of Chemistry and Physics.

(b) The density meter (Anton Paar K.G.). This instrument was developed by Kratky et al. (1969, 1973) and is based on the principle that the frequency of oscillation of a hollow glass tuning fork depends on the mass of any solution with which it is filled. It depends therefore on the precise electronic measurement of the change in frequency of the tube when it is filled with different solutions. Although expensive, the instrument has the merit that it can determine the density of less than 1 ml of a solution to four places of decimals very rapidly. The equipment must be calibrated by means of two liquids of accurately known density, but the calibration appears to remain very stable provided that the cell is kept scrupulously clean. Contamination of the tube can be detected by noting the instrument reading before and after use, when the tube is dry. If the two values are not identical, the tube must be cleaned, thoroughly washed and dried with acetone. Because the tube is narrow and U-shaped, it may be difficult to remove any persistent deposit and contamination should therefore be avoided. But if it should occur, the cell may be cleaned with any solvent or acid, even with 'chromic acid' mixture. With concentrated acids, a small amount of the cell wall may be etched away, but the resulting change in natural frequency necessitates only a re-calibration with fluids of known density.

It is worth noting that one version of this instrument can measure the density of solutions to six places of decimals and can therefore be used to determine the partial specific volumes of lipoproteins. For this purpose however, it is necessary to control the temperature of the measuring cell to within ±0.01 °C.

(c) Refractometry. This is a simple and speedy way of estimating the density of a solution of a single salt from its known correlation with the refractive index of the solution. This relationship is tabulated for NaCl and NaBr in Tables 2.2 and 2.3, and can be found for other substances in such compilations as the CRC Handbook of Chemistry and Physics. The method has the disadvantage that it requires special equipment and is not easily used when the solution contains a mixture of salts. Moreover, it is useless if the composition of the mixture is unknown. For the particular case of solutions of NaBr that contain a constant concentration of 0.196 molal NaCl, the refractive index has been tabulated by Lindgren (1975). In this case, the refractive index (n') of the mixed salt solution of density d differs from that of the isopycnic NaBr solution (n) by an increment (Δn) that is given by:

$$\Delta n = 0.00114 - 0.00102d + 0.00035d^2$$

The refractive index of a solution of NaBr in 0.196 molal NaCl can thus be found by looking up the value for a NaBr solution of the same density in Table 2.3 and adding the calculated value of Δn. For other NaCl/NaBr mixtures, the relationship between d and n' must be determined experimentally.

Details of the refractometric technique are given in Section 2.1.2.5.

(d) Other methods. Although methods based on the displacement principle e.g. the hydrometer and the Westphal balance, have occasionally been used, they are generally unsatisfactory and cannot be recommended for anything but the crudest preparative work.

The density of a salt solution that is to be used as a medium for the centrifugation of lipoproteins should be within ±0.0002 g/ml of the specified value. However, the density of a freshly weighed-up solution will often lie outside these limits and will need adjustment. If dilution is necessary, the volume of water to be used can be calculated by means of eqn. (1) of Section 2.1.1.1, while the amount of salt needed to raise the density can be determined by the method of Section 2.1.1.2. However, the amount of water or salt to be added can usually be estimated with sufficient accuracy by simple interpolation of the data from Tables 2.2 or 2.3 as appropriate. For example, if 1000 ml of a solution of NaCl of nominal density 1.093 g/ml is found to be 1.095 g/ml, the volume (V) of water to be added is given by:

$$V = 1000\left(\frac{1.095 - 0.9982}{1.093 - 0.9982}\right) - 1000 = 21.1\ \text{ml}$$

where 0.9982 is the density of water at 20 °C.

If, on the other hand, the density of the solution is found to be only 1.091 g/ml, the weight (M) of NaCl to be added is approximately

$$M = \frac{(144.91 - 137.06)\ (1.093 - 1.091)}{(1.095 - 1.090)} = 3.14\ \text{g}$$

The density of the solution must, of course, be checked after it has been re-adjusted in this way.

2.1.1.4. Choice of rotor and tubes The rotors most commonly used for the isolation of plasma lipoproteins are probably those of the Beckman-Spinco range that are detailed in Table 2.7. However, although these are the rotors to which we shall refer in this manual, the equivalent products of other manufacturers can also be used satisfactorily.

The actual choice of rotor is determined by several factors. It must be remembered that the purpose of the centrifugation is not only to float the required lipoproteins to the top of the tube but also to sediment the plasma proteins an adequate distance towards the bottom. In the case of VLDL, the density of both the lipoproteins and the plasma proteins differ from that of the solvent by enough to allow them to be almost completely separated by a few hours centrifugation at a relatively low speed. To float HDL however, the solvent not only has a density much closer to that of albumin, but is also more viscous than that used for the isolation of VLDL. A longer centrifugation, or a higher field force will therefore be needed and a 'low-speed' rotor that is adequate for the isolation of low-density lipoproteins may be unsatisfactory for the preparation of HDL. When lipoproteins are to be isolated from large volumes of plasma, it is desirable that the rotor

TABLE 2.7

This table summarises some operational details of some useful Beckman-Spinco fixed-angle centrifuge rotors. T is the time needed to achieve a practicable approximation to complete flotation of the relevant lipoprotein fraction, at the speed indicated by the subscript (in thousands of rev/min), at 20 °C. Δr = maximum radius − minimum radius.

	(RCF at r_{av}) · 10^{-3} at rotor speed			Min. time of flotation for LDL in hours			Min. time of flotation for HDL in hours		
	60	50 (rev/min × 10^{-3})	40	T_{60}	T_{50}	T_{40}	T_{60}	T_{50}	T_{40}
Rotor 60Ti Volume 8 × 38.5 ml Δr 5.3 cm Angle 23.5°	254.0	176.2	112.8	12	17	27	28	40	63
Rotor 50.2Ti Volume 12 × 38.5 ml Δr 5.5 cm Angle 24°		228.3	145.0		13	20		32	50
Rotor 50.3Ti Volume 18 × 6.5 ml Δr 3.2 cm Angle 20°		179.2	114.7		10	16		23	36
Rotor 50Ti Volume 12 × 13 ml Δr 4.3 cm Angle 26°		165.3	105.8		14	22		34	53
Rotor 40.3 Volume 18 × 6.5 ml Δr 3.2 cm Angle 20°			114.5			15			36
Rotor 45Ti Volume 6 × 94 ml Δr 6.7 cm Angle 24°	161 at 45 000 rev/min			$T_{45} = 24$			$T_{45} = 57$		

will accept the sample in the smallest number of tubes. This will not only minimise the cost of the tubes but also that of the labour involved in filling them. Although it is tempting to use the highest possible centrifugal force in order to effect a speedy separation of the lipoproteins, it can also cause the low-density lipoproteins to pack into a gel at the top of the tube. Because this gel can be awkward to recover, it may in practice be more satisfactory to use a high speed rotor at less than its maximum rate when isolating this fraction. This not only prolongs the life of the rotor, but can sometimes lead to the centrifugation being of a more practicable duration: it is usually more convenient to run for 18 hours overnight than to lengthen the working day to accommodate a 12 hour run. Note that, the longer the duration of the run at any selected speed, the better will be the recovery of the lipoproteins of low flotation rate but that this improvement must be paid for in terms of the life of the centrifuge drive and rotor. Fortunately, the HDL do not form gels so readily as the larger lipoprotein particles and full advantage can be taken of the available high-speed rotors to minimise the length of the run that is needed.

One of the most useful of the rotors listed is the 50.2Ti, which has a maximum capacity of 462 ml in 12 tubes of 39.5 ml each. Its maximum speed is 50 000 rev/min but, for the isolation of low-density lipoproteins, it need be run only at 40 000 rev/min when the centrifugal force at the top of the tube is 94 900 *g*. This will concentrate LDL at the top of the tube within 18 hours. For the essentially quantitative recovery of HDL however, it would be necessary to centrifuge for up to 48 hours at this speed, according to the temperature of the experiment. This time can be considerably reduced by running at 50 000 rev/min, when the field at the top of the tube amounts to 149 000 *g*. However, it is important to bear in mind that large volume, high speed rotors such as the 50.2Ti or 60Ti can only be used in centrifuges of the appropriate rating. If the machines available are not rated to handle these high energy rotors, it may be necessary to use others such as the Beckman 30 or 40 series. The Beckman 35 rotor (total volume 564 ml) can be used to prepare VLDL or LDL, but is inadequate for the isolation of HDL. For this purpose, the 40 rotor (total volume 162 ml) can be used. At the time of writing, little experience has been gained with the Beckman 45Ti rotor (nominal capacity 564 ml). It remains to seen whether the large capacity will offset the need to centrifuge for a long time in order to separate HDL.

When small amounts of lipoprotein are to be prepared, or if a quantitative recovery is important, the Beckman 30.2 or 40.3 rotors are the most suitable. These use narrow tubes which, with their skirt-less caps, simplify the recovery of the lipoproteins. (Tubes of more than 0.5 inches diameter are usually fitted with caps that have a skirt to support the wall of the tube during the centrifugation. This dips into the layer of lipoprotein that is formed at the top of the tube and will stir it up when the cap is removed. Alternatively, if the cap is left in place and the lipoproteins are recovered by tube slicing (Section 2.1.1.7), the skirt hinders the washing process.)

Clearly, the laboratory that aspires to carry on a range of studies on lipoproteins will need more than one type of fixed angle ultracentrifuge rotor, and the most useful compromise selection would seem to be a 50.2Ti, a 50.3Ti and a 30.2. If a large number of small samples must be centrifuged, as in some assays which involve radioactively labelled lipoproteins, or the estimation of HDL cholesterol, it is worth noting that Beckman market a rotor which will centrifuge 72 tubes of 175 μl capacity at 210000 *g*.

Whatever type or make of rotor is used, it is necessary to use tubes with parallel sides. Bottles are unsatisfactory because the floating lipoproteins cannot be properly recovered from below the shoulders. For the quantitative recovery of small samples in a fixed, small volume, it is also necessary to use transparent tubes. Cellulose nitrate, or the recently introduced Beckman Ultra-Clear tubes should be used. The latter have the advantage that they can be used in high speed rotors like the 50.2Ti. For large scale preparations when the final volume of the product is of lesser importance, the cheaper polyallomer tubes may be used. The fact that these are translucent is of no consequence when the lipoproteins are recovered by tube slicing, but makes it difficult to achieve a quantitative recovery by the pipetting technique (Section 2.1.1.7). Clear polycarbonate tubes are also available for many rotors. These are expensive but, like cellulose nitrate tubes, can be used several times. Moreover, they can be run without a cap and only partially filled, if the rotor speed is reduced. Under these circumstances, great care must be taken to ensure that the rotor lid seals properly or the samples will be lost when the chamber is evacuated.

All tubes should be be examined carefully for blemishes such as air bubbles or thin areas that may lead to leakage during the run. At the same time, check that the components of the tube cap are in good condition, paying particular attention to the gasket which should be flexible and free from nicks or cracks. The Donner group recommends the use of caps with incorrodible stainless steel stems, but others seem to be less scrupulous in this respect. In any event, the advice is technicaly difficult to follow if tubes of large diameter are used, since it would then be necessary to reduce the maximum speed at which the rotor is run. If steel components are used, take great care that they do not become mixed with duralumin ones, since this could lead to imbalance of the rotor.

Many of the problems that are inherent in the use of tube caps can be eliminated by the use of 'Quick-Seal' tubes (Beckman-Spinco). These are essentially polyallomer ampoules that can be filled throogh a narrow neck that is then hermetically sealed by heat. They are well suited to the bulk preparation of lipoproteins and the centrifugation of radioactive specimens, but may make the quantitative recovery of the lipoproteins more difficult.

2.1.1.5. Loading the tubes The routine for the adjustment of the density of the plasma or other lipoprotein solution will already have been decided, as in Section 2.1.1.1. First, label the tubes with a note of their contents, including the amount of plasma taken and the density of the run. Use a waterproof ink. Pipette the appropriate amount of plasma into each tube with a grade A pipette and follow this with the requisite volume of the specific diluent. This should fill the tube to a point slightly below the cap, e.g. a total of 9 ml in a 10.5 ml tube. Because the tube is not completely filled and there is accordingly a risk of it splitting during the run, it is advisable to use thick-walled tubes. This procedure is suitable for the treatment of small samples of plasma. For a large sample however, it may be simpler to adjust the density in bulk and then to add the appropriate volumes of the mixture to the centrifuge tubes.

Assemble a cap for each tube, with a slight smear of grease on the screw threads and the underside of the nuts. It is essential that the gasket and the top of the tube should be dry, or the cap will not seal when it is tightened. Fit a cap onto a tube and lightly tighten the nut with a box spanner, making sure that the cap is pressed fully home on the tube and is squarely seated. Clamp the tube in the tube vice and screw down the nut firmly. Be careful not to damage the cap by over-tightening the vice. Invert the capped tube 6 or 8 times to mix the contents.

If the density of the plasma is to be adjusted by the addition of solid salt, calculate the required amount as described in Section 2.1.1.3. Weigh this quantity into a centrifuge tube and add the relevant volume of plasma or lipoprotein solution from a pipette. Again, large volumes may be adjusted in bulk. Cap the tube as described above, and mix the contents thoroughly.

Note. The N.I.H. Manual of Lipid and Lipoprotein Analysis (1975) recommends that the tubes should be completely filled with liquid before centrifuging. This has the merit that the tubes are less likely to fracture, but the lipoproteins accumulate close under the tube cap and must be recovered by the tube slicing technique described in Section 2.1.1.7. It is also necessary to use tube caps with a central filling hole that can be closed with a screw plug. The centrifuge tube is filled to about 75% capacity with the density-adjusted plasma and a layer of salt solution of the same density floated on the surface to about 5 mm below the rim. The cap is then fitted and the tube completely filled with the salt solution through the filling hole, which is then plugged.

Quick-Seal tubes are filled as follows. Wet a syringe with salt solution of the same density as that at which the centrifugation is to be made, and fit a 1.5 inch, 18 gauge needle with a 9 inch length of fine polythene tubing (bore 1.2 mm) attached. After adjusting its density, take up a suitable volume of the sample into the syringe e.g. 30 to 35 ml for a 1″ × 3.5″ tube. Carefully expel the air from the syringe and tubing and then slowly deliver the sample into the bottom of the centrifuge tube. Avoid trapping any bubbles on the side of the tube. Finally, the sample is over-layered with salt solution of the same density until the required total volume is reached, which in this example is 39.5 ml. To do this, expel the salt solution very slowly from the syringe, with the tip of the polythene tube held in the meniscus at the wall of the centrifuge tube. Hold the syringe below the level of the meniscus and express the solution manually; gravity feed will be too rapid with this size of needle. There will be a small air-space below the opening to the Quick-Seal tube which allows it to be pierced after the centrifugation without the lipoproteins spurting out. After filling, the neck of the tube should be dried with a corner of paper tissue and sealed in accordance with the maker's instructions.

An alternative filling procedure uses a 4 inch, 16 gauge cannula fitted to a large syringe barrel, as a funnel with which to deliver the sample to the bottom of the Quick-Seal tube. The over-layer can then be added from a similar funnel made by attaching a 1 inch, 23 gauge needle to the barrel of a 10 ml syringe. The needle is bent to touch the wall of the centrifuge tube when the syringe is supported above the tube in a stand. The over-layering solution can be poured into the funnel and allowed to trickle slowly into the tube by gravity. Using this procedure, several tubes can be filled simultaneously.

Of course, the technique of over-layering with a solution of the same density as the sample can also be used to make up the required volume if the sample is too small to fill the centrifuge tube.

2.1.1.6. Loading the rotor It is of the utmost importance that the following precautions should be taken to ensure that the rotor is balanced at the start of the run. Care in this respect can materially increase the life of the centrifuge drive.

(a) If the rotor is not completely full, make certain that the tubes oppose one another in pairs. If there is an odd number of tubes, include a ballast tube that is filled with the salt solution of matching density.

(b) If tubes containing solutions of different density are centrifuged at the same time, make certain that they oppose each other in pairs of the same density. Include ballast tubes of the correct density where necessary. Make sure that the rotor gasket is clean and lightly lubricated; put a trace of grease on the screw thread and screw down the lid of the rotor in accordance with the maker's instructions. When transferring the rotor to the centrifuge, lower it gently onto the spindle. Clumsiness in this respect can materially shorten the life of the drive, particularly if large, heavy rotors are being used.

2.1.1.7. Unloading the rotor Because the density gradient at the top of the tubes after centrifugation is very shallow, it is necessary to handle the rotor and the tubes very smoothly. Take care not to jerk or shake the rotor when removing it from the centrifuge. If the rotor *must* be carried across the laboratory to the place where it will be unloaded, do so slowly and smoothly; do not bang it on the bench, or shake it when removing the lid. Use the tube extracting tool to remove the first tube from the rotor. Do this very smoothly and without haste. Gently return the tube to the vertical position.

There are two general ways of recovering the lipoproteins from the tube:

(a) By means of a fine-bore pasteur pipette.

This technique requires some skill and practice, but results in little dilution of the supernatant lipoproteins. It is probably the best way of collecting small amounts of lipoprotein especially when it is intended to estimate them quantitatively.

Many who attempt this method try to use commercial mass-produced pasteur pipettes. These are totally useless because their bore is too large. The internal diameter of the tip *must not exceed* 0.6 mm and, after a few trials, it is simple to draw satisfactory pipettes in the laboratory blow-lamp. First cut several 12 cm lengths of glass tubing of 0.7 cm external diameter. Draw the centre portion down to a capillary about 16 cm long by 1.0 mm external diameter at the mid-point. When cool, scratch transversely at the mid-point with a diamond pencil; the capillary will then break neatly, at right angles to the bore. Never try to use a pipette with a tip that is ragged, or is not normal to the axis of the capillary.

Note. These pipettes *must* be used with a bulb of heavy rubber, like those used with a medicine dropper. The ordinary laboratory teat is so thin that it responds to every involuntary twitch of the fingers, making it impossible to withdraw the lipoproteins smoothly.

Also needed for this method is a small lamp with which to project a

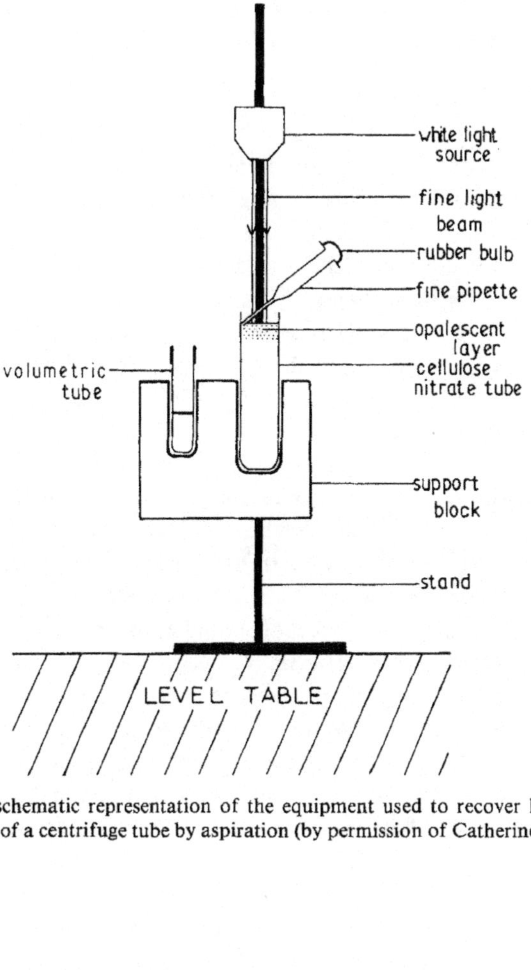

Fig. 2.2. A schematic representation of the equipment used to recover lipoproteins from the top of a centrifuge tube by aspiration (by permission of Catherine J. Briggs).

narrow beam of light vertically into the top of the centrifuge tube (Fig. 2.2). The light scattered by the (large) lipoproteins makes them easily visible against a dark background. The tube itself should be mounted at such a height that the seated worker can rest an elbow on the bench and comfortably introduce the tip of the pipette into the surface of the lipoprotein layer. This illuminated stand should be in a shaded place, as near to the ultracentrifuge as possible.

If the recovery of the lipoproteins is to be quantitative, as we shall assume, they are collected in small volumetric flasks. For volumes of plasma of about 2–3 ml, flasks of 1 ml capacity with a calibration at 0.5 ml are needed. For purely preparative purposes, volumetric ware is not necessary.

Before starting the operation, ensure that there is enough glassware to deal with all the lipoprotein samples and that the vials to which they are to be transferred are properly labelled. These vials should have a close-fitting cap with a plastic (e.g. teflon) liner and should accommodate the lipoprotein concentrate with a minimal air space above it. Vessels of water in which to soak the equipment after it has been used should also be at hand.

Fit a bulb to a clean, dry pipette. It is important that the assembly does not leak, so reject any bulbs that have frayed inner edges. Clamp the centrifuge tube into the tube vice. Hold the tube with one hand and loosen the nut with the box spanner until the cap is only lightly held (do not unscrew the nut completely). Carefully transfer the tube to the pipetting stand and remove the cap. Any material that adheres to the stem can be sucked up with the pipette (if it is very viscous, scrape it up with the tip) and transferred to the volumetric flask. Disassemble the tube cap and inspect the inside of the crown. If it contains lipoprotein, the gasket probably needs renewing: transfer the lipoprotein to the volumetric flask.

Slightly squeeze the bulb of the pipette (N.B. this should always be done with the tip in the volumetric flask and *never* with the tip in the centrifuge tube) and introduce the tip into the meniscus where it meets the wall of the tube. Support your elbow on the bench and keep the pipette at an angle of 50–60 ° above the horizontal. Now slowly release the bulb, at the same time withdrawing the pipette almost imperceptibly until a stream of bubbles begins to enter the tip. This will cause the lipoprotein layer to be skimmed off very effectively. However, if the suction is too rapid, some of the lipoprotein that is drawn across the surface fails to enter the pipette, strikes the side of the tube and is carried by its momentum 2 or 3 mm below the surface. It is therefore important not to aspirate too rapidly, and to rotate the tube so that suction is applied at several points on the circumference. Transfer the lipoproteins to the volumetric flask as they are removed.

When most of the supernatant lipoprotein layer has been recovered, hold the pipette vertically, with the tip in the meniscus and the capillary touching the wall of the centrifuge tube. Liquid will be drawn up between the wall and the pipette and, by moving the latter around the walls of the tube, any adherent lipoprotein can be washed down and recovered. At this stage, the presence of bubbles will probably make it impossible to read the position of the meniscus in the flask. Tubular flasks of 1 or 2 ml capacity may then be lightly centrifuged; alternatively the bubbles can be removed by sucking them into the pipette and gently expelling the liquid with the tip of the pipette touching the side of the flask some distance above the meniscus. Finally, fill the flask up to the mark with the infranatant solution from the centrifuge tube and transfer the contents to a labelled vial.

This technique works well when the lipoprotein layer is essentially fluid, recoveries of at least 95% being possible. But, if the concentration of VLDL in the sample is high, a layer with the consistency of vaseline may be formed which is difficult to pipette. Under these circumstances, use a pipette with a slightly larger bore than normal and scoop off as much as possible of the greasy layer with the side of the tip before starting suction. It is sometimes helpful to allow the tube to stand a while before beginning the recovery, to allow the layer to become less viscous. It is often necessary to centrifuge samples of this kind a second time and the volume in which the lipoprotein can be recovered is inevitably increased to 1.5 or 2 ml. If a thick, greasy layer is formed during the bulk preparation of lipoproteins, it can be removed with a spatula and the more fluid material beneath it taken off with the pipette.

If the centrifugation was performed in Quick-Seal tubes, these should be opened in the following way. Support each tube in a heavy rack such as that supplied by Beckman for sealing the tubes and cut away the spigot, together with some of the top of the tube, with a dissecting knife fitted with a short *concave* blade. The hole should be made large enough to allow the sample to be aspirated as described above. Note that the hole may need enlargement to allow the walls of the tube to be washed adequately.

Some workers have used a syringe and needle instead of a pasteur pipette to aspirate the supernatant lipoproteins after centrifugation. In practice, this technique is more clumsy and difficult to control than the one we have described. It should only be used for the non-quantitative collection of bulk preparations.

(b) By tube slicing

This procedure requires less skill than the aspiration method and can be essentially quantitative, but it leads to considerable dilution of the supernatant lipoprotein layer. It is best suited to the preparation of large amounts of lipoprotein; do not use it to recover lipoproteins that are to be determined by refractometric methods, or for experiments with radioactive lipoproteins.

Note. It is desirable, though perhaps not essential, to use tube caps of the kind that have a central filling hole that is closed with a screw plug.

Remove the tube from the rotor as described above but do not loosen the cap. Insert the tube into the slicer to an extent that will be determined by the depth to which the lipoprotein layer occupies the tube. Under normal circumstances it should suffice to have the bottom of the cap about 1.0 cm above the blade of the slicer. The dent that is often found in the side of the tube should face away from the point of the blade. Remove the plug from the filling hole. Cover the hole with a finger and slice the tube with a single smooth stroke. Use a syringe to withdraw the lipoprotein fraction as completely as possible through the hole. If it is to be used for quantitative estimations, transfer it to a small volumetric flask. Replace the screw plug and remove the cap, together with the top portion of the tube, from the slicer. With the syringe, recover the residual lipoprotein from the cap and the slicer by several washings with small amounts of 0.15 M NaCl. Add the washings to the volumetric flask and make the contents up to volume.

If the material in the bottom of the tube is to be retained, withdraw the slicer blade slightly, remove about 1 ml of solution with a Pasteur pipette and transfer it to a flask. Complete the withdrawal of the blade and remove the bottom portion of the centrifuge tube from the slicer. Mix up the contents of the tube, making sure that no sedimented material remains stuck to the wall. Transfer the mixture quantitatively to the flask with 0.15 M NaCl.

If Quick-Seal tubes were used, the slicing procedure must be slightly modified, as follows. Position the tube in the slicer so that the blade will strike through the clear region just below the lipoproteins. Thrust the blade firmly home and then, with a stout syringe needle, make two holes in the top of the tube, into the air bubble. With a syringe fitted with a second needle, withdraw the lipoproteins through one of these holes. If the lipoprotein layer extends beyond the reach of the needle, make a second hole further down the side of the tube.

It is important to understand that the success of the tube slicing technique depends in large measure on the condition of the slicing equipment. Keep the blade sharp and its tip in good condition, and do not use sealing rings that are damaged. The alloy of which the slicer is made is easily corroded by strong salt solutions and it is important to wash all parts of the apparatus thoroughly after use. Failure to do this may make it difficult, or even impossible to operate the blade. Judicious lubrication will also ease the working of the equipment and help to protect it against corrosion.

2.1.1.8. Specimen procedure for serial centrifugation As an example, we shall describe a representative procedure for the preparation of VLDL, LDL and HDL from human plasma. In the interest of generality, we have made the following assumptions when calculating the densities of the specific diluents required:

(1) That the centrifugation is to be performed at the same temperature as that at which the density is measured, i.e. at 20 °C.

(2) That the partial solvent volume of the plasma is 0.94. Note that this factor is irrelevant at the first stage of the preparation (separation of VLDL) and that the value will be changed by dilution of the plasma when the density is adjusted. Moreover, the value that applies to the infranatant after centrifugation will depend on the volume of solution in which the lipoproteins are removed. In the following example, the 'layered-load' technique is adopted and it will be assumed that the lipoproteins are recovered in a volume equal to that of the over-layer. Under these circumstances, the partial volume factor is dependent only on the amount of the specific diluent added. If this volume is large, e.g. twice that of the plasma, the partial volume becomes negligibly different from unity for subsequent adjustments.

(3) Re-distribution of salt will be ignored since this will depend on the rotor used and the speed and duration of the centrifugation.

Appropriate corrections for all these factors should be made in an actual experiment.

We shall also assume, in the interest of productivity, that the volumes of the diluents used to adjust the solvent density are to be kept small. Sodium bromide solutions of high density will therefore be used.

Materials. Note that all the following specific diluents should contain EDTA, NaN_3 and Thimerosal as detailed in Section 2.1.1.1.

0.196 molal NaCl solution, d = 1.0063 g/ml at 20 °C
0.844 molal NaBr solution, d = 1.063 g/ml at 20 °C
2.973 molal NaBr solution, d = 1.21 g/ml at 20 °C
4.778 molal NaBr solution, d = 1.3199 g/ml at 20 °C
7.593 molal NaBr solution, d = 1.4795 g/ml at 20 °C

Procedure. We shall assume that the plasma to be used is a fasting sample from a normal individual and that it contains effectively no chylomicrons. If this is not the case, the chylomicrons should be removed by a preliminary centrifugation of the native plasma under conditions that amount to a total of about 2.25×10^6 g·min. When the quantity of chylomicrons is very large, it may be necessary to re-suspend the infranatant and to centrifuge it a second time to ensure that the recovery is essentially complete. If the particles are to be retained as a separate fraction, they should be treated as described in Section 2.1.1.9.

Stage 1. Isolation of VLDL

Centrifuge native plasma or serum Because the very large lipoprotein particles of low density will float very rapidly, even in comparatively weak centrifugal fields, there is a temptation to use a low speed rotor of large capacity for the isolation of VLDL (e.g. the Beckman Type 35). But it must be remembered that the smaller VLDL particles have a density little more that that of the solvent and therefore float very slowly. To minimise the cross-contamination of the VLDL and LDL fractions it is therefore desirable to centrifuge at the highest practicable speed for the longest practicable time, for example, for 18–24 hours in the Beckman 50.2Ti rotor at 50 000 rev/min. However, because there is some evidence that very high centrifugal fields may lead to increased degradation of the lipoproteins, it may be prudent to refrain from using the fastest rotors at their maximum speeds, e.g. to restrict the use of the Beckman 60Ti to 55 000 rev/min. Unfortunately, there are few data from which to estimate the maximum acceptable rotor speed.

Having selected a rotor that is appropriate to the size of the preparation to be undertaken fill the tubes to about two-thirds capacity with plasma i.e. 20–25 ml in a 35 ml tube. Float on this a layer of 0.196 molal NaCl to fill each tube to just below the cap. Seal the tubes, centrifuge them overnight and then recover the lipoproteins as described in Section 2.1.1.7. If a large amount of VLDL is present, it can form a thick greasy layer at the top of the tube, that is difficult to remove with a pipette. It may then be necessary to recover as much as is possible, make up the contents of the tube to the original volume with 0.196 molal NaCl, and to repeat the centrifugation. (When tube-slicing is used, this second centrifugation will be less often needed.) When the VLDL have been recovered as completely as possible, retain the residue at the bottom of the centrifuge tubes for the preparation of LDL and HDL.

The crude VLDL must be washed to free them from residual plasma proteins. Re-suspend the pooled preparation, thoroughly but gently, in 0.196 molal NaCl. The final volume of this solution will depend on the amount of VLDL that has been isolated. If the concentration in the source plasma was normal, the volume of the wash solution need amount to only about 25% of that of the plasma. By contrast, if the source was rich in VLDL, it may be necessary to wash in a volume as great as, or even greater than the original. Distribute this solution between an appropriate number of centrifuge tubes and run overnight, as before. Recover the VLDL from the top of each tube, pool them and continue this washing process until the preparation satisfies the criteria of purity that have been chosen, e.g. until they no longer react with antiserum to serum albumin.

Stage 2. Isolation of LDL

Centrifuge this residue from Stage 1 at a solvent density of 1.063 g/ml (at 20 °C) As with all lipoprotein preparations that are made by serial centrifugation, the recovery of the LDL particles that have a density close to that of the solvent is never complete. The centrifuging times that are quoted in Table 2.7 are a compromise that will generally result in an acceptable recovery of the major part of these slow-moving particles, which will be only marginally improved by a longer and more expensive centrifugation. Nonetheless, it must be noted that these data refer to an experiment at 20 °C; for every 5 °C below this temperature, the duration of the run should be increased by about 12%.

Pool the infranatant solutions that result from Stage 1, making sure that the sticky residue at the bottom of the tubes is fully dissolved and is washed into the pool with 0.196 molal NaCl. Then make the volume up to a convenient total with 0.196 molal NaCl. To every 5 ml of this mixture, add 1 ml of 4.778 molal NaBr solution and mix thoroughly.

Fill the centrifuge tubes to two-thirds capacity with this solution, which now has a solvent density of 1.063 g/ml, and float 0.844 molal NaBr on the surface until the tubes are full. Cap the tubes carefully, to avoid disturbing the floating layer, and centrifuge them for 18 to 24 hours at the maximum speed of the rotor. Finally, recover the floating LDL according to Section 2.1.1.7, pool them, and retain the infranatant residues for the isolation of HDL.

The crude LDL preparation must be washed several times by centrifugation to remove the contaminating plasma proteins. Dilute the pool with 0.844 molal NaBr to a volume of 50 to 100% of that of the original plasma, according to the concentration of the lipoprotein, and centrifuge under the original conditions. This procedure must be repeated until the required degree of purity is achieved (Ch. 3 et seq.). The final product may then be dialysed against several changes of 0.196 molal NaCl solution in a stoppered flask that is filled to the brim, and then concentrated as necessary (Appendix 2).

Stage 3. Isolation of HDL

Centrifuge the residue from Stage 2 at a solvent density of 1.21 g/ml (at 20 °C) Because a large proportion of the HDL have an apparent flotation rate of less than 3.5 Svedbergs, a long centrifugation is necessary if a good recovery of this important fraction is to be obtained. Moreover, the sedimentation rate of plasma albumin is significantly reduced by the high density and high viscosity of the solvent. A prolonged centrifugation is therefore important if the top half of the centrifuge tube is to be adequately cleared of protein. As Table 2.7 shows, a minimum of 24 hours centrifugation will be needed, even at 20 °C in the most efficient rotor. The times shown must be increased by about 9% for each 5 °C by which the temperature of the run is reduced.

Thoroughly mix the infranatant residues from Stage 2, to ensure that the syrupy deposit at the bottom of the tubes is dissolved. Pool the contents and make up to a convenient volume with 0.844 molal NaBr solution. To every 2 ml of this mixture, add 1 ml of 7.593 molal

NaBr and mix well. Fill the centrifuge tubes with this solution to two-thirds capacity and then complete the process by carefully floating 2.973 molal NaBr solution on the surface. Cap the tubes and then centrifuge at the highest attainable speed (Table 2.7). After the appropriate time, recover the lipoproteins, pool them and wash by centrifugation in 2.973 molal NaBr until purity is established. Do not continue this process unnecessarily however, because there is evidence that the HDL are prone to undergo disproportionation on repeated centrifugation in high field strenghts.

In describing this procedure, we have assumed that the volume of the preparation will occupy several large centrifuge tubes. However, a small scale preparation may use only one or two small tubes and it is then possible to design the experiment in such a way that the plasma is never removed from its centrifuge tube. As each lipoprotein fraction is removed, the infranatant is stirred thoroughly in the tube, and the required volume of the next diluent added directly. This technique requires the use of capped tubes which must be very carefully selected for the absence of air bubbles, crazing etc., if they are to withstand the repeated centrifugations. With this precaution however, failures are unusual.

2.1.1.9. Subfractionation of chylomicrons and VLDL Chylomicrons, defined as particles of S_f greater than 400 svedbergs in NaCl solution of density 1.063 g/ml, can be recovered from plasma as described in Section 2.1.1.8. They should then be washed by re-suspending them in 0.196 molal NaCl solution and repeating the centrifugation. This washing process is often continued until the chylomicrons no longer give a precipitin reaction with antiserum to serum albumin. However, because repeated washings appear to 'destabilise' the particles and to cause some of them to disproportionate, the process must be performed with due care and attention.

As we briefly mentioned in Section 2.1.1, neither serial centrifugation in solvents that are increased step-wise in density, nor density gradient centrifugation can be used to sub-fractionate chylomicrons or VLDL. For this purpose, it is necessary to use a method that resolves them according to their different flotation rates i.e. by using a combination of mass and hydrated density as the discriminant. For preparative purposes a procedure such as that of Gustafson et al. (1965) is probably the most suitable but, because of the large amounts of sample that are used, its resolution is comparatively low. The technique developed by Lindgren (1975) is superior in this respect but is less practicable for preparative purposes because it has a smaller capacity and is far more cumbersome to set up. However, it probably has potential as an analytical method and it is described in this context in Section 4.3.

In the following section we shall describe a procedure based on that of Gustafson et al. (1965) which, in its original form, made use of the Beckman 40 rotor. If it is necessary to substitute a different fixed angle rotor, the speed and/or the duration of the centrifugation must be adjusted accordingly. The appropriate values can be determined by making use of the '*k*' value of the rotor, which is defined by:

$$k = \frac{\ln r_{max} - \ln r_{min}}{3600\ \omega^2} \times 10^{13}$$

The time taken to centrifuge all the particles of a given *F* value to the top of the tube (in hours) is then given by *k/F*. Note that the 'top' of the tube in this context is not equal to the minimum radius quoted for the rotor but will be at a point further down the tube which will be defined by the volume of lipoprotein solution that will be collected and also by the fact that the tube is never filled to the brim. Despite the greatest care in filling the tubes, there is always an uncertainty as to the exact position of the 'top' of the tube, which in turn degrades the precision with which the fractions can be defined. In the case of the 40 rotor, the effective *k* is probably about 80 at full speed. At a lower speed this will be increased in proportion to the square of the ratio of the maximum speed to the actual speed.

When making these calculations it is critically important to remember that the value of *F* must relate to the conditions in the preparative tube. It is not acceptable to use a value which is derived from the conventional scale because this is defined by analytical centrifugation at 26 °C in NaCl solution of density 1.063 g/ml. Thus, values that are referred to this scale (S_f) must first be corrected to the conditions of the preparative experiment by means of the equation:

$$F = S_f \frac{(1.0063 - d)\ 1.0173}{(1.063 - d)\ \eta}$$

where d is the density of the lipoprotein particle (cf. Table 4.2) and η is the viscosity of a NaCl solution of density 1.006 g/ml at the temperature of the preparative ultracentrifuge. The importance of making this correction can be seen from the fact that a particle of S_f 400 has a flotation rate of about 220 svedbergs in NaCl of density 1.006 g/ml at 26 °C, and only about 140 svedbergs at 4 °C.

Reagents. 0.196 molal NaCl solution, d = 1.0063 g/ml at 20 °C containing EDTA, NaN_3 and Thimerosal as detailed in Section 2.1.1.1.

Procedure. For the purposes of this example, we shall assume that the preparative centrifuge will be operated at a temperature of 20 °C, and that sub-fractions of chylomicrons and VLDL are to be prepared from plasma.

(1) Isolation of particles of S_f greater than 5000 Put 5 ml of fresh plasma into a 76 mm × 16 mm centrifuge tube and layer 5 ml of 0.196 molal NaCl on the surface. Cap the tube and centrifuge it in the Beckman Type 40 rotor at 12000 rev/min for 17 min (including the equivalent acceleration and deceleration times) at 20 °C. This is equivalent to 1.62 g·min at the mean radius of the rotor and will concentrate particles of S_f greater than about 5000 into the top 1 ml of the supernatant NaCl solution. Remove this layer (Section 2.1.1.7), resuspend the particles in 9 ml of 0.196 molal NaCl solution and repeat the centrifugation under the same conditions. Recover the chylomicrons and repeat the washing procedure twice more.

(2) Isolation of the S_f 400–5000 particles Thoroughly re-mix the infranatant from (1), make the total volume up to 10 ml with 0.196 molal NaCl and devide it equally between two centrifuge tubes. Float 5 ml of 0.196 molal NaCl onto each of these samples and centrifuge them at 20000 rev/min for 87 equivalent minutes (about 2.3 × 10^6

g·min) at 20 °C. Recover the supernatant 1 ml fractions from each tube, combine them and wash three times by centrifugation in 0.196 molal NaCl.

(3) Isolation of the S_f 100–400 particles Combine the re-mixed infranatants from (2) and adjust the total volume to 20 ml. Distribute this equally between 4 centrifuge tubes and layer each one with 5 ml of 0.196 molal NaCl. Centrifuge at 35 000 rev/min for 120 equivalent minutes (about 9.7×10^6 g·min) at 20 °C. Recover the top 1 ml fractions, combine them and wash three times by re-centrifugation.

(4) Isolation of the S_f 50–100 particles By analogy with the foregoing, distribute the infranatants from (3) between eight tubes. Centrifuge at 40 000 rev/min for 4.5 equivalent hours (about 28.5×10^6 g·min) at 20 °C. Wash the combined lipoprotein fractions by re-centrifugation as before.

(5) Isolation of the S_f 20–50 particles Replace the fractions removed from each of the tubes in (4) by the same volume of 0.196 molal NaCl solution and centrifuge for 24 hours at 40 000 rev/min at 20 °C. Recover the supernatant lipoprotein fractions, combine them and wash three times by centrifugation for 24 hours at *d* 1.0063 g/ml.

Note that the resolution of this process is low because slow-moving particles that start from the top of the sample layer will reach the top of the tube at the same time as faster particles that travel from the bottom of the tube. There is consequently a considerable contamination of each fraction with smaller particles and the apparent distribution of material between the fractions is distorted. Moreover, if the whole of the infranatant from one stage is not carried over to the next, the yields are substantially reduced and the distortion of the apparent distribution is even greater. The contamination of the fractions is reduced by the washing process, which is the more effective if it is done by the layering technique. However, it is almost impossible to layer 0.196 molal NaCl onto the recovered lipoprotein fractions unless their density is first increased. But, if this is done, a macromolecular solute such as albumin or high molecular weight polyethylene glycol must be used.

2.1.1.10. Serial centrifugation at different temperatures In principle, the difference in the rates of thermal expansion of the lipoproteins and the solvent solution can be used to sub-fractionate the lipoproteins. The technique would be subject to certain limitations, for example, because the maximum usable range of temperature is about 15–20 °C, the density limits of the fraction that is isolated cannot exceed a differential of about 0.01 g/ml. Moreover, both the resolution and recovery of the lipoproteins would suffer from the low mobility of particles that have a density close to that of the solvent. Nonetheless, the method could be of some value for the production of relatively large amounts of LDL of narrow density range.

The principle of the method can be illustrated by an example: Prepare a solution of LDL in NaCl of density 1.03 g/ml at 20 °C. Centrifuge this (Section 2.1.1.8) at 5 °C, at which temperature the density of the solvent is 1.0334 g/ml. Remove the supernatant lipoproteins which consist of the particles that have a hydrated density less than 1.0334 g/ml at 5 °C i.e. of density less than 1.0233 g/ml at 20 °C. Mix the infranatant solution, make up to the original volume with NaCl d_{20} 1.03 g/ml and centrifuge again, this time at 20 °C. The floating layer will then consist of the lipoproteins of hydrated density less than 1.03 g/ml at 20 °C i.e. the fraction of d_{20} 1.0233–1.030 g/ml.

2.1.2. Density gradient centrifugation

If the plasma lipoproteins were distributed between a few discrete classes, in each of which the particles were nearly of the same density, it would be possible to separate them completely by centrifugation to isopycnic equilibrium on a density gradient. However, since the distribution is actually continuous over a wide range of densities, this is not strictly possible. No matter how well the major lipoprotein classes are separated into bands, there will be some low concentration of material between them that can only be sub-divided on a purely arbitrary basis. Nonetheless, in practice, the conventionally defined VLDL, LDL and HDL can be adequately separated in a long tube. To resolve HDL_2 and HDL_3, on the other hand, requires a gradient that would cramp the VLDL and LDL into an inconveniently short space at the top of the tube. Moreover, if plasma is the starting material, it may be difficult to resolve the lipoprotein classes adequately and at the same time to separate the HDL completely from the plasma proteins. It is also a disadvantage of the method that it cannot be used to fractionate VLDL or chylomicrons, which have a density less than that of water. However, isopycnic centrifugation has the merit that it should, in theory at least, lead to less cross-contamination of the fractions than serial centrifugation. Unfortunately, few studies have been made of the 'purity' of the fractions obtained, or of the precision with which they can be defined.

Two practical considerations arise in connection with the loading of the gradient. If the load is small, it may be difficult to detect the positions of the lipoproteins in the tube, and the amounts of the separate fractions may be too small to study by the available physical or chemical techniques. The usual response to this challenge is to increase the load until the separated lipoproteins from visible bands. But, if this is carried too far, the density gradient itself may be significantly distorted which will lead to errors in the exact identification of the density of the fractions. In practice, a careful compromise on the size of the load is necessary. Unfortunately, there seem to be no theoretical analyses of this problem that would help to make this decision.

Many different gradient systems have been described. Those that are formed by the continuous mixture of two different solutions in a 'gradient maker' are well suited to the analysis of isolated lipoprotein fractions, e.g. the sub-fractionation of radioactively labelled LDL. Such gradients are usually linear over the greater part of their length, although there will be some curvature at the top and bottom due to the re-distribution of salt during the centrifugation. For the separation of more complex mixtures, for example the isolation of all three major lipoprotein classes from plasma, it is desirable for the gradient to change along its length in such a way as to cause the different lipoproteins to equilibrate at the most advantageous positions in the tube. Gradients of this kind are usually built up by the superimposition of several solutions of decreasing density. At the start therefore, the gradient has several plateaux on which lipoproteins of the appropriate density could sit. It seems likely that these plateaux may persist in the form of inflections in the gradient for long enough for bands of lipoprotein to accumulate thereon. The final resolution of these bands would then take place during the subsequent diffusion of the inflections. It is most important therefore that the centrifugation should be continued until a continuous, smooth gradient has been established. If this is not done, resolution will be seriously diminished, or could even lead to the spurious isolation of 'new' lipoproteins. One of the most widely used of the non-linear gradients was designed by Redgrave et al. (1975). This is of fairly general application and can easily be modified to suit special requirements. One modification has been proposed by Chapman et al. (1981) with the object of improving the resolution of the HDL when all the lipoprotein classes are to be isolated from plasma in a single run. This gain is partially offset by the fact that the VLDL layer at the top of the tube includes substances of density as high as 1.016 g/ml, i.e. it overlaps the conventional LDL density class. Nonetheless, this is probably the best characterised gradient system at present available for the simultaneous isolation of LDL and HDL.

No matter what method is used to form the gradient, two factors must be decided when the experiment is designed:

(*1*) *The temperature.* It should be remembered that the gradient must be unloaded and its profile determined at the temperature at which it was centrifuged. Suppose the run is carried out at 5 °C, and that the density profile is then determined by measurements at 20 °C. Since the different parts of the gradient will not have the same temperature coefficient of expansion, the measured profile will not be the same as that prevailing during the centrifugation. Moreover, it cannot be argued that the density measured at 20 °C would also be that of the equilibrated lipoproteins at 20 °C, because the solvent and the lipoproteins do not generally expand at the same rate. Consequently, it is only at 5 °C that the density of the particle and its solvent would be the same. The differential expansion of the lipoproteins and the solvent that will occur as the system warms up, will also tend to cause it to become unstable. In practice however, this does not seem to lead to a significant loss of resolution provided the gradients are unloaded without delay. In general, the technically easiest course is to perform the centrifugation at a temperature close to room temperature.

(*2*) *Duration of the run.* It is important that the lipoproteins should reach isopycnic equilibrium at the relevant point on the density gradient and that the centrifugation should continue for long enough to achieve this. Because the viscosity of the solvent is greater at low temperatures, the duration of the run will be dependent on the temperature at which it is carried out.

However, a distinction must be made between the equilibrium between the lipoprotein particle and the solvent, and the establishment of an equilibrium gradient. To achieve the latter condition may take a much longer time than is needed to cause the lipoproteins to reach their equilibrium positions, and small changes in the positions of the lipoproteins during the latter stages of centrifugation probably reflect alterations in the shape of the gradient as it tends to approach the equilibrium distribution of salt. Whether the original gradient was 'continuous' or 'discontinuous', it is unlikely to approximate to the equilibrium gradient and the re-adjustment towards this condition will continue for a considerable time. Since the system may therefore be operating under dynamic conditions, it is essential that the acceleration time, the period of centrifugation at full speed and the deceleration time are always exactly the same for both the experimental and the calibration tubes. As mentioned above, it is also important that gradients made by the 'layering' method are centrifuged for long enough to establish a smooth concentration gradient that is free from inflections. The time required to do this must be established before the experiment proper is begun, by centrifuging gradients to which a dummy load has been applied.

Lipoproteins that are centrifuged on a density gradient are subjected to a substantial hydrostatic pressure that varies with their position in the gradient. If the compressibilities of the lipoproteins and the solvent are different, an equilibrium can be established between them which is isopycnic only during the centrifugation. Unfortunately, little is known about the compressibility either of the lipoproteins or of the relevant solutions. The significance of the error introduced by equating the density of the lipoprotein with that of the solvent under bench conditions after the centrifugation is therefore difficult to assess.

The gradients used for the fractionation of lipoproteins are often made from solutions of NaCl or NaBr that are prepared as described in Section 2.1.1.1. Sodium chloride can be used for gradients in which the maximum density does not exceed 1.18 g/ml, i.e. for work on low-density lipoproteins. Sodium bromide must be used if the maximum density is to be sufficient for work with HDL, i.e. up to about 1.23 g/ml. Buffered sucrose solutions have also been used but have the disadvantages that they are more viscous than salt solutions, and undergo a greater re-distribution of solute during the centrifugation. Because of the higher viscosity, the centrifugation will be rather longer than is necessary in salt solutions but, on the other hand, the gradient may be somewhat more stable.

Before passing to a more detailed consideration of the density gradient technique, it is worth to remarking that lipoproteins which are isolated in this way are, per se, no less likely to be contaminated with plasma proteins than those prepared by serial centrifugation. In principle, there seems to be no reason to suppose that the gradient technique will be a more effective way of removing adsorbed plasma proteins than a conventional centrifugation. If it is considered necessary to centrifuge the products of the latter process several times to purify them, it is likely that lipoproteins from a gradient will require similar treatment. Moreover, if VLDL, LDL and HDL are all isolated from plasma on a single gradient, the relatively close proximity of the HDL to the plasma protein residue may lead to their gross contamination. Fractions that are prepared from plasma by gradient centrifugation should therefore be submitted to the same purification procedures that are used on fractions made by serial centrifugation.

2.1.2.1. Choice of rotor and tubes Most work with density gradients has been done with swing-out rotors, on the grounds that wall effects could lead to significant errors in the conventional fixed angle variety, and that the re-orientation that take place in the latter must lead to some loss of resolution. Recent experience with vertical-tube rotors suggests that this last difficulty may have been exaggerated but, at the time of writing, the advantages claimed for this type of rotor have yet to be fully confirmed with lipoproteins (but see Chung et al., 1980). Moreover, the full benefits of the vertical-tube rotor can only be obtained by the use of a specially designed or modified centrifuge.

Not only does the swinging-bucket rotor avoid the problem of re-orientation, but some examples can offer a longer tube length than is available in fixed-angle rotors of similar speed rating. As with the fixed-angle type, the most widely used are those made by Beckman-Spinco and we shall assume the use of these. However, the equivalent products of other manufacturers can also be used with only minor modifications to the technique. The Spinco SW-41Ti rotor has been used in many laboratories for the fractionation of lipoproteins because it has the capacity to centrifuge a long tube (89 mm; 3.5 in.) in a powerful field (286 000 g_{max}). The SW-40Ti rotor uses a slightly longer tube (95 mm), albeit at a slightly lower speed (285 000 g_{max}), a combination which may offer advantages in experiments for which the length of the gradient is critical. An even longer tube can be used in the SW-27.1 rotor, but the maximum g force generated by this is only 135 000 and the centrifugation must therefore be very protracted by comparison with those in the titanium rotors. Rotors capable of higher speeds can establish isopycnic equilibrium more quickly than those listed above, but use shorter tubes (50–60 mm). It is therefore necessary to use either a shorter gradient, or a steeper one, with consequent limitations on its capacity to separate lipoproteins.

The choice of tubes is normally between polyallomer and cellulose nitrate, both of which can be used satisfactorily. However, the transparent cellulose tubes allow the lipoprotein bands to be seen more easily, and may also somewhat easier to puncture than the polyallomer variety. The latter, on the other hand, are not wetted by the solvent and may therefore drain with less disturbance of the bands.

2.1.2.2. Making the gradient The density of each lipoprotein fraction that is isolated on the gradient is estimated by measuring the density of the solvent at the same point in a dummy gradient that has been centrifuged at the same time as the one carrying the load. The gradients must therefore be reproduced with high precision and the greatest care taken to avoid mechanical or thermal disturbance. It will be evident that a mechanical gradient former is essential for the production of a smooth, continuous gradient and many commercial instruments are available. Most of these can be used satisfactorily, provided that they can handle the relatively small quantities of solution required. However, the simple and inexpensive device illustrated in Fig. 2.3 will produce perfectly adequate linear gradients, provided that the following precautions, which may also apply to other instruments, are observed:

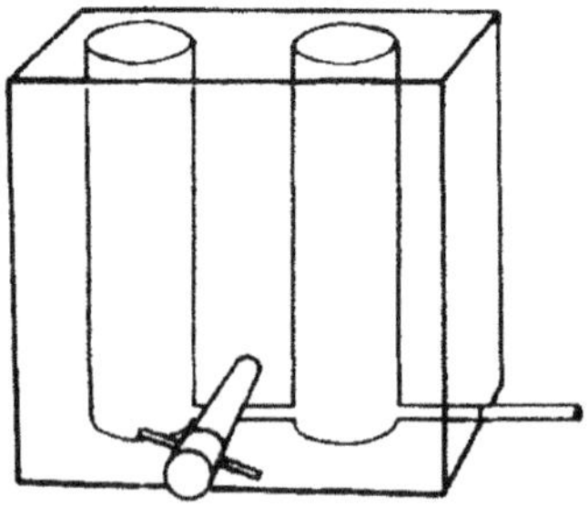

Fig. 2.3. This simple gradient former can be made by drilling two holes (ca. 12 mm diam.) into a block of plastic of the appropriate size. The right-hand chamber is the mixing compartment and must be fitted with a small stirrer. The method of use is detailed in Section 2.1.2.2.

(1) The tube that conveys the gradient solution into the centrifuge tube should be of very fine bore and, like the bore of the stop-cock that separates the two reservoirs, should be filled with the dense solution before the operation begins.

(2) The volumes of the solutions in the chambers of the mixer should be inversely proportional to their densities, otherwise there is a flow between the chambers when the connecting cock is opened.

(3) The exit tube should touch the side of the centrifuge tube and be kept a short distance above the meniscus. This distance must be carefully adjusted to reduce to a minimum any disturbance of the meniscus. If it is either too great or too small, the incoming solution 'jets' into the surface. If the centrifuge tube is mounted on a jack, the position of the meniscus relative to the end of the inlet tube can be easily and smoothly maintained. It is also advantageous to keep the pressure head that drives the flow from the gradient maker as low as possible. Note that this assumes the use of cellulose nitrate tubes, which are wettable. If non-wettable polyallomer tubes are used, the tip of the filling tube must be kept just in the meniscus. Since this is a difficult operation it may better, in this case, to load through a fine metal or glass tube that dips to the bottom of the centrifuge tube. The solution of lowest density is then loaded first. This procedure may reduce mixing while the centrifuge tube is filled, but it is difficult to avoid some stirring as the 'dip-leg' is removed.

The preparation of a gradient in this way can be exemplified by the formation of a linear gradient of buffered sucrose from about *d* 1.01 g/ml to about 1.08 g/ml in an 89 mm × 14 mm centrifuge tube (3.5 in. × 0.563 in.).

The following solutions are required. *Phosphate buffer containing:* 0.1 M NaCl, 0.025 M Na_2HPO_4, 1 mM EDTA, 1.5 mM NaN_3, the mixture adjusted to pH 7.5. '*Heavy solution*': phosphate buffer + approximately 20.5% (w/v) of sucrose to give a final density of 1.080 ± 0.0005 g/ml. '*Ligth solution*': phosphate buffer + approximately 1% (w/v) sucrose to give a density of 1.0093 ± 0.0005 g/ml.

Procedure: Put 0.132 g of sucrose into the bottom of a centrifuge tube. Add 0.5 ml of the solution to be analysed (or 0.5 ml of NaCl solution *d* 1.006 g/ml for a control gradient) and carefully dissolve the sucrose. Fill the bore of the stop-cock that connects the two reservoirs of the gradient maker (Fig. 2.3), and the exit tube, with 'heavy' solution. Ensure that no bubbles remain, clamp the exit tube and close the stop-cock. Then put 6 ml of 'heavy' solution into the mixing chamber and 6.42 ml of 'light' solution into the other reservoir. Mount a small stirrer in the mixing chamber, open the stop-cock and the exit tube and allow the solution to run slowly through a fine plastic tube (diam. 0.5 mm) down the side of the centrifuge tube, as described above.

When the gradient is made by the 'layering' technique, it is vital that the successive solutions are superimposed without mixing. If this can be achieved, the subsequent diffusion of the boundaries takes place in a system that starts with the same configuration in each tube. If mixing occurs before the centrifugation is begun, it will be random in character and will result in a gradient of slightly different shape in each tube is prepared. The best way of loading the tube is to add each successive solution slowly through a fine tube by means of a smooth-acting pump that does not give a pulsatile flow. If the loading has to be done manually, a pipette with a fine, drawn-out tip that is bent through 90° should be used. Additional control of the flow may be had by connecting a short length of fine capillary tube to the upper end of the pipette to restrict the flow of air. It is advisable to hold the centrifuge tube in a clamp, and some authorities believe that it should be inclined to the vertical to simplify the manipulation of the pipette, the tip of which should be allowed to touch the side of the tube just above the meniscus. This modification probably makes it easier to control the filling process but introduces the hazard of mechanical disturbances as the tube is re-orientated.

As an example of the construction of a gradient by the discontinuous method, we shall describe the technique of Chapman et al. (1981). This will resolve the plasma lipoproteins into four fractions in a single run, viz. 'VLDL' ($d < 1.016$ g/ml), LDL (*d* 1.028–1.050 g/ml), HDL_2 (*d* 1.066–1.10 g/ml) and HDL_3 (*d* 1.10–1.153 g/ml) on a gradient of NaCl/NaBr that ranges from

1.016 to about 1.23 g/ml.

The following solutions are required (Sections 2.1.1.1 and 2.1.1.2).

(A) 11.46 g NaCl + 0.372 g EDTA + 0.13 g NaN_3 + 1.0 kg H_2O (d = 1.006 g/ml).
(B) 34.11 g NaBr + 100 ml of solution A. Final density, 1.24 g/ml.
(C) 29.63 g NaBr + 100 ml of solution A. Final density, 1.21 g/ml.
(D) 7.60 g NaBr + 100 ml of solution A. Final density, 1.063 g/ml.
(E) 1.66 g NaBr + 100 ml of solution A. Final density, 1.019 g/ml.

Check the density of each solution according to Section 2.1.1.3, and adjust as necessary.

Procedure: Adjust the solvent density of the plasma to 1.21 g/ml by the addition of 0.276 g NaBr to each ml. Put 2 ml of solution B into the bottom of an 89 mm × 14 mm centrifuge tube and carefully layer onto it 3 ml of the plasma/NaBr mixture. This is a particularly delicate operation because the total density of the adjusted plasma is very close to that of the salt solution beneath it. Next, layer 2 ml of solution C onto the plasma, followed successively by 2.5 ml of solution D and 3 ml of solution A, in that order. Transfer the tubes to their buckets and instal them in the rotor with the greatest care to avoid mechanical disturbance. Finally, begin the centrifugation as soon as is practicable. The originators of this system continued the centrifugation for 48 hours at 40000 rev/min, at 15 °C. If a lower temperature is used, a longer run may be needed and this should be checked by trial. In the dummy tubes that are used to establish the shape of the gradient, 3 ml of solution E are substituted for the plasma/NaBr mixture.

2.1.2.3. Loading and unloading the rotor The precautions to be taken when using fixed-angle rotors (Section 2.1.1.6) also apply to swinging-bucket rotors. In addition however, greater care must be taken not to squeeze the loaded tubes when handling them since, as they have no caps, they can easily be distorted. The lack of cap also means that the tubes must be withdrawn from the buckets with the help of forceps. Scrupulous care must be taken to ensure that this does not disturb the surface of the gradient (where lipoproteins may be floating), or cause scratching and bruising of the metal at the top of the bucket. Moreover, the tube must be slid smoothly from the bucket and must not pop out with a jerk as it finally comes free.

2.1.2.4. Unloading the gradient There are innumerable ways of performing this operation, none of which is entirely beyond criticism. In general, there seems to be no advantage in using highly elaborate methods. At the same time, there is little point in performing a long and expensive fractionation and then losing the potential resolution of the method by crudely collecting the fractions with a pipette or syringe. In this context, it is also advisable to keep the gradients at the temperature at which they are centrifuged, until the unloading can be commenced.

The fractionation of the gradient can begin either at the top or at the bottom of the tube. Of the two, it is technically easier to puncture the bottom of the tube with a hypodermic needle and to allow the contents to run out by gravity. However, if the needle is left in situ, the volume below its tip is stagnant and is not collected. It is therefore important to have a dense solution at the bottom of the centrifuge tube, on which even the densest lipoproteins will float. It is also vital to ensure that there is no leak where the needle penetrates the tube, since this will either lead to a loss of liquid or to the entry of air bubbles that will destroy the gradient. The equipment needed for this method is shown in Fig. 2.4 and can be prepared as follows:

Snap the socket off a 1.5 inch, 19 gauge hypodermic needle (score the needle with a file first) and attach a length of fine-bore plastic tubing. Then fit the needle with a collar of thick-walled plastic capillary tube so that it can be gripped in a small clamp. Cut off the end of a rubber teat to form an approximately hemispherical cap of such a size that it will fit closely to the base of the centrifuge tube being used. Lightly smear the inside of this cap with grease and press it carefully into place. Assemble the apparatus as shown, with the centrifuge tube lightly grasped in the clamp and centred above the needle. To puncture the tube, press it onto the needle with a cork until it meets the collar surrounding the needle. Tubes that are punctured in this way rarely leak and the rubber cap may be used several times.

When the gradient is fractionated from the top, it is not only

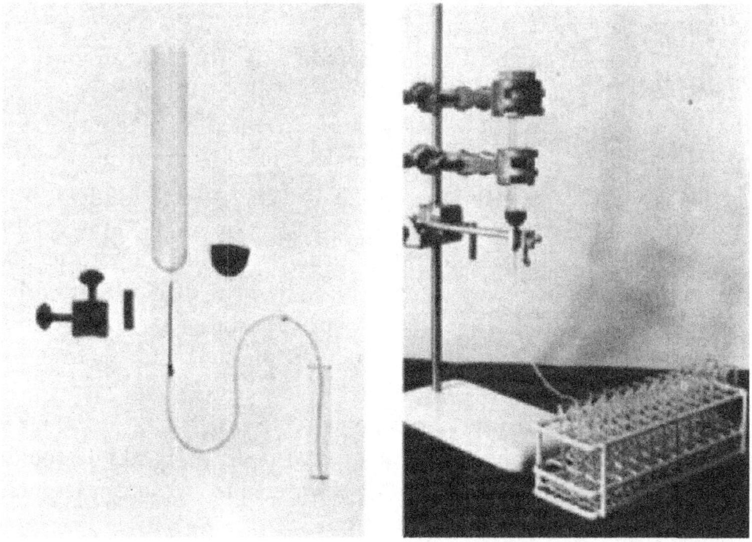

Fig. 2.4. The assembly of the equipment needed for recovering density gradients by the method described in Section 2.1.2.4.

necessary to puncture the bottom of the tube, with the attendant hazards, but also to construct a special cap through which the fractions are syphoned off. This introduces an element of engineering complexity which, although it tends to discourage the average bench-worker from making his own apparatus, is attractive to the manufacturers of equipment and many commercially available fractionators work on this principle. A sophisticated example is the ISCO Model 185 (Instrumental Specialties Co.), in which a very dense, inert liquid (Fluorinert FC-40, *d* 1.85 g/ml; manufactured by the 3M Co.) is slowly and continuously pumped into the bottom of the centrifuge tube. The displaced gradient passes through a fine tube in the special cap and is conducted to a small fraction collector.

2.1.2.5. The profile of the gradient If the control gradient is collected in fractions that exceed about 0.8 ml in volume, their densities can be measured directly and with great accuracy by means of the density meter (Section 2.1.1.3). Alternatively, the density of smaller fractions can be determined indirectly from measurements of their refractive index (*n*). With an Abbe type refractometer, this can be done on one drop of solution. However, it is then advisable that the gradient be formed from a single solute, or from mixtures of solutes that will re-distribute in the centrifugal field in a similar way, to avoid the uncertainties that may otherwise arise from differential re-distribution. This consideration usually becomes important only when plasma is being centrifuged on a gradient of NaBr. It is then desirable to construct the gradient from salt solutions that contain the same molal concentration of NaCl as plasma, i.e. 0.196 molal (Section 2.1.1.1).

The refractometer must be capable of measuring *n* to $\pm$ 0.00002 and requires a monochromatic source of light, preferably a sodium lamp. The instrument must also be connected to a constant temperature bath that can be controlled to at least $\pm$ 0.05 °C.

A calibration is needed to relate the density and refractive index for each type of gradient. To obtain this, first make several different mixtures of the 'heavy' and 'light' solutions from which the gradient is to be prepared. Use one of the methods described in Section 2.1.1.3 to determine the density of each mixture at the temperature at which the gradient is to be centrifuged. Then measure the refractive index of each solution as follows:

Switch on the apparatus and allow it to reach thermal equilibrium.

Open the prism.

Take up 0.2–0.3 ml of solution in a pasteur pipette and deposit a bubble-free drop in the centre of the lower prism.

DO NOT TOUCH THE SURFACE OF THE PRISM WITH THE PIPETTE.

Immediately close the prism and observe the position of the boundary between the light and dark fields through the eyepiece.

Turn the scale adjustment to bring the boundary close to the cross-wires (N.B. the boundary should always approach the wires from the same direction). Then turn the adjusting head slowly and smoothly until the boundary just bisects the cross-wires and take the scale reading. To minimise the possibility of losses by evaporation, the interval between the closing of the prism and the setting of the cross-wires should be brief and should always be the same. About 1 min is usually sufficient.

Having taken the reading, clean both prisms thoroughly with distilled water from a wash-bottle wit a *polythene* jet. Dry the surfaces with a soft tissue. Take care over this; an abrasive tissue or over-vigorous rubbing may scratch the prisms.

By plotting the refractometer scale values against the corresponding density measurements, a calibration curve can be obtained that is usually linear. The two variables need not be determined at the same temperature, provided that both temperatures are constant. If either of them are changed, the calibration must be repeated.

The refractive index of each fraction that is taken from the control gradient can be measured in the same way, and its density estimated by interpolation of the calibration curve.

2.1.2.6. Rate-zonal ultracentrifugation The value of conventional gradient centrifugation for the preparation of lipoproteins is limited by the relatively small capacity of the centrifuge tubes. This can be overcome, in part at least, by the use of zonal rotors (e.g. the Beckman Ti-14 or Z-60) which have a total capacity of 600 ml or more, with the added advantage that the isolation can be completed much more rapidly than is possible in a conventional rotor. Moreover, the lipoprotein distribution can be oberved and measured concurrently with the fractionation, thus providing a much more detailed understanding than can be obtained by sequential centrifugation at fixed densities.

Despite these cogent advantages, the zonal rotor has not been widely used for isolating lipoproteins, perhaps because it requires a fairly heavy capital investment in the rotor, in equipment to fill it with a density gradient and in apparatus to measure and record the lipoprotein distribution as the rotor is unloaded. In addition, the zonal rotor can only deal with one sample at a time and it is not an appropriate instrument for small-scale preparations, although the short duration of the centrifugation (e.g. 30–45 min for VLDL, according to Danielsson et al. (1978)) allows more frequent use for the machine. There is also a considerable dilution of the lipoprotein fractions that may necessitate re-concentration (Appendix 2).

The greatest potential use for the zonal rotor would seem to be for the large scale preparation of VLDL, or for the separation of HDL_2 and HDL_3. Both Patsch et al. (1974) and Danielsson et al. (1978) have claimed that the two HD-lipoproteins can be satisfactorily isolated in this way from relatively large volumes of plasma. However, whereas Patsch et al. found it necessary to use a three-step, non-linear gradient to achieve this separation, Danielsson et al. (1978) claim that it can be done on a simple linear gradient. In neither of the original papers are the criteria used to judge the effectiveness of the separation fully described but later work by Patsch et al. (1980) not only separated HDL_2 and HDL_3, but also subfractions of these lipoproteins which were comprehensively characterised.

A useful bibliography on the separation of lipoproteins by zonal ultracentrifugation can be found in the paper by Danielsson et al. (1978).

2.2 Isolation by precipitation

These methods allow the plasma lipoproteins to be divided into two fractions that approximate to the α- and β-lipoproteins, or to the high-density (*d* 1.063–1.20 g/ml) and the low-density ($d < 1.063$ g/ml) classes. However, as we shall show later, the fractions obtained by precipitation differ somewhat from those isolated by ultracentrifugation and seem to be more closely akin to the fractions that are separated by electrophoresis. We shall therefore refer to them as α- and β-lipoproteins respectively.

It has been said by some that the low-density lipoproteins that are isolated by precipitation can be substantially pure, but this claim will not survive the most critical tests. It is certainly difficult to remove the last traces of plasma proteins by these methods, particularly from the α-lipoproteins. For critical work, it is necessary to use an additional stage of purification, and ultracentrifugation is the method usually chosen. Nonetheless, precipitation is useful as a means of concentrating the lipoproteins from a large amount of plasma while, at the same time, effecting a considerable degree of purification. Moreover, when properly done, it can be an effective way of separating the α- and β-lipoproteins. The value of this for the measurement of the lipoprotein classes in plasma will be discussed in Section 7.3.

Three well-established precipitation methods are available:

(a) The low-temperature ethanol fractionation methods developed by Cohn and his collaborators. These can, in principle, be used to purify the lipoproteins to a considerable degree. However, several stages are needed, in each of which it is necessary to control the pH, ionic strength, dielectric constant, protein concentration and temperature. Acetate and carbonate buffers of different metals are used to control the pH and ionic strength. Ethanol is used to adjust the dielectric constant, and all the operations are carried out at temperatures of about − 5 °C. In practice, the method is more suited to the preliminary separation of the α- and β-lipoproteins into separate fractions than to their purification. However, it can be useful for the isolation of the so-called LP-X from the plasma of patients with cholestatic jaundice (Section 2.4).

When properly performed on the bench scale, the cold ethanol procedure brings about no detectable degradation of the lipoproteins. It is therefore interesting to find that it often results in extensive break-down of the α-lipoprotein when used on the industrial scale. It is presumably to this that the contamination of some commercial preparations of albumin with apo-lipoproteins is due (Fainaru and Deckelbaum, 1979; Ostlund-Lindquist and Boberg, 1979).

(b) Methods that are based on the interaction of lipoproteins with macromolecular polyanions like the sulphated polysaccharides heparin and dextran sulphate. This subject was comprehensively reviewed by Cornwell and Kruger (1961) and by Burstein and Scholnick (1973).

The fact that sulphated polysaccharides will react with plasma proteins at physiological pH to form soluble complexes has been known at least since the work of Chargaff et al. (1941). However, their affinity for the low-density lipoproteins is considerably greater than for other proteins. If the concentration of the polysaccharide is low, and the pH between 7.5 and 8.6., complexes with lipoproteins are the sole product of reaction (Bernfeld et al., 1957, 1960). Under the appropriate conditions, all three of the main classes of lipoprotein can be precipitated by a mixture of sulphated polysaccharide and one of the divalent cations Ca^{2+}, Mg^{2+}, Mn^{2+} or Co^{2+}. Each lipoprotein class is precipitated by a different concentration of the reagents which, in turn, depends on the nature of the polyanion and the cation used. Moreover, the solubility of the compplexes is influenced by the ionic strength of the solution and by the presence of plasma proteins.

All these lipoprotein complexes will dissolve if the ionic strength is high enough, the critical value depending on the nature of the lipoprotein, the polyanion and the cation. For example, heparin-Mg^{2+} will not precipitate LDL at the ionic strength of normal plasma, but will do so if the plasma is dialysed against 0.02 M Tris–HCl buffer at pH 7.7. By contrast, heparin-Mn^{2+} will form a complex with LDL in native plasma, and dextran sulphate will do so even in the presence of Mg^{2+}. Comparative studies of this kind have shown that dextran sulphate is a more reactive precipitant than heparin, and that Mn^{2+} (or Co^{2+}) is a more effective co-reagent than Mg^{2+} (or Ca^{2+}). But it is important to note that this property of dextran sulphate is a function of its molecular weight: a preparation of molecular weight 2×10^6 will precipitate LDL in the absence of divalent cations (albeit not quantitatively), whereas a fraction of molecular weight 15 000 will not do so unless Mg^{2+} or Ca^{2+} are present. These properties are summarised in Table 2.8.

The presence of plasma proteins tends to inhibit the formation of insoluble polyanion–lipoprotein complexes (Burstein et al., 1970). The reasons for this are not clear and the effect seems to be of little practical importance. By contrast, the presence of sucrose facilitates complex formation (Burstein, 1962). For example, heparin-Mg^{2+} will precipitate LDL from plasma in the presence of 1 g/ml of sucrose, but will not do so if the sugar is absent.

The foregoing discussion refers only to reactions at physiological pH. At values near the isoelectric point of the lipoproteins, the polyanions will precipitate LDL almost completely, without the addition of metal ions. The products of precipitation above and below pH 7.0 are therefore quite different. At pH 7.5, heparin-Mg^{2+} will precipitate chylomicrons and VLDL but not LDL, whereas at pH 5.0 heparin alone will precipitate only the LDL. Burstein and Scholnick (1973) showed that, under standard conditions of reaction in an acidic buffer, the pHs at which the different lipoprotein classes were precipitated formed the series LDL > VLDL > chylomicrons > HDL. Hitherto, these observations have not found any practical application.

In 1955, methods for the precipitation of low-density lipoproteins from plasma were reported by three laboratories. Bernfeld (1955) used sulphated amylopectin as the reagent, and Oncley et al. (1955) used dextran sulphate. Burstein and Samaille (1955) however, chose to use heparin, which has a relatively low molecular weight and needs the addition of a divalent cation to form an insoluble complex. All of these methods were subsequently developed into methods of preparative fractionation, but those based on the Burstein procedure have become the most widely used (Burstein and Scholnick, 1973).

(c) Precipitation with sodium phosphotungstate. This method, which was also developed by Burstein (1970), uses an inorganic precipitant of relatively high molecular weight that, like heparin, has anticoagulant properties and will release clearing factor (Bragdon and Havel, 1954). It has the merit that it is much cheaper than the organic polyanions and, unlike them, has constant properties. The method almost duplicates the results obtained with heparin or dextran sulphate but, as explained below, it requires even more careful control than the heparin procedure. In the presence of Mg ions, sodium phosphotungstate, like high molecular weight dextran sulphate, will precipitate all the lipoproteins including the HDL. However, the reaction differs in two respects from that with organic polyanions: (i) high ionic strengths favour the formation of the lipoprotein complexes, (ii) the reaction is more markedly inhibited by plasma proteins.

It is important to be aware that all these precipitation methods are extremely exacting and must be performed precisely, and with the greatest care if good results are to be obtained. It should also be observed that the fractions isolated by precipitation are not exactly the same as those obtained by ultracentrifugation at the traditional densities. On average, the HDL fraction ($1.063 < d < 1.21$ g/ml)

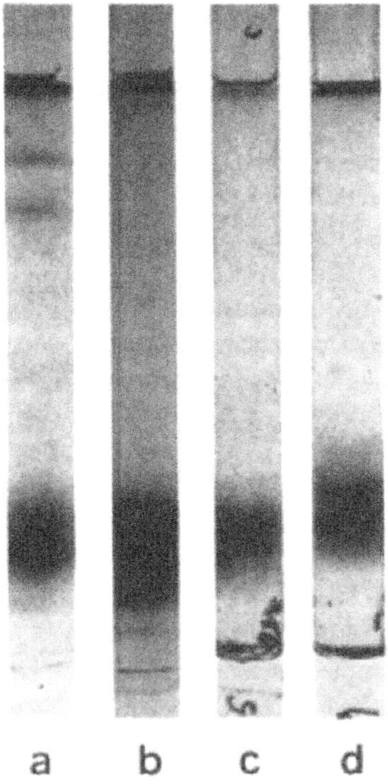

Fig. 2.5. Analysis of α-lipoprotein preparations by electrophoresis in polyacrylamide gradient gel (Section 5.5.3.3). (a) HDL ($d = 1.063$–1.21 g/ml) isolated by ultracentrifugation. (b) α-Lipoproteins isolated by precipitation with ethanol at low temperature. (c) α-Lipoproteins isolated by the phosphotungstate precipitation method. (d) α-Lipoproteins isolated by the heparin precipitation method.

contains about 17 mg more cholesterol per 100 ml of plasma than the corresponding (α-) fraction that is isolated by precipitation. When the fractions are analysed by electrophoresis on poly-acrylamide gel gradients (Fig. 2.5), it can be seen that the HDL includes a small proportion of high molecular weight particles that are absent from the precipitated fractions. These large particles have a density between 1.063 and 1.090 g/ml and can be shown by immunodiffusion analysis to contain apo-lipoprotein B. They are therefore chemically related to the lipoproteins of density less than 1.063 g/ml, which also contain apo-lipoprotein B, and differ from the principal components of HDL in which this apo-lipoprotein is absent. It is presumably the presence of apo-lipoprotein B that causes these particles of density 1.063–1.090 g/ml to be precipitated in the β-fraction, together with the lipoproteins of lower density.

It is more difficult to be certain that the precipitation methods themselves all yield the same products. Electrophoretic analyses like those of Fig. 2.5 suggest that all three methods can produce α-lipoproteins that are completely free of β-lipoproteins. However, the reverse is not always true. The precipitation of Cohn fraction I + II + III (Section 2.2.1) does not completely remove the α- from the β-lipoproteins (Fig. 2.6a) even though the precipitate is washed and re-precipitated. The selectivity of this procedure is evidently not good enough to achieve a complete separation of the two lipoprotein classes. By continuing the sub-fractionation of I + II + III by the Cohn procedure, a purer preparation of β-lipoprotein can be obtained, albeit at the cost of much labour and a diminished yield.

By contrast, a single, properly performed precipitation with heparin can be relied upon to produce β-lipoproteins that are free from α-lipoproteins (Fig. 2.6c). This may be less consistently true of the phosphotungstate method, which occasionally yields a β-fraction that contains traces of α-lipoproteins. This is probably due to the greater affinity of phosphotungstate for α-lipoproteins, which causes these to be precipitated by high concentrations of the reagent. Thus, whereas ten times the amount of heparin required for complete precipitation of the β-lipoprotein can be added without bringing

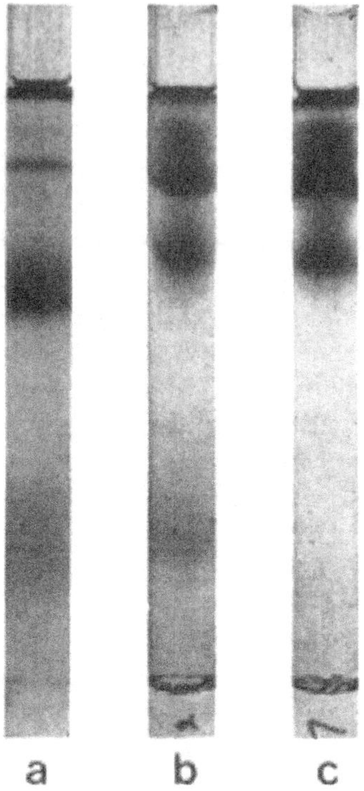

Fig. 2.6. Analysis of β-lipoproteins by electrophoresis in polyacrylamide gradient gel (Section 5.5.3.3). (a) Cohn fraction I + II + III. (b) β-Lipoprotein isolated by precipitation with phosphotungstate. (c) β-Lipoprotein isolated by precipitation with heparin.

about the precipitation of α-lipoprotein, excess sodium phosphotungstate causes the precipitation of progressively larger quantities of α-lipoprotein as well (McTaggart et al., 1978).

It is important to appreciate that the effectiveness of these precipitation methods may depend on the criteria by which they are judged. As the gradient gel analyses show, the heparin method can separate plasma lipoproteins cleanly into two fractions. But these are heterogeneous in almost every respect. The β-fraction contains most (perhaps all) of the apo-lipoprotein B, while the α-fraction contains most (though not all) of the apo-lipoprotein A. Both however will contain varying proportions of all the other apo-lipoproteins. An adjustment of the conditions of the precipitation may change these proportions. For example, it may be possible to increase the amount of apo-lipoprotein A that is recovered in the α-fraction but there is a high probability that apo-lipoprotein B will then be precipitated as well.

(d) Precipitation with non-ionic macromolecules. Non-ionic polymers like polyvinylpyrrolidone (PVP) or polyethyleneglycol (PEG) have been proposed as reagents for the precipitation of certain classes of lipoproteins. For example, PVP has been said to bring about the flocculation of chylomicrons of intestinal origin, while PEG was used by Kostner and Alaupovic (1972) to separate LP-A from other lipoproteins. At present however, it is not possible to separate the conventional lipoprotein classes quantitatively by means of these reagents, and they have not come into general use.

(e) Salting out. Ammonium sulphate was used as a precipitant in the early work of Machebœuf (1929) and has recently been re-investigated in this context by Parra et al. (1976, 1977) who have devised a method that is said to allow the lipoproteins to be fractionated without the need for an ultracentrifuge. It is reported to yield 'pure' lipoproteins by precipitation first with ammonium sulphate and then with sodium phosphotungstate. However, the details of the method are not easily accessible and it is impossible to assess its value in the absence, for example, of critical information on the effect of pH on the course of the precipitation. While it is certainly possible to bring about a partial separation of the lipoprotein classes by salting out, the method's only real value at present appears to be for the easy concentration of lipoproteins from large volumes of dilute solutions. Virtually all the plasma lipoproteins are found in the fraction precipitating between 35 and 70 % saturation with ammonium sulphate at room remperature and at pH 7.

2.2.1. Choice of method

All the precipitation methods suffer from the disadvantage that it is difficult to maintain exactly reproducible fractionations of the highly polydisperse starting material. Moreover, the problem is probably the greater when the heparin or dextran sulphate reagents are used, since these can vary in properties according to their source. It is therefore desirable, when setting up any of these procedures or when changing the source of a macromolecular reagent, to check their working by another method, preferably by ultracentrifugation. When an effective fractionation has been confirmed, great care must be taken to ensure that the technical details of the method are continually reproduced.

Of all the methods available, the cold-ethanol procedure not only requires an accurately controlled low-temperature bath but is by far the most tedious to perform if isolated lipoproteins are the sole objective. For this reason it has been supplanted by methods that use polyanion precipitants. Of these, heparin has come to be favoured for the separation of the β- from the α-lipoproteins, despite its cost and some variability in properties, because it gives a clean and reliable fractionation. When it is necessary to precipitate the α-lipoproteins, either dextran sulphate or the phosphotungstate reagents must be used. Of these, dextran sulphate is considerably the more expensive and it is important to use a product of known and reproducible molecular weight (e.g. the Pharmacia product). On balance therefore, phosphotungstate (Section 2.2.3.2) appears to offer slight practical advantages over dextran sulphate, but some workers have found lipoproteins prepared by this method to be unsatisfactory for their purpose. We have not had this experience and it is not clear whether the effect is merely due to inadequate removal of the reagent, or to some degradative process such as a catalysed oxidation. If the latter is found to be the case, it may be necessary to use the dextran sulphate method (Section 2.2.3.3).

Note that the following technical sections deal only with the precipitation of β-lipoproteins (i.e. the total 'low-density' lipoproteins) and the α-lipoproteins. Although methods have been described whereby chylomicrons and 'VLDL' can be separated from 'LDL' (Burstein and Scholnick, 1973) they are, in our experience, far inferior to the centrifugal methods.

2.2.2. Re-solution of precipitated lipoproteins

It is most important to use the correct technique when re-suspending or re-dissolving lipoprotein precipitates. The following procedure is of general application, but when the precipitate is the result of the cold-ethanol process, all the operations should be carried out as near to −5 °C as is practicable until the residual ethanol has been adequately diluted. In all cases, it is important that the solvent is added very slowly, with continuous stirring with as glass rod. The first objective should be to reduce the precipitate to a smooth cream, without lumps. Either 0.5 M sodium citrate, 0.5 M Na_2CO_3 or 1.0 M NaCl can be used as the solvent, but sodium citrate is recommended when the cold-ethanol fractionation is being used.

To dissolve a precipitate that has been obtained by the treatment of 10 ml of plasma, proceed as follows:

Add 0.5 ml of 0.5 M sodium citrate dropwise to the precipitate, stirring carefully between each drop to disperse lumps. When a uniformly smooth cream has been produced, add 2.0 ml of 0.15 M NaCl, stir and dialyse in the cold against several changes of NaCl/azide (Section 2.2.3.1).

Note. Lipoproteins that have been re-dissolved in this way may be immediately dialysed against a salt solution of low ionic strength. This observation also applies to lipoproteins that have been precipitated with heparin or dextran sulphate, once these reagents have been removed by reaction with barium. However, in our experience, low-density lipoproteins show less signs of aggregation and may be somewhat more stable if they are first dialysed against 9 % NaCl before they are transferred to 0.15 M NaCl.

2.2.3. Precipitation with polyanions

Since fibrinogen is also precipitated in these methods, the starting material should be serum or defibrinated plasma unless the product is to be further purified by other pethods. If serum is used, the main source of contamination of the precipitated β-lipoproteins appears to be the mechanical transfer of proteins on the precipitate. By contrast, the conditions needed to form insoluble complexes with α-lipoproteins also seem to precipitate small amounts of other serum proteins.

TABLE 2.8

The concentration of reagents needed to precipitate lipoproteins from human serum. The values given are those required to complete the precipitation of the classes that approximate to VLDL, LDL and HDL (taken from Burstein and Scholnick, 1973).

Reagents	Chylomicrons plus 'VLDL'	'LDL'	'HDL'
Heparin Mg^{2+} (Ca^{2+})	0.25% 0.1 M	No pptn.	No pptn.
Heparin Mn^{2+} (Co^{2+})	0.01% 0.05 M	0.1% 0.05 M	No pptn.
Dextran sulphate[a] Mn^{2+} (Co^{2+})	0.007% 0.025 M	0.05% 0.025 M	0.65% 0.2 M
Dextran sulphate[a] Mg^{2+} (Ca^{2+})	0.01% 0.1 M	0.1% 0.1 M	No pptn.
Dextran sulphate[b] Ca^{2+}	—	0.05% 0.025 M	0.55% 0.2 M
Sodium phosphotungstate Mg^{2+}	0.05% 0.1 M	0.2% 0.1 M	2.0% 0.2 M
Sodium phosphotungstate Mn^{2+}	—	0.08% 0.05 M	0.6% 0.2 M

[a] Dextran sulphate mol. wt. 15 000.
[b] Dextran sulphate mol. wt. 2×10^6.

2.2.3.1. Precipitation of β- and pre-β-lipoproteins with heparin. This procedure precipitates a fraction that approximates to VLDL plus LDL. The α-lipoproteins remain in solution. This is probably the best method for separating these fractions on an analytical scale, but it must be performed with great attention to detail. The slightest carelessness can lead to unreliable results.

Reagents: 4% heparin in 0.15 M NaCl. This should be made up from sodium heparin of activity about 160 units/mg. Since different batches may vary in their properties, it is advisable to test each one to ensure continuity of performance (cf. Burstein and Scholnick, 1973).

1.0 M $MnCl_2$; 2.0 M $MgCl_2$; 5 % $BaCl_2$; 0.5 M Na_2CO_3; 1.0 M NaCl;
Tris–HCl buffer, 0.02 M, pH 7.7;
Neutral NaCl/azide: 8.77 g of NaCl
0.13 g of NaN_3
0.372 g of disodium ethylene-diamine tetra-acetic acid
Water to 1.0 litre. Adjust to pH 7.0–7.5 with $NaHCO_3$

Procedure: To 10 ml of serum in a 25 ml centrifuge tube, add 0.5 ml of 4 % heparin solution and 0.5 ml of 1.0 M $MnCl_2$, with stirring. Continue to stir for at least 15 min, then centrifuge at about 15 °C for 10 min at 6000 *g*.

Note. Some workers perform this precipitation at 0 °C, but Burstein and Scholnick (1973) remark that is can be less selective in the cold. We find however that the method may perform badly if the ambient temperature is unduly high and some cooling may be desirable if the room temperature exceeds 20 °C.

Remove the supernatant, which contains the α-lipoproteins, and retain it for further treatment if required. Re-dissolve the precipitate by the addition of 0.5 ml of 0.5 M Na_2CO_3 as described in Section 2.2.2, and restore to a volume of 10 ml with 0.02 M Tris buffer, pH 7.7. Centrifuge off the precipitated manganese carbonate, add 0.25 ml of 2.0 M $MgCl_2$ and stir for 15 min. Centrifuge for 10 min at 6000 *g*. Discard the supernatant and re-dissolve the precipitate as before, but use 0.5 ml of 1.0 M NaCl instead of Na_2CO_3. Make up to 10 ml with Tris buffer, pH 7.7, and re-precipitate with $MgCl_2$ a second time.

Finally, dissolve in 0.5 ml of Na_2CO_3, add 2 ml of neutral NaCl/azide and dialyse against 5 % $BaCl_2$ for 48 hours at 4 °C. It is most important that this reaction be allowed to go to completion, or lipoproteins will re-precipitate when the ionic strength of the solvent is lowered. Centrifuge off the precipitate of barium heparin salt and dialyse the solution against several changes of neutral NaCl/ azide for 48 hours.

This procedure is designed to yield a product of high purity. When it is used as a preliminary to the quantitative estimation of α- and β-lipoproteins, the repeated re-precipitations may not be necessary.

2.2.3.2. Precipitation of α- and β-lipoproteins with sodium phosphotungstate. This procedure allows the α-lipoproteins to be precipitated after the β- and pre-β-lipoproteins have been isolated. As with the heparin method, the most careful technique is needed if good results are to be obtained. Moreover, since the efficiency with which the two lipoprotein classes are separated depends on the concentration of the reagent, it may be advisable to dilute hyperlipoproteinaemic sera to a lipoprotein level that approaches the normal values for which the procedure was designed.

Reagents: 1.0 M NaCl; 2.0 M $MgCl_2$; 0.5 M Na_2CO_3;
Tris–HCl buffer, 0.02 M, pH 7.7
Neutral NaCl/azide solution (Section 2.2.3.1)
Sodium phosphotungstate solution: 40 g of dodeca-tungstophosphoric acid
160 ml of 1.0 N NaOH
Water to 1.0 litre. Adjust to pH 7.6

Procedure: Burstein recommends that this method is carried out at room temperature. In our experience it works satisfactorily at 0 °C, but there may be less adsorption of foreign proteins at the higher temperature. It may therefore be better to work at room temperature if the products are not to be further purified, e.g. by ultracentrifugation.

Put 10 ml of serum into a 25 ml centrifuge tube and cool it in ice. Add 1.0 ml of sodium phosphotungstate and 0.25 ml of $MgCl_2$ with stirring. Continue to stir for 15 min at 0 °C and then centrifuge for 30 min at 3000 *g*. Remove the supernatant (SI), which contains the α-lipoproteins, and set it in ice to be worked on later. Dissolve the precipitate in 0.5 ml of 0.5 M Na_2CO_3 (see Section 2.2.2) and dilute to 10 ml with 0.02 M Tris buffer. Then repeat the precipitation by adding 1.0 ml of sodium phosphotungstate solution and 0.25 ml of $MgCl_2$. After stirring for 15 min, centrifuge as before and discard the supernatant. Dissolve the precipitate in 0.5 ml of 0.5 M Na_2CO_3, add 2 ml of neutral NaCl/azide and dialyse against several changes of neutral NaCl/azide for at least 36 hours at 4 °C. The resulting solution contains a mixture of lipoproteins that approximates to the VLDL plus LDL. Although much purified, they are nonetheless still contaminated with small amounts of serum proteins.

To complete the isolation of the α-lipoproteins, take the supernatant Sl and add to it 9 ml of cold sodium tungstate solution. Stir for 1 hour at 0 °C and then centrifuge for 30 min at 3000 *g*. The precipitate should contain only traces of lipoproteins and may be discarded. To the supernatant, add 0.75 ml of 2 M $MgCl_2$. Stir for 1 hour and allow to stand at 4 °C for 18 hours (this long resting period leads to higher recoveries of lipoprotein). Centrifuge for 30 min at 3000 *g* and discard the supernatant. Re-dissolve the precipitate in 0.5 M Na_2CO_3 (Section 2.2.2), dilute to 10 ml with 0.02 M Tris buffer and re-precipitate by the addition of 10 ml of sodium phosphotungstate solution and 2 ml of 2 M $MgCl_2$. Stand the mixture at 4 °C for 18 hours before recovering the precipitate by centrifugation at 3000 *g*. Dissolve this precipitate and dialyse it against neutral NaCl/azide as described for the β-fraction.

Good yields of α-lipoprotein can be obtained by this procedure but the product is contaminated with small amounts of plasma proteins which are difficult to remove by repeated precipitations. If this contamination cannot be tolerated, the preparation must be purified by ultracentrifugation (Section 2.1.1.8) or by gel filtration chromatography (Section 2.3.1), although the latter is probably the less efficient. The fact that the ultracentrifuge is needed to complete this preparation underlines the precipitation method is more useful as a concentration process than as a preparative method in its own right.

2.2.3.3. Precipitation of α- and β-lipoproteins with dextran sulphate. The following directions apply only to the use of a dextran sulphate of high molecular weight e.g. about 5×10^5. Preparations of much lower molecular weight can also be obtained, e.g. 15 000, but the conditions under which these will precipitate lipoproteins are quite different. Even when the molecular weight and reproducibility of the dextran sulphate can be relied on, it is advisable to give the system an adequate test before putting it to serious use. If the provenance of the reagent is uncertain, this precaution is indispensable.

Reagents: A 10% solution of dextran sulphate (Pharmacia Fine Chemicals; Sigma Chemical Co.) in water;

1.0 M $MnCl_2$; 2.0 M $MgCl_2$; 1.0 M NaCl; 1.0 % $BaCl_2$ in 0.15 M NaCl; 10 % $NaHCO_3$; 0.5 M Na_2CO_3;
Tris–HCl buffer, 0.02 M, pH 7.7 in 0.15 M NaCl
Tris–HCl buffer, 0.02 M, pH 7.7 (Section 2.2.3.1)
Neutral NaCl/azide (Section 2.2.3.1).

Procedure: Pipette 10 ml of plasma into a 25 ml centrifuge tube that is equipped with a stirrer. With continuous stirring, add 0.05 ml of 10 % dextran sulphate, followed by 0.5 ml of 1.0 M $MnCl_2$. Allow to stand for 10 min and centrifuge at 6000 *g* for 10–15 min at 15 °C. Transfer the supernatant, which contains the α-lipoproteins, into a second 25 ml centrifuge tube. Dissolve the precipitate in the first tube in 0.5 ml of 10 % Na_2HCO_3 followed by 5 ml of Tris buffer and stand for 10 min. Remove the precipitate of manganese carbonate by centrifugation and discard it. To the supernatant, add 45 ml of Tris buffer and 0.25 ml of 2 M $MgCl_2$ and mix well. Stand for 10 min and then centrifuge as before. Re-dissolve the precipitate in 0.5 ml of 1.0 M NaCl followed by 10 ml of Tris–NaCl, and re-precipitate the lipoproteins by adding 0.5 ml of 2 M $MgCl_2$. Centrifuge off the lipoproteins and repeat this washing procedure in the same way. Finally, dissolve in 0.5 ml of 1.0 M NaCl, add 2 ml of NaCl/azide and dialyse against 1.0 % $BaCl_2$ for 48 hours at 4 °C. Centrifuge off the precipitate of barium salt and dialyse the solution of β-lipoproteins against several changes of NaCl/azide for 48 hours at 4 °C.

To complete the isolation of the α-lipoproteins, add to the solution that was reserved from the first stage, 0.6 ml of 10 % dextran sulphate and 1.5 ml of 1.0 M $MnCl_2$. Stir thoroughly and stand for at least 2 hours, though better results are obtained if the rest period is 18 hours. If the rest period is short, the precipitate should be centrifuged down at 20 000 *g* for 30 min but, if the precipitate is allowed to mature overnight, 30 min at 6000 *g* should be sufficient. Pour off the supernatant and suspend the precipitate in 5–10 ml of Tris–NaCl buffer which contains 0.1 % dextran sulphate and 0.1 M $MnCl_2$. Recover the washed precipitate by centrifugation at 6000 *g* for 10 min, re-dissolve it in 0.5 ml of 0.5 M Na_2CO_3 and dilute with 2 ml of NaCl/azide (Section 2.2.2). Note that the precipitate will not dissolve unless the pH is at least 8.0. After 30 min, centrifuge off the precipitated manganese carbonate at 6000 *g* for 10 min. The resulting solution of α-lipoproteins can be dialysed against $BaCl_2$ as for the β-lipoproteins but the preparation will be contaminated with plasma proteins that must be removed by ultracentrifugation (Section 2.1.1.8) or by gel filtration chromatography (Section 2.3.1). However, if an ultracentrifuge is available, the simplest course is to proceed immediately to raise the density of the crude lipoprotein preparation to 1.21 g/ml by the addition of NaBr (Section 2.1.1.2; note however that allowance must be made for the partial volume of the dextran sulphate) and then to ultracentrifuge for 24–48 hours (Table 2.6). The HDL can then be recovered from above the sedimented impurities in the usual way (Section 2.1.1.7).

2.2.4. Procedure for low-temperature ethanol precipitation

This is essentially the method 10 of Cohn et al. (1950), which is very suitable for bench scale preparations.

Reagents: Use analytical grade materials wherever possible.
0.5 M Na_2HPO_4; 0.5 M NaH_2PO_4; 0.5 M Trisodium citrate; 1.0 M acetic acid; 1.0 M sodium acetate;
0.2 M NaCl (stock for dilution as required)
Sodium acetate buffer: 100 ml of 4 M sodium acetate
200 ml of 10 M acetic acid
Water to 1.0 litre
The pH of this buffer should be 4.0 after dilution 1:80 with water
Reagent 'A': 250 ml of 95% ethanol (prepared from absolute ethanol immediately before use)
2.5 ml of sodium acetate buffer
Water to 1.0 litre

Also required are a cold bath (or room) at −5 °C, a refrigerated centrifuge and a small mechanical stirrer.

Method: All grossly lipaemic plasma should be rejected if possible, since they cause great difficulties in the centrifugation of the precipitates.

Cool the reagent 'A' to −5 °C before starting the fractionation.

Cool 10 ml of plasma to 0 °C in a plastic centrifuge tube of 100 ml capacity which is immersed in the bath at −5 °C. Do *NOT* allow any part of the plasma to freeze. Ice crystals will form very rapidly if the plasma is stirred, but this can be avoided in the following way. Plunge the tube deeply into the cooling bath, clamp it lightly, and arrange a small stirrer to rotate just above the surface of the plasma. Draw up the pre-cooled reagent 'A' into a pipette. Begin to add this reagent to the plasma and, at the same time, raise the tube until the stirrer is submerged in the plasma.

Add 40 ml of reagent 'A' at the rate of 20 ml/min, with continuous stirring which should be maintained for 15 min after all the reagent has been added. The pH of the mixture should be 5.8–5.9. Finally, centrifuge at 3000 *g*, at −5 °C for 30 min. The supernatant consists of Cohn fractions IV + V + VI and contains the α-lipoproteins; the precipitate consists of fractions I + II + III and contains the β-lipoproteins. Although 'pure' lipoproteins can be isolated from these fractions by several further stages (Cohn et al., 1950), the process is uneconomically laborious by comparison with the precipitation with polyanions described in the preceeding section. However, the technique given here is useful for the preparation of LP-X (Section 2.4.2) or for the analytical separation of the α- and β-lipoproteins when it is not necessary to remove the other plasma proteins, e.g. as a preliminary to their estimation by cholesterol estimation. Under these circumstances, the precipitate of I + II + III should be washed by suspension in cold reagent 'A' to remove adherent α-lipoproteins.

Dissolve the precipitate in 0.5 M sodium citrate as in Section 2.2.2 and dialyse both this solution and the fraction IV + V + VI exhaustively against NaCl/azide if they are to be stored.

2.3. Isolation by chromatography

The selectivity of most of the available methods does not permit the complete separation of the lipoproteins from plasma by chromatography. Moreover, it is difficult to define the products as precisely as those obtained by ultracentrifugation and, unlike the centrifuge, chromatography usually produces a diluted product rather than a concentrated one. Nonetheless, chromatography has been used by some workers for the fractionation of lipoproteins that have been isolated by other methods and can be of great value in the making of certain preparations that are described in detail in the following sections.

2.3.1. *Gel filtration chromatography*

The large size of the plasma lipoproteins restricts the use of this very mild technique to columns of weak agarose gels. Moreover, until recently, even the most porous gels were unable to handle the largest lipoprotein particles. With the advent of Sephacryl S-1000 however, it may become possible to extend the range of the method to include lipoprotein particles of up to at least 100 nm in diameter. But, at the time of writing, experience with this system is lacking and established technique is limited to the separation of lipoproteins whose diameter is less than about 30 nm (i.e. mol. wt. about 10×10^6) on more rigid gels such as Sepharose 4B or Biogel A-15M. Unfortunately, the resolution obtainable with these gels tends to be low. Moreover, the fractionation is achieved on the basis of particle size, which is an inconvenient parameter to measure, which in turn makes it difficult to classify the isolated fractions in terms of their behaviour on the column. As Fig. 2.7 shows, the gel-filtration chromatogram of human plasma lipoproteins on Sepharose or Biogel has three peaks. These bear a rather ill defined relationship to the main density classes and are therefore often referred to as 'VLDL', 'LDL' and 'HDL', but it must be remembered that these may not correspond exactly with the fractions that would be isolated by centrifugation. This disparity is at least partly due to the difficulty of deciding where the boundaries between the conventional fractions would lie on the chromatographic elution curve. It should also be noted that, although gel filtration can be successfully used to remove free contaminating proteins from lipoprotein preparations, it cannot be assumed that the product will be free of adsorbed proteins, particularly albumin.

As was said in Chapter 1, there tends to be an inverse relationship between the diameter of a lipoprotein particle and its density. Gel

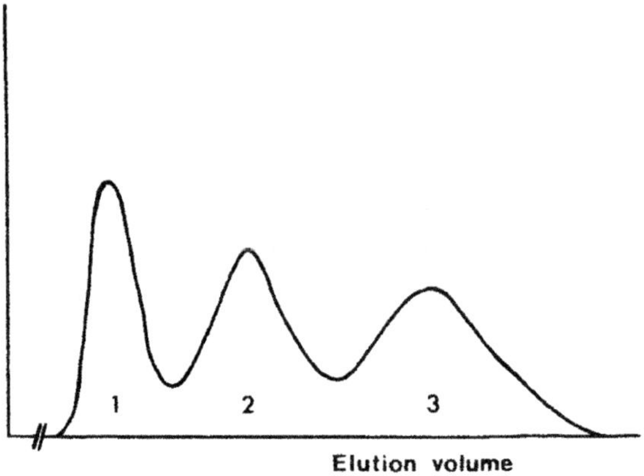

Fig. 2.7. The distribution of lipoproteins obtained by filtration chromatography of plasma on gel of 4–6% agarose content (Section 4.4).

filtration is an especially useful method of analysis in cases where this relationship does not hold. It has been used for example, to detect the aggregation that is brought about by the mishandling of LDL during radio-iodination (Schonfeld et al., 1974; Weech et al., 1978), or as the result of other procedures (Margolis, 1967). It may also be used to isolate Lp(a) and LP-X, both of which are larger than would be expected for particles of their density. The preparation of these substances is described in Section 2.4.

The method of Rudel et al. (1974) is a good example of the way gel chromatography can be used to separate 'VLDL', 'LDL' and 'HDL' from an ultracentrifugally prepared total plasma lipoprotein fraction. These authors claimed that, by selecting appropriate fractions of the eluate, products could be obtained that were free from cross-contamination. They substantiated this view by comparing the chemical composition of the lipoproteins isolated by chromatography with the equivalent fractions that had been obtained by ultracentrifugation. The agreement was, in general, good. However, the dilemma is clear. If an ultracentrifuge is necessary to prepare the specimen that is to be chromatographed and to produce the fractions with which the column is to be calibrated, why use chromatography at all? The reply to this criticism seems to be as follows: provided that the calibration of the column can be assumed to remain constant and that the lower resolution of the method can be tolerated, chromatography is quicker and less expensive than ultracentrifugation and can also be used to fractionate lipoproteins that have been isolated by precipitation.

Sometimes, the nature of the experiments will allow plasma to be applied directly to the column. For example Johnson et al. (1981) used chromatography on Sephacryl S-300 to show that the 'HDL' in plasma from the inferior vena cava and from the hepatic vein probably differ in the proportions of free and esterified cholesterol that they contain. Note that, because the molecular weight of HDL is only about 0.5×10^6, it is feasible to use a comparatively dense gel in this experiment. Moreover, the fact that the column would not separate the lipoproteins cleanly from the plasma proteins was unimportant since the only variables to be measured were cholesterol and cholesteryl ester.

Procedure: This is essentially the method of Rudel et al. (1974). The general procedure for preparing the column and for applying the sample may be found in such reference books on chromatographic technique as that of Fischer (1974). Rudel et al. found that both Bio-Gel A-5M and Bio-Gel A-15M could be used, but the resolution of 'VLDL' from 'LDL' was greater on A-15M, whereas 'LDL' and 'HDL' were separated better on A-5M.

A column of 1.5–2.0 cm diameter is packed to a height of 100 cm with Bio-Gel A-5M (200–400 mesh) that has been equilibrated with 0.15 M NaCl containing 0.01 % EDTA (pH 7.0). The column should be exhaustively purged with the solvent to remove any soluble substances that may be derived from the matrix material. Up to 40 mg of lipoprotein cholesterol in a volume of not more than 2 ml may be loaded per square cm of column cross-section. Some authors have recommended phosphate buffered saline for the elution, but Rudel et al. found 0.15 M NaCl to be equally satisfactory. The best results are probably obtained with elution rates of 3–5 ml/cm^2/hour. The profile of the eluted lipoproteins can be determined by spectrophotometric measurement of the absorption at 280 nm, or by the chemical estimation of cholesterol or phosphorus. No matter which method is used, it must be remembered that none of these components is a constant proportion of all the lipoproteins and that the profile observed in this way cannot be directly interpreted in terms of lipoprotein mass.

2.3.2. Affinity chromatography

This method has not been widely used, but its specificity makes it a valuable tool for the isolation of lipoproteins that are defined by a chemical rather than a physical property. It follows from their lack of popularity that relatively little work has been done to refine the various methods that use this principle. Those described below may therefore perform fairly well but may also be open to considerable improvement.

2.3.2.1. Immunoadsorbents These have been used to adsorb lipoproteins that contain a particular protein moiety of interest. There is little to be gained however, in using this technique to isolate the lipoproteins that contain either the A (i.e. A-I + A-II) or the B apo-lipoproteins, since there are simpler ways of doing this when the concentration of the lipoprotein is comparable with that in plasma. If the concentration is low, as for example in a liver perfusate, or because the apo-lipoprotein of interest is a minor one, the adsorber can be used to advantage. Thus, Gibson et al. (1982) have used this technique to isolate the lipoprotein particles that contain apo-lipoprotein E, and Kunitake et al. (1982) have prepared lipoproteins that contain apo-lipoprotein A-I in a similar way.

The first step is to obtain an antiserum that is monospecific for the protein of interest. In the case of apo-lipoprotein B, it may be possible to use an antiserum to an LDL preparation of density 1.02–1.04 g/ml. In general however, it will be necessary to isolate a small amount of the protein in a pure state by one of the methods described in Chapter 8.

Unfortunately, this is a long and tiresome process that seriously limits the utility of the immunoadsorption technique. Once the protein has been prepared, it is used to raise an antiserum in a convenient animal. Although the whole globulin fraction can be used to prepare an immunoadsorbent that is adequate for many purposes (McConathy and Alaupovic, 1976), a product with greater capacity will result from the use of the specific immunoglobulin. Several ways of isolating this by dissociation of the immune complex have been reported, some of which are conflicting in their recommendations. For example, Kostner and Holasek (1969) dissociate the complex at pH 3.4 in a solvent which they stress must be of low ionic strength. By contrast, Picard and Veissiere (1970a) used 4 M NaCl acidified with HCl. Burstein (1964) proposed the use of 0.9 % NaCl buffered to pH 1.6–1.8 with veronal–acetate–HCl but the composition and ionic strength of this buffer were not stated: presumably the recipe of Michaelis (1931) was used. The immune complex can also be dissociated at a high pH (e.g. Chung and Nishida, 1967) but this technique has been less often used.

As an example, we shall describe the procedure used by Kostner and Holasek (1969) for the isolation of immune globulins to LDL. As with most of the other methods, the globulins are precipitated with a suitable lipoprotein fraction from which they are later separated by ultracentrifugation at a density which allows the lipoprotein to float while the globulins sediment.

Reagents: 0.9 % NaCl, 0.1 M HCl
17 % sucrose–0.9 % NaCl, pH 7.4 (density approx. 1.07 g/ml)
17 % sucrose adjusted to pH 3.4 with 0.1 M HCl

Note that it may be necessary to change the density of these sucrose solutions if a high density lipoprotein fraction is used as the precipitant.

Procedure: Precipitate the antibody–antigen complex at the equivalence point with LDL (Clausen, 1974) and allow the mixture to stand for 2 hours at 37 °C, followed by 18 hours in the refrigerator. Centrifuge off the precipitate, wash it three times with 0.9 % NaCl and once with distilled water. Suspend it in a volume of water equal to about 10 % of that of the original antiserum and cautiously adjust to pH 3.4 with 0.1 M HCl. This operation is made more difficult by the poor buffering capacity of the solution but, according to Kostner and Holasek (1969), the presence of buffer salts other than glycine will prevent the dissociation of the complex.

After the adjustment of the pH, centrifuge off any undissolved material and float the supernatant on a gradient prepared in an ultracentrifuge tube by layering 17 % sucrose (pH 3.4) on top of an equal volume of 17 % sucrose (pH 7.4). Centrifuge the tube at a minimum of 50 000 *g* (at the top of the tube) for 24 hours at 10 °C. Dialyse the bottom fraction, which contains the antibodies, against 0.9 % NaCl and concentrate by ultrafiltration if necessary. If need be, the globulins can be further purified by chromatography on DEAE-cellulose (Peterson, 1975).

For those who do not have easy access to an ultracentrifuge, the method of Burstein (1964) may be worth considering. It is based on the adsorption of the lipoprotein liberated from the re-dissolved immuno-precipitate on $BaSO_4$. The original description of the method referred to the preparation of anti-β-lipoproteins but it was claimed that it could be used to isolate other immune globulins. However, we have no experience of this procedure and it is not clear whether all lipoproteins are equally well adsorbed by $BaSO_4$.

To make an adsorbent suitable for chromatography, the immunoglobulins (or antiserum) must be linked to a suitable matrix. Kostner and Holasek (1969) attached the protein to bromacetyl-cellulose by the method of Jagendorf et al. (1963), but it is now probably more convenient to use the commercially available cyanogen bromide–activated agarose as matrix (e.g. CNBr–activated Sepharose 4B: Pharmacia Fine Chemicals AB). Alternatively, the activated gel can be prepared in the laboratory by the method of March et al. (1974).

Reagents: 0.1 M $NaHCO_3$ in 0.5 M NaCl
1.0 M ethanolamine in 0.1 M $NaHCO_3$–0.5 M NaCl
1.0 M NaCl–0.1 M acetate buffer, pH 4
0.15 M NaCl–0.05 M Tris–HCl (pH 7.5) to which is added 0.01 % NaN_3

Procedure: Dialyse the globulin solution against 0.1 M $NaHCO_3$–0.5 M NaCl and adjust to a final concentration of 2–3 mg/ml. If a commercially activated gel is used, wash it thoroughly with 0.001

M HCl to remove the stabilisers that are present. Add 5 ml of the globulin solution to each 1.0 ml of swollen, activated gel and stir very gently for 2 hours at room temperature, or overnight at 4 °C. Wash the product with $NaHCO_3$–NaCl and incubate it for 2–3 hours in the ethanolamine–$NaHCO_3$–NaCl solution to block any remaining active groups. Filter off the gel and wash alternately with three cycles of pH 4 acetate and $NaHCO_3$–NaCl. Finally wash exhaustively with the pH 7.5 Tris buffer.

The use of the immunoadsorbent in a chromatography column is generally conventional (Lowe, 1979). The lipoprotein sample is applied in the pH 7.5 Tris buffer and the column is washed until no further protein is eluted. The adsorbed material may then be eluted either with 0.2 M glycine–HCl at pH 3.2 (Kostner and Holasek, 1970), or with 3 M NaCNS (McConathy and Alaupovic, 1976), after which the column is exhaustively washed with pH 7.5 Tris buffer. The eluted lipoprotein is dialysed and concentrated by any convenient method (Appendix 2).

In order to reduce the period of contact between the column and the thiocyanate solution as much as possible, McConathy and Alaupovic (1976) used a stratified column that was packed as follows:

Prepare a suspension of Sephadex G25 in pH 7.5 Tris–HCl buffer. Pour sufficient of this into a column of 2.5 cm diameter to form a packed layer 15 cm high. Next, in the same buffer, pack a 40 cm layer of the immunoadsorbent, followed by 4 cm of Sephadex G25. Condition the column by running through it 50 ml of 3 M NaCNS followed by exhaustive washing with the NaCl–Tris buffer pH 7.5. Apply the sample in the same buffer and wash until no further protein is eluted. Then displace the bound lipoprotein with 50 ml of 3 M NaCNS, followed by the pH 7.5 Tris buffer. Unfortunately, there is no published evidence that this practice is really superior to the simpler one. Moreover, although much of the lipoprotein will appear in the Tris buffer, as much as 25 % will also be present in the thiocyanate fraction.

2.3.2.2. Immobilised heparin This adsorbent has been used, along with other immobilised glycosaminoglycans, in studies of the binding reaction that is thought to contribute to the process of atherogenesis (Iverius, 1972) but has been little used for preparative purpose until recently. The discovery that the presence of apo-lipoprotein E will, like the B protein, enable a lipoprotein to form a complex with heparin has led to the use of this reagent for the concentration and partial purification of lipoproteins that carry the E protein (Shelburne and Quarfordt, 1977; Hay et al., 1978; Marcel et al., 1980). Conversely, the ability to adsorb the B and E proteins can be of value for the isolation of lipoproteins from which they are absent. It has to be remembered however, that other proteins (e.g. fibrinogen) may also be adsorbed by a column of immobilised heparin and that plasma will therefore not yield a product that is free of other proteins if it is applied directly to the column. Moreover, since the gel matrix is highly permeable, any proteins that are present in the material that is applied will be eluted comparatively slowly. The column must therefore be thoroughly washed to ensure the removal of contaminating but non-adsorbed proteins before the elution of the lipoproteins is begun. In general, immobilised heparin is best used as a means of fractionating lipoproteins that have previously been isolated in some other way.

Procedure: The heparin is coupled with Sepharose 4B that has been activated with cyanogen bromide, as described in Section 2.3.2.1. Adsorbents of adequate capacity may be made from the best commercial preparations of heparin, which have an activity of 160–170 units/mg. If need be, other preparations can be purified by precipitation with cetyl-pyridinium chloride as described by Lindahl et al. (1965). Note however that there is now evidence that the adsorptive properties of immobilised heparin can differ considerably according to the manufacturer or even the batch of heparin that is used.

There is a critical ionic strength of solvent for which the degree of adsorption of a lipoprotein by the immobilised heparin is a maximum. For LDL this is between $I=0.05$ and 0.15. Desorption takes place as the ionic strength is raised and, for LDL, is essentially complete at $I=0.45$.

For other lipoproteins the conditions may be slightly different. Usually however, it will be satisfactory to prepare a column of heparin–Sepharose in a solvent such as 0.05 M NaCl containing 0.002 M sodium phosphate, pH 7.5, and to apply the sample in the same solvent. After a thorough washing to remove the non-adsorbed substances, the bound lipoproteins can be eluted by raising the ionic strength, either stepwise or by means of a concentration gradient. The effect of adding the ions of magnesium or calcium has apparently not been systematically studied but there is some evidence for an enhancement of the binding effect.

2.3.2.3. Immobilised lectin Apo-lipoprotein B is a glycoprotein that contains about 5 % of carbohydrate by weight, including about 2 % of mannose. Lipoproteins that contain this apolipoprotein will therefore complex with lectins like concanavalin A that react with terminal α-D-mannopyranosyl reidues. By contrast, concanavalin A does not react with any of the A or C apo-lipoproteins, or with native apo-lipoprotein D (McConathy and Alaupovic, 1974). The immobilised lectin therefore provides an easy way of recovering lipoproteins that carry the B protein, which can be particularly useful if their concentration is low. For example, Fainaru et al. (1977) used concanavalin A to trap particles containing apo-lipoprotein B from perfusates of rat livers. The lectin will also complex with apo-lipoprotein D if this is first treated with neuraminidase. This manoeuvre was used by McConathy and Alaupovic (1976) to facilitate the isolation of D-containing lipoproteins. However the lectin, like heparin, will react with other components of plasma such as the immunoglobulins (Weinstein et al., 1972) and is therefore of limited use for the direct extraction of lipoproteins from plasma.

The active site of concanavalin A requires the presence of both calcium and manganese ions, which form a stable complex with the lectin above pH 7. Since these ions are easily dissociated at a pH below 5, McConathy and Alaupovic (1974) proposed that a 0.05 M Tris buffer, pH 7.0, containing 0.001 M Ca^{2+} and Mn^{2+}, should be used for the preparation of the column and the sample. However,

Tris–HCl is a poor buffer at this pH and it is difficult to prevent the slow precipitation of manganese hydroxide. Moreover, provided that the presence of the ions is ensured during the preliminary equilibration and that the pH during use is not allowed to fall below about 7.0, their inclusion in the elution buffer appears to be unnecessary. Thus, Shore and Shore (1973) successfully chromatographed VLDL on columns of concanavalin A that were eluted only with 0.2 M NaCl in 0.1 M sodium acetate buffer at pH 6.8. It may also be observed that other plasma proteins have been successfully chromatographed on this adsorbent, using solvents that do not contain divalent cations. Note however, that there has been no systematic study of the effects of different solvents on the adsorbtion of apo-lipoprotein B.

Reagents. 0.2 M NaCl in 0.1 M sodium acetate, adjusted to pH 6.8 with acetic acid
0.2 M methyl α-D-glucopyranoside in NaCl–acetate pH 6.8

Procedure. If desired, concanavalin A can be coupled to agarose (e.g. Sepharose 4B) in the laboratory by the cyanogen bromide method (Section 2.3.2.1) but the same product is also available commercially (Pharmacia Fine Chemicals AB). The latter is supplied as a suspension in 1.0 M NaCl–0.1 M acetate buffer, pH 6, which also contains calcium, magnesium and manganese ions at 0.001 M concentration. If the adsorbent is home-made, it should also be equilibrated with this mixture to ensure that the lectin is fully activated.

Dialyse the sample of lipoprotein that is to be chromatographed against the NaCl–acetate buffer, pH 6.8. In the meantime, wash the adsorbent gel with the same buffer and prepare a column 30 cm × 2 cm also in this buffer. Apply the lipoprotein to the column and wash with NaCl–acetate until the eluate is free of unretained lipoprotein. Then elute the retained material with the methyl glucopyranoside solution. The retained (or unretained) fractions should be dialysed and then concentrated as required (Appendix 2).

Finally, wash the column thoroughly with NaCl–acetate buffer to prepare it for further use. In order to ensure that the adsorbent maintains its full activity, it should be periodically equilibrated with the calcium–magnesium–manganese solution.

2.3.3. *Adsorption chromatography*

Although there are several adsorbents that can be used to separate the various classes of plasma lipoproteins, e.g. glass beads (Carlson, 1960), the only one to have been used effectively is hydroxyapatite. Hjerten (1959) showed that this substance adsorbs 'β-lipoprotein' strongly from solvents of low ionic strength and that selective elution is possible as the ionic strength is raised. Later work has shown that it has some ability to distinguish between lipoproteins that contain the apo-lipoproteins A, C and D. But it must be borne in mind that the material to be fractionated (usually HDL) does not consist of a few, discrete lipoprotein species of distinctly differing properties, but is composed of a wide range of particles, each of which differs only slightly from its neighbours in the series. Consequently, every increase in the ionic strength of the eluting buffer will displace some of the adsorbed lipoprotein and there is no value at which increments will fail to elute material from the column. The 'separation' achieved is therefore purely arbitrary and can be adjusted to produce whatever result is desired. Kostner (1977) has used this technique to obtain an HDL fraction in which apo-lipoprotein A was almost the only protein present, and another which was likewise rich in apo-lipoprotein C. In its most successful application, hydroxyapatite chromatography was used in conjunction with other methods of purification to isolate lipoprotein particles containing a single protein component, namely LP-A, LP-C and LP-D (Kostner and Alaupovic, 1972; McConathy and Alaupovic, 1976). The method has not yet found any more general use.

Hydroxyapatite also has the practical disadvantage that different preparations can vary quite widely in their properties and may need individual characterisation before use. Moreover, commercially prepared hydroxyapatite may form columns with a rather low flow rate. Some users have overcome this difficulty by using a mixture of two volumes of hydroxyapatite with one volume of microcrystalline cellulose. However, the following procedure, due to Kostner (1977), yield an adsorbent with satisfactory flow properties:

Reagents. Prepare a stock solution of 1.0 M potassium phosphate by neutralising 98 g of *ortho*-phosphoric acid with 30 % KOH to pH 6.8, in a final volume of 1.0 litre. 0.5 M $CaCl_2$ (1.0 litre); 0.5 M Na_2HPO_4 (1.0 litre); 40 % NaOH (50 ml)

Procedure. Add 1.0 litre of 0.5 M $CaCl_2$ to 1.0 litre of Na_2HPO_4 drop by drop, over a period of 24 hours with gentle stirring. Allow the coarse particles of the precipitate to settle, decant off the turbid supernatant and wash the residue four times by suspension in 2 litres of distilled water. Next, re-suspend the precipitate in 1.5 litres of water, add 50 ml of 40 % NaOH and heat on a boiling water bath for 2 hours. Allow to cool and decant off the supernatant. Finally, wash the precipitate three times with 2 litres of distilled water, twice with 2 litres of 0.01 M potassium phosphate and twice with 2 litres of 0.001 M potassium phosphate (both these solutions diluted from the 1.0 M stock). This preparation will retain its properties for several months if it is stored at 4 °C in the presence of sodium azide.

For use, the hydroxyapatite is poured into a column in dilute phosphate buffer, pH 6.8. The exact concentration of this buffer may need to be established for the batch of adsorbent that is used, and will be determined by the separation that is desired. As a general guide, LP-D is not adsorbed even from 0.001 M phosphate, LP-A is eluted at a concentration of 0.05–0.1 M, while LP-C requires a concentration of about 0.5 M phosphate to displace it (N.B. Kostner and Alaupovic, 1972, used 0.1 M sodium citrate buffer, pH 7.0, to elute LP-C). The material to be applied to the column is first stripped of lipoproteins that contain apo-lipoprotein B, preferably by means of an immunoadsorbent (Section 2.3.2.1). It is then dialysed against the buffer with which the hydroxyapatite has been equilibrated, applied to the column, and washed with the same buffer. Elution is then carried out by such stepwise increments in the concentration of the buffer as are found by experiment to achieve the desired fractionation.

2.4. *Isolation by preparative electrophoresis*

This technique has little to recommend it for the isolation of the main lipoprotein classes, but it can be used to advantage for the separation of particles that have the same density but differ markedly in electrical charge. One of the first successful applications was the resolution of two low-density lipoproteins by Mahley et al. (1973). Subsequently, the method was used to resolve the so-called HDL_c (produced by feeding cholesterol) from LDL (Mahley et al., 1975). In these experiments, the electrophoresis was carried out in a mixture of equal parts of powdered Geon (polyvinyl chloride) and Pevikon (vinyl chloride–vinyl acetate co-polymer). Later experience has shown however, that good results can often be obtained with Pevikon alone. With this modification, the following description of the technique is closely based on that of Mahley and Weisgraber (1974).

Equipment. The mixture of Pevikon and electrolyte is contained in an acrylic tray of about 29 × 38 × 1.5 cm (total volume 1300 ml). This is supported between two electrode vessels of essentially conventional design, with three baffle plates. Each is 30 cm wide and is capable of holding 3–4 litres of electrolyte. (Note that drawings for the construction of the cell can be obtained from the original authors.) The power supply must deliver a constant current of up to 50 mA. Electrical contact between the Pevikon and the electrode vessels is by means of thick wicks about 20 cm long by 30 cm wide. Mahley and Weisgraber (1974) recommend the use of Telfa surgical swabs for this purpose (Kendall Surgical Supplies). These are 7.5 × 20 cm and are 2–3 mm thick. Sintered glass filters will also be needed, as specified below.

Reagents. Pevikon C-870; this is a co-polymer of vinyl chloride with 13% of vinyl acetate. It is obtainable from some laboratory suppliers (e.g. Shandon Scientific; Mercer Chemical Corp.).

Electrolyte: about 8 litres of barbital buffer, pH 8.6, containing 2.07 g of diethyl barbituric acid and 11.5 g of sodium diethyl barbiturate per litre.

Procedure. Dialyse the crude lipoproteins against the electrolyte buffer to give a solution containing up to 15 mg of protein/ml. A sample of the lipoprotein equivalent to about 1 mg of protein must be pre-stained with Sudan Black (Appendix 4) for use as a marker.

Stir 600 g of Pevikon into 1 litre of distilled water. Allow the polymer to settle for 10 min and then suck off the supernatant and the suspended fines. Filter off the polymer, under light suction, on a 125 mm, coarse grade sintered-glass filter. Wash twice with water, by re-suspension on the filter. Finally, wash in the same way with 800 ml of the electrolyte buffer. Suck the polymer almost dry, transfer to a beaker and re-suspend in 500 ml of electrolyte.

Line the bottom and sides of the acrylic tray with Parafilm and set it on a level surface. Soak the wicks in the electrolyte, squeeze out the excess and position them across the ends of the tray. If Telfa swabs are used, arrange four at each end so that they extend about 2.5 cm along the bottom of the tray, leaving some 15 cm outside. If other materials are used for the wicks, several layers may be needed to build up a thickness of 2–3 mm. Stir the slurry of Pevikon and pour it into the tray, taking care not to disturb the wicks, nor to entrap air bubbles. At this stage, the consistency of the medium will be unusably thin and the excess buffer must be allowed to drain off, through the wicks, until a stable slit can be cut in the polymer with a spatula. This surplus electrolyte can conveniently be absorbed by wads of towelling. When the correct consistency is achieved (which may take about 10 min: be careful not to over-dry), stop the drying process by folding the wicks over onto the top of the bed of polymer.

To load the sample, use a spatula to make a slit in the Pevikon, about 1 mm wide, at a distance of 7 cm from the cathode end of the tray. This slit must penetrate to no closer than 2 mm from the bottom of the tray, or 5 mm from the sides. Pass the sample slowly into this slit from a syringe fitted with a 19 gauge needle. The bed will accept about 0.4 ml of sample/cm of slit, which should be equivalent to not more than 10 mg lipoprotein/cm. Make a small, separate slit at each side of the tray and load these with the pre-stained marker specimen. Finally, close the slits by carefully pressing the sides together with the spatula. A total of up to 100 mg of lipoprotein protein can be loaded onto the polymer bed.

In the meantime, cool the electrolyte vessels, each containing 3 litres of electrolyte, to about 5 °C in the cold room. When the loading of the polymer bed is completed, position the tray between the electrolyte vessels. Note that it is vital for the tray to be horizontal and for the surfaces of the polymer and of the electrolyte in each reservoir to be at the same level. Submerge the wicks in the electrolyte, cover the tray with a sheet of acrylic to reduce evaporation, and pass a current of 50 mA for about 18 hours.

After the electrophoresis, remove the plastic cover (be careful that condensate does not drip onto the bed), set the tray on the bench and allow excess buffer to drain off, through the wicks, for about 10 min. If the tray is then illuminated from below, it is often possible to see the lipoprotein bands directly. If this is not possible, use the dyed markers as a guide to their position. Excavate the bands with a spatula and transfer the polymer to a sintered glass funnel (40–45 mm diam.) that is mounted on a small Buchner flask or a large sidearm test-tube. Wash the polymer four times with 10 ml volumes of 0.15 M NaCl solution, using light suction to assist draining. This procedure will give an essentially complete recovery of fractions containing up to 50 mg of lipoprotein protein. More extensive washing may be necessary if the load is greater. Finally, centrifuge at about 7000 g for 15–20 min to remove fine particles of Pevikon, and concentrate the product (Appendix 2).

2.5. *The isolation of Lp(a) and LP-X*

The isolation of these two lipoproteins will be described in detail because they are thought to have a pathological significance and are the only lipoproteins that are widely prepared in a 'pure' state. The first, which was rather unfortunately called Lp(a), was originally thought to be a genetically determined polymorph of lipoprotein B, but is now known to contain a characteristic protein, apo-Lp(a), in addition to apo-lipoprotein B and small amounts of other apo-lipoproteins. It has attracted interest by reason of an alleged involvement in atherogenesis. LP-X is a substance of abnormal shape and composition (Seidel et al., 1969; Hamilton et al., 1971) that has come to notice through its association with liver disease. Both Lp(a) and Lp-X have unusual combinations of properties that enable them to be separated from other lipoproteins comparatively easily.

The substances isolated by Alaupovic et al. (Kostner and Alaupovic, 1972; McConathy and Alaupovic, 1976) and called by them LP-A, LP-C and LP-D were also 'pure' in the sense that they contained only one apo-lipoprotein each but they have not been widely studied and are, at present, of little practical interest.

2.5.1. Lp(a)

Ehnholm et al. (1972) have shown by polyacrylamide gel electrophoresis that Lp(a) breaks down on storage at 0 °C to give a lipoprotein that seems to be indistinguishable from LDL, together with apo-Lp(a) and serum albumin. Since the characteristic apo-lipoprotein is also dissociated from the Lp(a) by freezing, it is important that the lipoprotein should be prepared promptly, from fresh material. According to Kostner (1976a), Lp(a) becomes progressively less stable as the purification proceeds and has a pronounced tendency to aggregate during storage. Lp(a) has a particle weight of about 5×10^6 and is therefore rather bigger than the average LDL but, because it contains some 40% protein, it is considerably more dense (approx. 1.09 g/ml). This combination of properties has been made the basis of one method of isolation. Alternatively, advantage may be taken of the fact that Lp(a) is adsorbed relatively strongly by hydroxyapatite. Berg (1963) adsorbed the lipoprotein on to hydroxyapatite directly from serum, but later workers have preferred to start by isolating a fraction of density 1.06–1.12 g/ml, using the ultracentrifuge. The Lp(a) may then be separated from the other lipoproteins that are present by chromatography.

Since the amount of Lp(a) in plasma rarely exceeds 80 mg/dl, with an average of only about 15 mg/dl, it is necessary to start with a large volume of blood. If possible, it is desirable to use plasma from subjects who are known to be 'Lp(a) positive'. The following method is that recommended by Kostner (1967a).

Reagents. 4% (w/v) sodium phosphotungstate; 2 M $MgCl_2$; 9% NaCl (several litres); 0.15 M NaCl containing 0.4% (w/v) sodium phosphotungstate and 0.05 M $MgCl_2$; 0.1 M sodium citrate; NaCl solution of density 1.055 g/ml (several litres);

0.15 M NaCl (several litres); 0.25 M potassium phosphate buffer, pH 6.8; 0.65 M potassium phosphate buffer, pH 6.8; 0.15 M NaCl containing 0.1 M potassium phosphate buffer pH 7.5 and 0.01% EDTA.

Procedure. Plasma or serum is obtained from young, fasting, healthy donors of either sex. To each ml of plasma add 1 mg of NaN_3 and 0.1 mg of Na_2EDTA. Mix 1 litre of the pooled plasma with 100 ml of 4% phosphotungstate and 25 ml of 2 M $MgCl_2$. Allow to stand for 1 hour at room temperature and centrifuge off the resulting precipitate. Wash once with the NaCl–$MgCl_2$–phosphotungstate mixture and then dissolve the precipitate in 0.1 M sodium citrate (Section 2.2.2). Make up to a volume of about 50 ml and dialyse against 9% NaCl (Section 2.2.2) for at least 36 hours. Next, adjust the density of the crude lipoprotein concentrate to 1.055 g/ml by dialysis and ultracentrifuge it at 100 000 *g* for 20 hours at 16 °C. Remove and discard the supernatant layer of VLDL and LDL.

The fraction in the lower third of the tubes, which contains most of the Lp(a), may be treated in either of two ways:

(a) raise the density to 1.21 g/ml and ultracentrifuge again for 20 hours at 16 °C. It is desirable to use a high speed rotor to effect the most complete sedimentation of the contaminating plasma proteins.

(b) dialyse against 0.15 M NaCl and then apply to a column of hydroxyapatite (about 1 cm × 30 cm; Section 2.3.3). Wash the column with 0.25 M phosphate buffer, pH 6.8, to remove contaminating proteins and lipoproteins. The Lp(a) is then eluted with 0.65 M phosphate buffer, pH 6.8.

Note. Since hydroxyapatite is a material of variable properties, the concentrations of these buffers may need slight adjustment.

Concentrate the fraction produced by either of these methods to a volume of about 20 ml (Appendix 2). Apply this concentrate to a column of Biogel A-15M (2.5 cm × 100 cm) that has been equilibrated with NaCl-phosphate buffer, pH 7.5, and elute with the same buffer (Section 2.3.1). The Lp(a) is eluted in front of the residual contaminants. Check the purity of the product as described in Chapters 4–6.

2.5.2. LP-X

This lipoprotein is a large, vesicular particle of unusual composition, that behaves in the ultracentrifuge like a normal LDL. In some cases of obstructive jaundice, it may even be possible to detect the presence of LP-X by analytical ultracentrifugation, as a peak moving with an S_f of 12–15 svedbergs. In the heparin precipitation procedure (Section 2.2.3.1) it behaves like a β-lipoprotein (LDL), but in the cold-ethanol process (Section 2.2.4) it behaves like an α-lipoprotein (HDL). These properties are the basis of the classical isolation procedure of Seidel et al. (1969).

The use of gel chromatography to isolate immunologically pure LP-X directly from plasma would be appear to be limited by the low resolution of the method, to which reference has already been made. This seems to be borne out by Magnani's (1976) observation that LP-X prepared in this way contained appreciable amounts of normal LDL that could not be removed by re-chromatography but which could be precipitated immunologically. Nonetheless, Picard and Veissiere (1970a,b) claim to have used gel chromatography on agarose gel to prepare LP-X that did not react with antiserum to albumin or antiserum to normaL LDL. Moreover, Agorastos et al. (1978) successfully used this method in their study of the relation between the presence of LP-X and the level of lecithin: cholesterol acyl transferase activity in plasma. They were using the method qualitatively however, and were not attempting to isolate 'pure' fractions. At present, it appears that gel chromatography can play a useful though limited role in the isolation of LP-X.

Methods which make use of the zonal ultracentrifuge have also been described (Danielsson et al., 1973; Patsch et al., 1976) and these may be of interest when there is easy access to one of these machines.

We shall describe a modification of the method of Seidel et al. (1969), which is probably still the simplest and most reliable. Burstein and Scholnick (1973) have proposed a method that does not require the use of an ultracentrifuge but, since the normal low-density lipoproteins are removed by immunoprecipitation, it is suitable only for small scale work.

Procedure. Serum or citrated plasma should be obtained from patients with obstruction of the bile duct. The VLDL may be removed by ultracentrifugation (Section 2.1.1.8) but, as the level of these substances in cases of liver disease is usually low, this step is not always necessary. Use the cold-ethanol process described in Section 2.2.4 to precipitate fraction I + II + III from the serum (or the infranatant from the ultracentrifugation). This residue can be centrifuged off and discarded. The supernatant, which contains the LP-X, may be treated in two ways: (a) it may be concentrated to about one third volume at the lowest practicable temperature, in a rotary evaporator, or (b) pour the solution into a dialysis tube of 6–12 mm diameter and immerse it in ice-cold 0.15 M NaCl which contains 1.0 mM EDTA and is adjusted to pH 7 with $NaHCO_3$. Gently agitate the dialysing bath and change the NaCl solution frequently during 24 hours. The first method is quicker but requires more attention than the latter. In either case, dialyse the resulting solution for 24 hours at 4 °C against 0.075 M NaCl–1.0 mM EDTA, pH 7, and then precipitate it with heparin–$MnCl_2$ as described in Section 2.2.3.1. Allow this reaction mixture to stand for 30 min at room temperature before centrifuging off the precipitate of LP-X. Dissolve the precipitate in sodium citrate solution (Section 2.2.3.1) and re-precipitate with $MgCl_2$; repeat this twice. Finally, remove the heparin by dialysis firstly against $BaCl_2$ and then successively against 9% NaCl and 0.15 M NaCl–1.0 mM EDTA (Section 2.2.2).

Note. In the original method, the heparin precipitation precedes the fractionation with cold ethanol. This produces the LP-X in a relatively large volume of solution, which may be inconvenient. In the procedure described above, the heparin precipitation leads to a concentration of the LP-X. Both methods appear to be equally efficient. The use of immobilised heparin in this preparation does not seem to have been investigated.

After a thorough dialysis, adjust the density of the lipoprotein solution to 1.063 g/ml and centrifuge at 100 000 *g* for 24 hours at 4 °C (Section 2.1.1.8) to remove residual plasma proteins. The LP-X forms the supernatant layer. This procedure is surprisingly effective and the product is rarely contaminated with albumin or substances containing apo-lipoprotein B. However, if these are detected they can be removed either by chromatography on agarose gel (e.g. Bio-Gel A-50M; Section 2.3.1) or, which is probably better, by treatment with the appropriate immunoadsorbent (Section 2.3.2.1).

2.6. The preservation of isolated lipoproteins

While the lipoproteins are dissolved in the plasma, the presence of the other plasma components apparently helps to protect them against oxidation and the effects of freezing. When isolated however, they are more susceptible to oxidative degradation and, like many emulsions, they will not withstand freezing. Once prepared therefore, they should be used as soon as possible.

It is not clear whether the storage of isolated lipoproteins brings about changes in composition like those to which they are subject during the storage of plasma, but the ultimate degradation product is a similar insoluble aggregate. The early stages of this aggregation can be detected by a progressive increase in the amount of light that is scattered by the preparation. In the case of the low-density lipoproteins this can be observed in the simple nephelometer described by Thorp et al. (1967). The tendency to aggregate is important when the lipoproteins are examined by electron microscopy, since old preparations form clumps or chains of particles on the grids. This is very noticeable with VLDL, and it is impossible to make good observations on anything but fresh preparations. Once degradation has become fairly well advanced, it can also be detected through the broadening of the schlieren pattern that is observed in the analytical ultracentrifuge.

The loss of integrity of the lipoproteins also leads to changes in their electrical properties, of which the increase in electrophoretic mobility is the most easily observed. There is also an increase in the conductivity of the solution, the cause of which is unknown. This would provide a convenient way of monitoring the early degradation of a lipoprotein preparation, were it not for the disadvantage that it can only be used in solvents of low ionic strength, since the magnitude of the effect is small.

At present, the only steps that can be taken to protect the chemical integrity of lipoproteins are the maintenance of sterility and the prevention of oxidation. If the recommended procedures are followed, the preparations will contain EDTA and sodium azide, which have a stabilising effect. But, if a period of storage is unavoidable, it would be advisable to de-aerate the solution and to store it under nitrogen. Schuh et al. (1978) found that the addition of the anti-oxidants butylated hydroxytoluene (2 mM) or propyl gallate (5 mM) to be beneficial and this is supported by the later work of Lee et al. (1981). These workers focussed their attention on the properties of apo-lipoprotein B and concluded that the anti-oxidant effect of a mixture of butylated hydroxytoluene (0.002%) or butylated hydroxyanisole (0.01%) with reduced glutathione (0.05%) was superior to that of any of these substances used separately. On the evidence of these observations it may be possible to preserve LDL in this mixture of anti-oxidants for several months in a sealed tube, under nitrogen at 4 °C.

Another potentially valuable but virtually unexplored way of maintaining the stability of lipoproteins derives from the observation by Lovelock (1957) that substances like glucose, sucrose and glycerol can protect lipoproteins against denaturation. It has been known for many years that sugars or polyhydric alcohols will stabilise solutions of simple proteins but the mechanism of the action is still not entirely clear. Recently however, Lee and Timasheff (1981) have found evidence that, in a protein–sucrose solution, the sugar is excluded from the protein domain and they propose that the resulting increase in free energy of the system leads to an increase in the apparent activation energy of denaturation. Remarkably enough, little seems to have been done to pursue Lovelock's observation, although Triplett and Fisher (1978) have claimed that LDL can be frozen in 20% sucrose solution without apparent harm. Our own observations confirm this. Moreover, we have found that LDL can be freeze-dried in 20% sucrose and will retain its solubility properties. However, although a solution of LDL in sucrose can be kept at 4 °C for many months without evident change in its gross physical properties, the extent of any concurrent oxidation has not been investigated. More detailed studies of the lipoprotein–sucrose system would clearly be of great interest. On present evidence, the best system in which to store lipoproteins for an extended period is an oxygen-free solution of 20% sucrose that contains bacteriostatic agents and anti-oxidants, the mixture being sealed under N_2 at 0 °C.

The characterisation of plasma lipoproteins: introduction to principles and methods

As we remarked in Chapter 1, every lipoprotein preparation is a mixture of particles that differ with respect to size, density, electric charge and chemical composition. Moreover, no two preparations will be exactly alike (e.g. Fig. 1.3) and each must be identified by several physical and chemical measurements that will enable it to be adequately compared with similar preparations made on other occasions, or in different laboratories. This process of characterisation has three elements. Firstly, there must be evidence for the purity of the preparation. Secondly, it must be defined in terms of its mean physical and chemical properties. Thirdly, an estimate of the dispersion of its properties should be made. In this chapter we shall discuss the general principles of this process of characterisation and the merits of the various methods that are available. Working details of the recommended procedures will be given in the chapters that follow. It will become evident that a complete characterisation can be both protracted, and consuming of material. It is therefore necessary for each worker to adjust his effort in this direction to suit the nature and calibre of his main experiment; a delicate judgement in which it is easy to stop short of the optimum.

3.1. The purity of lipoprotein preparations

As commonly used, the term 'pure' would indicate an absence of substances other than lipoproteins. These will normally be plasma proteins, of which albumin is probably the most troublesome, as it can be tightly adsorbed to the lipoprotein. At present, the best way of detecting these impurities is by immunological assay, either with a specific antiserum, or with antiserum to whole human plasma. But it is important that the immunological balance of the system is such that it can detect a small amount of the suspected impurity. It is not enough to test a preparation against a single concentration of antiserum unless it has already been shown that this is capable of detecting the impurity at the low levels expected. Moreover, the lower limit of detection must be quantitatively known. It is unsatisfactory (though nonetheless common) merely to say that 'there was no reaction with antiserum to...'.

In practice, it may be difficult to decide whether a minor protein component of a lipoprotein preparation is an impurity that is adsorbed to the particle, or is a genuine apo-lipoprotein. At the time of writing, for example, there is some controversy over the role of β_2-glycoprotein-1. Furthermore, a lipoprotein may be contaminated with enzymes that are present at levels that can be detected only by their characteristic activity. Of these, proteinases and phospholipases could play a part in the progressive degradation of the lipoprotein during storage, even though they may be present at too low a concentration to be detected in the relatively brief reaction time that is used in most conventional assays. Care should therefore be taken to add appropriate inhibitors at the outset of the isolation.

3.2. Physical and chemical characterisation

All lipoprotein preparations should be identified by the determination of some characteristic physical or chemical properties. This is particularly important if the isolation technique is one, like gel filtration, precipitation or electrophoresis, that defines the product poorly, if at all.

Unfortunately, there are few properties that are suitable for the purpose and are also easy to measure accurately. In practice, many lipoproteins are identified by the limiting densities at which they are centrifuged during their isolation (Section 2.1.1). In principle, lipoproteins that are isolated by electrophoresis or gel filtration can be defined in an analogous way, in terms of relative mobility or particle size. However, there has been a general disinclination to do this, which may only partly be due to the technical reasons that we shall discuss later. In any event, there is no justification for the quite common malpractice of indiscriminantly applying the terms HDL, LDL and VLDL to lipoproteins that have not been isolated by centrifugation. For example, the assumption that α-lipoprotein is identical with HDL is often wrong, as is well illustrated by the work of Laplaud et al. (1980). It cannot be too strongly emphasised that it is improper to use the density terminology for lipoproteins that have not, at some stage in their isolation, been subjected to preparative centrifugation. In principle, of course, it is equally undesirable to apply the terms 'α-' and 'β-lipoproteins' to fractions that have not been prepared by electrophoresis. However, since these names are quantitatively imprecise, their application to the products of precipitation or gel filtration methods may be tolerated for lack of anything better.

3.2.1. Physical methods

In effect, the limiting densities of a lipoprotein preparation are an approximate measure of the hydrated density of the particles, although it would be unwise to assume that this is equal to the mean of the limits unless these are very close together. A better estimate of the hydrated density is obtained when the preparation is made (or is purified by) equilibrium centrifugation on a density gradient. However, density is an unsatisfactory way of characterising VLDL because these are often less dense than water and their lower limiting density cannot be established by centrifugation in aqueous solution. It is therefore usual to refer to VLDL as having a density less than 1.006 g/ml (i.e. less than that of the protein-free ultrafiltrate of plasma). This leaves the lower threshold density indeterminate although, in the limit, it cannot be less than about 0.92 g/ml. A single-ended identification of this kind is unsatisfactory and it is better to use a method that is based on differentiation by flotation rate, such as the preparative method described in Section 2.1.1, or analytical ultracentrifugation (Section 4.1).

As we have already pointed out in Chapter 2, the limiting densities quoted for a preparative centrifugation are of little definitive value if they are not corrected to take account of the effects of temperature, and the re-distribution of salt that occurs during the run. In any event, although the limiting densities may be a sufficient characterisation for lipoproteins that are used in biological experiments, those that are intended for more precise work should be identified with greater care. The most accurate determinations of hydrated density (or partial specific volume) are made by analytical ultracentrifugation, or by pycnometric methods. The method of choice is the speedy and accurate oscillating-tube density meter (Kratky et al., 1973) which is described in Section 2.1.1.3. Unfortunately, the equipment is expensive. The more conventional pycnometric methods are both tedious and experimentally exacting, and may sometimes require the use of special equipment. Ultracentrifugation, on the other hand, is technically simple but is time consuming since it requires the measurement of flotation or sedimentation rates in several different solvents (Section 4.2).

The three other physical characteristics that can be used as determinants are: flotation rate, relative particle weight and particle diameter. The diffusion constant is also required for the accurate estimation of particle weight from measurements of flotation rate, but is not itself of much value for identifying lipoproteins. All of these parameters can be estimated in the analytical ultracentrifuge, which is still one of the lipoprotein chemist's most powerful analytical tools despite the disadvantages that are discussed in Section 3.3.2.2. Of the four properties mentioned, the flotation (sedimentation) rate is the least laborious to estimate (Section 4.1.2.1). As will be seen in Section 4.1.2.2, it requires considerably more work to determine the relative particle weight with even modest pretensions to accuracy. This smaller demand for time makes the flotation rate the more attractive characteristic for routine estimation, particularly since its measurement can be combined with observations of heterogeneity. However, it must be remembered that the particle weight and hydrated density are intrinsic properties of the lipoprotein, whereas the flotation rate is a function of the conditions of observation, which must be carefully specified if the measurement is to be of value.

The diameters of lipoprotein particles can also be estimated by direct observation in the electron microscope, in which they take the form of spheres. Moreover, it has been shown that the diameter measured in this way is close to the Stokes diameter of the lipoproteins as determined by hydrodynamic methods. Microscopy may therefore be a useful and speedy method of characterisation where the equipment is available (Section 4.5), subject to the following two considerations: (a) the bodies that are observed are not the native lipoproteins, and (b) satisfactory observations can only be made if the lipoprotein preparation is very fresh.

To those who are denied the use of an analytical ultracentrifuge, the prospect of estimating the diameter and relative weight of particles of HDL, LDL and the smaller VLDL by filtration chromatography (Margolis, 1967) may appear attractive. However, apart from the fact that the original description of the method (Ackers and Steere, 1967) is in some respects obscure, the technique suffers from a practical handicap that limits its general use. In essence, a column of agarose gel beads is calibrated with substances of known particle size. Because the lipoprotein particles are almost spherical and are large in diameter, the standard substances should be of similar shape and of a comparable range of sizes. But there are few readily available, well characterised substances of a particle size that is comparable with that of VLDL (Margolis used bushy stunt virus, which has a Stokes diameter of 30.9 nm). Until this difficulty is overcome, the method is unlikely to be widely used. It also has to be said that the resolution of the method is comparatively low and is, of course, fixed by the porosity of the gel. In this respect it lacks the flexibility of the analytical ultracentrifuge. Moreover, it requires more material than the centrifuge. Nonetheless, as judged by the values quoted by Margolis (1967) for the diameter (25.8 nm) and particle weight (3.5×10^6) of LDL, the method appears to give results that are roughly comparable with those obtained by centrifugation. In general however, filtration chromatography seems to be more suitable for the detection of heterogeneity than for the estimation of particle size.

Analysis by zone electrophoresis in stabilising media has the great merit of speed and economy of sample, as well as requiring relatively inexpensive equipment. It has therefore found great favour with clinicians, for the routine semi-quantitative analysis of the distribution of lipoproteins in plasma. However, the migration of the lipoproteins is often strongly influenced by such factors as the properties of the stabilising medium, the conditions of the electrophoresis and the extent of adsorption to the medium. It is therefore difficult to make measurements of mobility that can be reproduced in different laboratories. These problems have been extensively discussed by Opplt and Opplt (1980). Even the use of a 'relative mobility' that is measured by reference to either an added standard, or to another plasma protein does not entirely overcome the difficulty. This disadvantage, coupled with its generally low resolving power, makes electrophoresis unsuitable for the quantitative characterisation of lipoproteins.

It is perhaps surprising that light scattering techniques have not been more often used to determine the size and relative weight of lipoprotein particles. Thanks to the large size of the particles, the intensity of the light they scatter is comparatively high. This, taken with the introduction of laser light sources would seem to make it a method worthy of further study. However, considerable care would be needed to ensure that the tendency of lipoproteins to aggregate did not falsify the results.

3.2.2. Chemical characterisation

The well established precept that a substance should not be identified solely by its physical properties is especially pertinent in the case of the lipoproteins because their physical properties are not uniquely related to their chemical composition. For example, it is possible for particles of different composition to have the same density. A good characterisation should therefore include an analysis for protein, esterified and unesterified cholesterol, triglyceride and phospholipid. Of course, these do not account for the whole particle, since lipoproteins contain small amounts of free fatty acids, as well as traces of colouring matter and other fat-soluble substances. However, the errors introduced by ignoring these components are small. On the other hand, it is important that all five main components should be included in the analysis. Analyses that omit triglyceride, for example, or that give a combined figure for esterified and unesterified cholesterol are useless. Incomplete analyses are often met with in reports on clinical experiments of the kind that investigate the effect of drugs on 'lipoproteins'. A typical example of this genre states that a certain treatment caused 'an increase in VLDL triglyceride but no change in cholesterol'. But the many different particles present in VLDL and their non-stoichiometric nature will not allow us to find a unique interpretation of this observation and its value is therefore severely limited. The same problem can arise with the LDL and HDL, but is slightly less acute for these fractions because they are usually somewhat less heterogeneous than VLDL. Nonetheless, it is worth noting that although a determination of the lipoprotein profile may make it possible to interpret chemical analyses of these traditional lipoprotein classes more exactly, it is usually better to design experiments around the use of well-defined sub-fractions that have a narrow range of densities and can be regarded as approximately homogeneous (cf. Lee and Alaupovic, 1970, 1974).

Mixtures of lipids are usually analysed by chromatography, followed by the estimation of the separate components (Kates, 1975). Unfortunately, although the conventional procedures are simple (Ch. 6), they are also laborious. Attempts to overcome this by using non-specific methods of estimation, e.g. by charring with sulphuric acid on thin-layer plates, suffer from the defect that the reproducibility between laboratories is low. Recently however, more precise techniques have been devised that use the flame ionisation detector as the means of quantification. With these it is possible to analyse microgram amounts of lipoprotein relatively quickly, but it is important to note that all methods that depend on non-specific assays require that each component of the mixture is completely separated from all the others, and it is sometimes difficult to be certain that this has been achieved. In the method of Kuksis et al. (1975) the phospholipids are first hydrolysed with phospholipase C. The total lipid fraction is then extracted, silylated, and analysed by conventional gas-phase chromatography. The apo-lipoprotein must be estimated separately. The need for separate determinations of protein and lipid is avoided in the method proposed by Mills et al. (1979), in which all five major components of the lipoprotein are separated by chromatography on silica-coated quartz rods that are then scanned directly in the flame ionisation detector (Section 6.7). This technique is quick and economical of material, but requires relatively expensive equipment and, in its present form, has comparatively low precision.

The proportions of the apo-lipoproteins are no more fixed than those of the lipid components of the lipoprotein. Moreover, recent studies have shown that many of them play a part in the regulation of lipoprotein metabolism that, in some cases, may be more important than their role in stabilising the lipoprotein particle. A quantitative analysis of the apo-lipoprotein distribution may therefore play an important part in the chemical characterisation of the lipoprotein. Although the determination of total protein content is quite simple (Section 6.4.1), it is much more difficult to estimate the individual apo-lipoproteins. For the best combination of specificity, sensitivity and technical simplicity, immunochemical analysis must be used (Section 8.9). However, the need for purified apo-lipoproteins and mono-specific antisera makes it difficult for any but the most specialised laboratories to determine all of the known apo-lipoproteins in this way.

The many laboratories that lack a complete set of specific antisera have to settle for a method of analysis that physically isolates each apo-lipoprotein, of which electrophoresis is the best available (Section 8.9.1). However, the tendency of the apo-lipoproteins to associate is so strong, even in solvents like 8 M urea or 6 M guanidine hydrochloride, that none of these techniques will yield pure proteins. Consequently, analyses that are made in this way do not have the specificity of the immunochemical methods. Electrophoresis in polyacrylamide gel is the most economical of material and has the highest resolving power, but has the defect that quantification depends on the staining of the protein bands which are then estimated by photometric densitometry. This is a procedure that requires rigorous technical control and validation with pure apo-lipoproteins. Since the latter are not often available, it follows that electrophoretic analysis as it is usually practiced is no better than a semi-quantitative method. Moreover, it may be particularly susceptible to the presence of artifacts that are formed in samples that are not fresh (Herbert et al., 1977). But, when the polymorphic forms of apo-lipoproteins A and E are analysed by iso-electric focussing (Section 6.4.2.6), the polymorphs may be expected to stain in a comparable way and reliable estimates of their relative proportions may be obtained even without the use of standards.

Before the apo-lipoproteins can be analysed by any non-specific physical method, they must be separated from the lipids. This is usually done by solvent extraction, which is a technique that has to be performed with great care if denaturation of the proteins is to be avoided (Appendix 3). A convenient and widely used development of this procedure was introduced by Kane (1973). In this, a differential extraction with tetramethyl urea is used to separate apo-lipoprotein B and the lipids from the smaller apo-lipoproteins (Section 6.4.1.2). The total of the latter can then be determined by the Lowry method, and their relative proportions estimated by electrophoresis in gels containing 8 M urea. To ensure that the results are reliable, control gels should be run with authentic pure apo-lipoproteins.

3.3. Heterogeneity and polydispersity

No strictly homogeneous lipoprotein preparation i.e. one in which all the particles are identical in every respect, has yet been made. Moreover, some preparations will be more heterogeneous than others, even if they are made in the same way (Fig. 1.3), and anyone who is proposing to perform experiments with isolated lipoproteins is well advised to determine what kind of mixture is at his disposal before embarking on a detailed, laborious and expensive study. The characteristics on which the analysis of heterogeneity is based are often simple properties such as particle diameter, flotation rate or protein content, but may be more complex parameters that are defined, for example, by the conditions under which the particle is eluted from a chromatography column. The examples given in Fig. 1.3 show the distribution of flotation rate in several conventional lipoprotein fractions that were isolated by centrifugation and were therefore defined by the densities at which this operation was carried out. The complex chemical heterogeneity of the lipoproteins is exemplified in Fig. 3.1, which shows how various apo-lipoproteins are distributed with respect to the size of the particles as determined by gel filtration chromatography. It is clear from these curves that both distinguishable groups of lipoproteins are polydisperse with respect to each apo-lipoprotein. Moreover, the maximum ordinate for the distribution of each protein occurs at a different particle size. As in the case of conventional macromolecules this effect is evidence of heterogeneity but, because the lipoproteins do not conform to the laws of stoichiometry, it is extremely difficult to determine exactly what form this heterogeneity takes.

The dispersion in properties that is illustrated in these two figures may be relatively unimportant in biological experiments, in which precision is usually low. Even in these circumstances however, it may be advisable to use lipoprotein fractions that have been isolated within limits that are closer than those conventionally used. In more precise physico-chemical experiments, dispersion can cause a significant loss of reproducibility and it may be desirable to establish its magnitude. This can be done by analysing in numerical terms the shape of the distribution curve that plots the concentration of lipoprotein as a function of the parameter of interest (e.g. Fig. 1.3). Unfortunately,

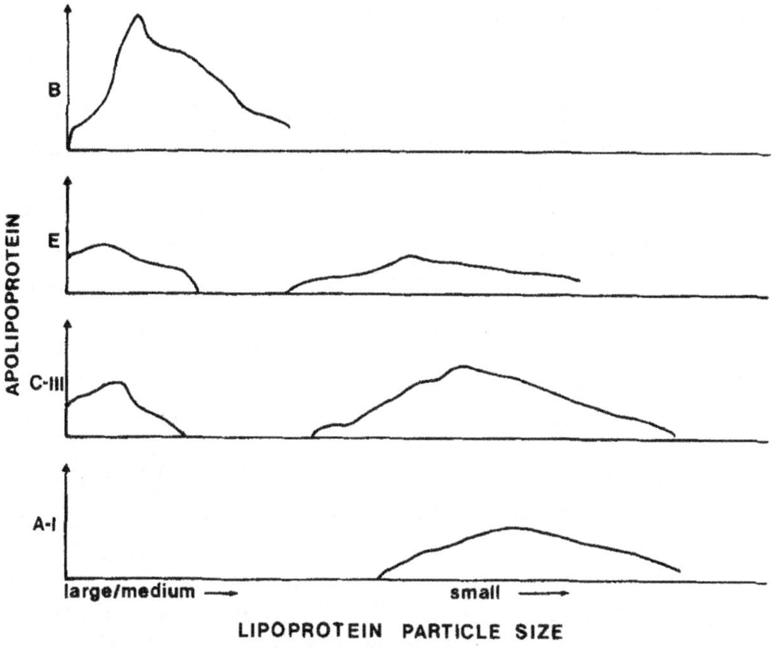

Fig. 3.1. The protein composition of the lipoprotein fractions resolved by filtration chromatography on a 6% agarose gel (Bio-Gel A-5m). Each fraction of eluate was separately assayed for each apo-lipoprotein by electroimmunoassay. The terms 'large', 'medium' and 'small', correspond approximately with VLDL, LDL and HDL. Note that, whereas the E and C-III proteins elute together in the large particles, they tend to precede the B protein. In the small particles the peaks of E, C-III and A-I elute in that order. (These data are made available by the kindness of Dr. P. Alaupovic.)

the procedure is discouragingly laborious, and the usual course has been to present the curve itself for visual appraisal, as in Fig. 1.3, or to compare the areas on each side of the maximum ordinate. With the advent of the cheap micro-computer however, routine mathematical analysis should become a more realistic proposition.

3.3.1. Contamination with lipoproteins that lie outside the nominal bounds of the preparation

Although lipoproteins are often prepared by methods that define, more or less precisely, the limits within which one physical property should lie, e.g. density or particle size, they will in practice be contaminated with lipoproteins that lie outside these limits. We shall regard this as an aspect of heterogeneity. An important analysis of the boundary indeterminacy that leads to the cross-contamination of ultracentrifugally prepared lipoproteins was made by Lindgren et al. (1972a,b). However, it has the significant drawback that it is based on an idealised model system. It therefore estimates the minimum and not the actual cross-contamination. The latter (for example, the small amount of VLDL that is normally present in a preparation of LDL) is very difficult to measure. In principle, it can be done either by ultracentrifugal or by electrophoretic analysis, but these methods are not sensitive enough to detect low orders of contamination. Perhaps the most sensitive way of detecting the presence of unduly large or unduly small particles is by electron microsocopy (Section 4.5), which can also be helpful if the lipoprotein particles are of different shapes, e.g. discoid LP-X particles in the presence of normal spherical LDL. Of the more accessible techniques, immunoelectrophoretic analysis probably offers the best combination of sensitivity and resolving power (Sections 5.6, 5.7). However, this can only be used when the possible contaminant carries an apo-lipoprotein that is absent from the principal lipoprotein. For example, LDL can be detected in LP-X through its reaction with antiserum to apo-lipoprotein B, which is not a component of LP-X. In general, it is difficult to estimate the degree of cross-contamination by chemical or immunochemical analysis because, (a) it is rare for any lipoprotein to be entirely devoid of one of the normal constituents, (b) the available methods of analysis are not sufficiently precise, and (c) the composition of what is ostensibly the same lipoprotein fraction varies from one blood sample to another.

3.3.2. Heterogeneity within the nominal bounds of the preparation

Dispersion within the nominal confines of the preparation can be estimated in several ways. In the most frequently used methods, the mass distribution of the lipoprotein particles is determined after they have been separated into sub-fractions by centrifugation, chromatography or electrophoresis, on the basis of the relevant physical property. Alternatively, the fractions obtained by a separation of this kind can be chemically analysed to give the distribution of the individual components. Although the chemical analysis of many fractions can be laborious, it can often lead to valuable information about the inhomogeneity of a lipoprotein preparation, as is illustrated in Fig. 3.2 or by the work described by Mills et al. (1976).

3.3.2.1. Electron microscopy Because the observations can be made quickly, on very small samples of material, it might appear that a size distribution obtained from measurements of representative electron micrographs would be a good way of characterising the

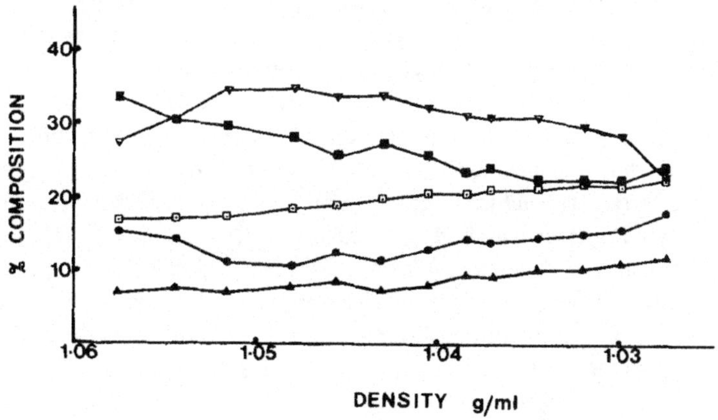

Fig. 3.2. Chemical heterogeneity of LDL exposed by analysis of fractions obtained by density gradient centrifugation. ▽, cholesteryl ester. ■, protein. □, phospholipid. ●, triglyceride. ▲, unesterified cholesterol.

heterogeneity of a lipoprotein preparation, and this method has been used in several laboratories (see for example Bierman et al., 1965; Kane et al., 1975; Chapman et al., 1981). However, it is important to bear in mind that the bodies observed are not the native lipoproteins and that the particles have a strong tendency to aggregate in preparations that are not fresh. Moreover, variation in size is technically more exacting to measure accurately than the dispersion of particle weight, which is proportional to the third power of the radius. Nonetheless, microscopy can be a valuable way of detecting the presence of abnormal discoid lipoprotein particles such as lipoprotein X, or the newly-secreted, nascent particles of HDL. For this purpose, it is probaly the most sensitive procedure available.

3.3.2.2. Ultracentrifugation By contrast with the electron microscope, the analytical ultracentrifuge allows the lipoproteins to be studied in their native state and is particularly useful because it allows polydispersity to be studied concurrently with the determination of sedimentation rate, relative particle weight, hydrated density and diffusion constant. Despite this versatility it has been generally ignored in favour of methods that are quantitatively inferior. Several misconceptions seem to have contributed to this neglect, notably the idea that analytical centrifuges are uncommon and difficult to use. In fact, many universities have an analytical centrifuge that spends much of its time standing idle, and there can be few lipoprotein laboratories that are denied all access to one. Moreover, the design of modern instruments is such that any competent person can easily be trained in their use, and the measurements are made by relatively simple routine procedures (Chervenka, 1969). Above all, the speed of the method is not sufficiently well appreciated. For example, it takes substantially less than two hours to obtain a record of a lipoprotein profile that will immediately yield qualitative information on polydispersity and particle size.

Nonetheless, it is true that the extraction of quantitative data from the ultracentrifuge records is time consuming, although the wide publicity given to the methods used at the Donner laboratory (Ewing et al., 1965; Lindgren et al., 1972; Lindgren, 1974) may have given potential users an exaggerated impression of the difficulties involved. These methods are based on techniques that were originally developed for the routine estimation of lipoprotein levels in large numbers of sera and are not well suited to the needs of the worker who only has to do an occasional analysis. Indeed, it is arguable that these complex procedures, with their emphasis on the use of digital computers, have discouraged the use of the analytical centrifuge as a tool in lipoprotein research, which might have been better served by a greater emphasis on technically simpler and less capital-intensive methods of treating the observations. It is notable that more conventional techniques have been used to great effect by, for example, Lee and Alaupovic (1973) and by Fisher and his colleagues (Fisher, 1970; Mauldin and Fisher, 1970; Hammond and Fisher, 1971; Fisher, Granade and Mauldin, 1971; Fisher, Hammond and Warmke, 1972).

Whatever the procedural details, the hydrodynamic analysis of lipoproteins is subject to certain limitations. In the first place, the theoretical bases for the analysis of highly inhomogeneous mixtures are uncertain. Secondly, some of the information that is required is still not well established. In 1953, Toro-Goyco remarked that 'although β-lipoproteins from human plasma have attracted considerable attention, much work on their physical chemistry remains to be done'. Over thirty years later, this is still true. For example, the published data on the relationship between the flotation rate and concentration of the lipoprotein are incomplete. Neither is it clear how this relationship changes with different experimental conditions, or whether it is the same for particles of all sizes. The whole question may be further complicated by evidence that the flotation rate may be a function of the intensity of the centrifugal field. The reason for this is unknown, but may be related to the distortion of the lipoprotein by hydrostatic pressure, or to a pressure sensitive association phenomenon. In fact, little is known about any aspect of the association of lipoprotein particles and the effect this may have on their behaviour in the centrifuge.

When used in the conventional way, the ultracentrifuge detects polydispersity in terms of rate of flotation (or sedimentation), which is a function of particle weight and the difference between the densities of the particle and the solvent. There is a well-established theoretical basis for the estimation of polydispersity by the analysis of boundaries produced by ultracentrifugation, which has been discussed by Schachman (1959). The technique was applied to lipoproteins by Toro-Goyco (1958) and, more recently, by Oncley (1969). It is the most sophisticated quantitative analysis available but is too laborious for routine use unless the extensive calculations can be done by computer. However, even without this help, it is possible to obtain useful information on the dispersion of the flotation rate by the comparatively simple methods that are described in Section 4.1.1.

The method of equilibrium centrifugation on density gradients that was devised by Meselson et al. (1957) has also been applied to lipoproteins, with apparent success, by Adams and Schumaker (1969a, 1970). Their procedure can give a quantitative estimate of the number of 'lipoproteins' that are present (i.e. a measure of heterogeneity), their mean densities and particle weights, and also their polydispersity with respect to density. However, the centrifugations are time-consuming and, since the computational procedures are described in only the barest detail, the would-be user faces a considerable task in familiarising himself with the mathematics of the model and in writing his own computer programmes. Like the flotation-rate method, this technique is of no value for the routine estimation of dispersion unless appropriate computing facilities are available.

Conventional equilibrium centrifugation can also be used to detect polydispersity through the non-linearity of the plot of log C against r^2 (Chervenka, 1969). Moreover, since it is possible to estimate both the weight average and the Z average particle weights by this technique, it provides a quantitative measure of the magnitude of the dispersion. In practice however, this method has not often been applied to lipoproteins.

In principle, many of these hydrodynamic measurements can also be made in the preparative ultracentrifuge and, as with the analytical centrifuge, there are two general ways of assessing polydispersity. In the first, a sequence of sub-fractions is isolated according to their flotation rates (Lindgren et al., 1972a,b; Gustafson et al., 1965). This is the only method that is at present available for the analysis of chylomicrons. The same procedure can be used for VLDL but, both for these and the denser lipoproteins, the analytical centrifuge will give results of greater resolution with less labour. In the second technique, the lipoproteins are centrifuged to equilibrium on a density gradient and then collected in a series of sub-fractions. From these it is possible to estimate the dispersion of the lipoprotein particles with respect to hydrated density. Moreover, each fraction can be subjected to chemical analysis and this combination of physical and chemical analyses gives the best available estimate of the polydispersity of a lipoprotein preparation. Technically, this is a simple procedure but, because of the need to analyse a large number of samples, it is time-consuming. Nonetheless, it is probably rather less laborious than the flotation-rate method and has the merit that, by judicious choice of gradient (Section 2.1.2), it can easily be adjusted to deal with mixtures that range from total plasma lipoproteins to sub-fractions thereof. However, it cannot be used to study chylomicrons, or VLDL that are less dense than water. When used to analyse lipoproteins that are radioactively labelled, it is potentially very sensitive and is the method of choice for analysing by density the small amounts of lipoprotein that are often the outcome of metabolic experiments in intact animals or, for example, liver perfusions.

3.3.2.3. Gel filtration chromatography As we have already pointed out (Section 3.2.1), this technique has low resolution and suffers from a dearth of suitable standards with which to calibrate the columns. Moreover, the analysis of the elution curves in terms of inhomogeneity is less well developed than that of the boundaries formed during analytical ultracentrifugation. Only if the dispersion in size becomes evident through a lack of symmetry in the distribution can gel filtration detect it with certainty, and even then it is difficult to quantify. However, as with preparative centrifugation, the simple determination of a 'size' distribution curve does not fully exploit the potential of the filtration method. By measuring the distribution of other parameters, such as chemical composition, a more penetrating analysis of inhomogeneity can be made (e.g. Fig. 3.1). This additional information can be particularly useful when the interpretation of the size distribution curve is uncertain, but it can only be obtained at a considerable cost in labour. In this context, the technique of quantitative microscale thin-layer chromatography (Section 6.7) offers great potential advantages.

Commercially available gels will distinguish three main groups of lipoproteins that are approximately equivalent to the VLDL, LDL and HDL. The diameter of the largest particles is close to, or beyond the upper limiting pore size of the gels and they therefore cannot be resolved by gel filtration. However, it is possible to achieve a modest resolution of the smaller lipoproteins and, despite its inferiority to centrifugal analysis, chromatography has sometimes been used to detect re-distribution of these classes during in vivo experiments. It is also a simple and inexpensive way of detecting any aggregation or disaggregation that is the result of handling the lipoproteins, for example, during radioactive labelling (Schonfeld et al., 1974), or demonstrating the presence of lipoproteins of abnormal size, as in patients with obstructive jaundice (Agorastos et al., 1978).

3.3.2.4. Adsorption chromatography This technique has not been widely used for the quantitative analysis of lipoproteins. However, there are two methods that seem to offer a practicable assessment of inhomogeneity. The adsorption of lipoproteins to hydroxyapatite (Section 2.3.3) is probably largely determined by the nature of their protein moieties, although it may be modified by other factors. By making appropriate chemical analyses of fractions eluted from a column of hydroxyapatite by a salt gradient, it is possible to estimate the polydispersity of the sample in terms of chemical composition, albeit only with respect to the arbitrary parameter that is defined by the conditions under which the lipoproteins elute from the column. Despite this restriction, the analytical potential of this procedure appears to warrant its being given more attention than it has hitherto received. By contrast, the greater specificity of affinity chromatography (Section 2.3.2) makes this a method more suited to the detection of heterogeneity than the estimation of polydispersity. It is a very versatile technique, since heparin, lectins and immunoglobulins can all be used to prepare adsorbents with different specificities. The most commonly used lectin is Concanavalin A (Section 2.3.2.3), which is specific for apo-lipoprotein B. Immobilised heparin, on the other hand, will adsorb either apo-lipoprotein B or E (Iverius, 1972; Hay et al., 1978; Marcel et al., 1980). A still greater range of specificities can be obtained by the use of immobilised immunoglobulins, but this advantage is to some extent offset by the need for antisera that are monospecific for the apo-lipoproteins of interest (Section 2.3.2.1). Since these are difficult to prepare and are, in most cases, not available commercially, this method of analysing heterogeneity can only be fully exploited by a few specialised laboratories.

3.3.2.5. Electrophoresis Electrophoretic methods are more speedy than those we have just discussed, and can be used to analyse small amounts of material. The orthodox methods are capable of resolving the plasma lipoproteins into three major classes that approximately correspond with the VLDL, LDL and HDL, but a simple mass distribution analysis will not usually detect polydispersity within these classes. By contrast, electrophoresis on a polyacrylamide gel gradient (Section 5.5.2.2) is exceptional in its ability to resolve both LDL and HDL into several apparently discrete components, thus demonstrating the heterogeneous nature of these classes. Unfortunately, these components have not been studied separately and their degree of polydispersity is unknown.

Isoelectric focussing, which has a very high resolving power for apo-lipoproteins, has not so far proved to be as good for lipoproteins. Nonetheless, the few results that have been published suggest that the method has a potential for the analysis of polydispersity that should warrant its further investigation. At present however, better results are given by two-dimensional immuno-electrophoresis (Section 5.7) which, through its combination of physical and chemical discrimi-

TABLE 3.1
Summary of physical techniques for the characterisation of lipoproteins

Procedure	Properties that can be estimated	Can polydispersity be analysed?	Comments
Analytical ultracentrifugation			
(a) velocity method	Flotation rate Particle weight Hydrated density	Yes, by estimating the deviation of these properties from a Gaussian distribution	The method is quick, but requires expensive equipment. Also the behaviour of lipoproteins in the cell is not fully understood
(b) density gradient	Particle weight Hydrated density	Density dispersion can be determined	Time consuming and the computational procedures are unpublished
Preparative ultracentrifugation			
(a) serial method	Limits of hydrated density only	Serial fractionation gives an estimate of density dispersion at low resolution but fractions can also be analysed by other methods to give dispersion of particle wt., chemical composition and size	Tedious unless combined with the preparative stage. Inaccurate as a way of estimating hydrated density but fair for studying dispersion when combined with other techniques
(b) density gradient	Mean hydrated density	As for (a) but resolution can be much better	Also time consuming, but generally a better method than (a)
Filtration chromatography	Particle size	Low resolution but fractions can be analysed by other methods	Simplicity is the main attraction. Can be of value when used in conjunction with other methods of analysis

Electrophoresis	—	At low resolution with respect to size or electrophoretic mobility	Simple and economical but essentially non-quantitative
Electron microscopy	Particle size	Yes, with respect to size	Requires only a very small sample and, with effort, can give detailed estimates of size distribution
Pycnometry	Partial specific volume	No	The equipment is simple but the technique is exacting

nants, has a considerably better resolution than methods that depend on differences in only one property (Fig. 5.10). Moreover, this method requires only a very small amount of material and provides a permanent record from which it is possible, at least in principle, to estimate the proportions of the distinguishable particles that are present in the sample. But the patterns of precipitin lines are often complex and can be very difficult to interpret unless a full array of monospecific antisera are available. This method is therefore subject to the same restrictions as affinity chromatography on immobilised immuno-globulins, few laboratories being equipped to do a complete analysis routinely.

At the technical level, it is important to note that the results of all these electrodynamic procedures are influenced by the age and manipulative history of the lipoprotein preparation that is under study.

Physical characterisation of lipoproteins

This chapter is concerned with physical methods for the characterisation of lipoproteins, but excludes electrophoretic techniques which are dealt with in Chapter 5, and the use of the centrifuge to determine lipoprotein concentration, which is explained in Chapter 7.

4.1. The analytical ultracentrifuge

At the outset, it must be pointed out that there are theoretical limitations to the use of the analytical ultracentrifuge as a means of characterising lipoproteins. Firstly, lipoproteins are heterogeneous, or at least polydisperse substances and sedimentation theory for systems of this kind is more complex and less well developed than that for homogeneous molecules. Moreover, there is an exchange of components between lipoprotein particles and possibly association of particles also. Because of the complexities they introduce, these effects are usually ignored and the lipoprotein fraction treated as though it is essentially a homogeneous, ideal solute. Estimates of sedimentation constant or molecular weight will therefore be approximations of an accuracy that will depend on the truth of this premise. Nonetheless, even approximate values of these properties can be used to characterise a lipoprotein if they are measured under defined conditions.

It must also be accepted that the most refined methods of analytical ultracentrifugation are comparatively time-consuming and are therefore not well adapted to the routine characterisation of lipoprotein preparations. For studies to the highest standards of accuracy, conventional procedures should be used as described, for example, in the Manual of Methods for the Analytical Ultracentrifuge (Chervenka, 1969). In the following sections we shall confine ourselves to the description of simplified sedimentation velocity methods which, though not capable of great accuracy, are relatively speedy and are adequate for the definition of preparations for descriptive purposes. It may be argued that a sacrifice in accuracy is less of an objection when the solute is inhomogeneous, but it must be remembered that, because the results obtained by these methods are subject to some uncertainty, they must not be made the basis of detailed or profound theories of lipoprotein structure.

For the sake of simplicity, we shall discuss the different centrifugal procedures separately but it will be evident that, in practice, several properties can often be determined from one properly designed exper-

TABLE 4.1
The physical characterisation of lipoproteins

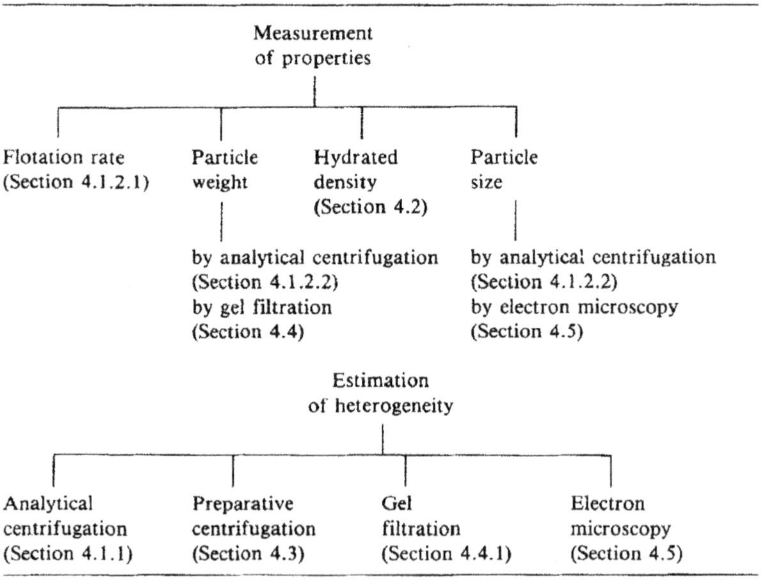

iment. Before opening these descriptive sections, it may be worth making the following three general comments:

(1) Unlike most of the methods that can be used to characterise lipoproteins, centrifugal procedures are non-destructive and the lipoprotein can easily be recovered.

(2) With all the methods based on analytical ultracentrifugation, it is helpful to have some mechanical assistance to lighten the drudgery of calculation. Fortunately, the advent of the desk-top micro-computer has brought this potential within the grasp of almost every laboratory.

(3) When reporting results obtained by analytical ultracentrifugation, it is important to quote the concentration of lipoprotein in the cell, as well as the temperature and speed of the rotor.

4.1.1. Estimation of polydispersity in the analytical ultracentrifuge

When the dispersion of a sample is minimal, as in the case of a sub-fraction of LDL, it may be investigated by a variety of ultracentrifugal techniques that are designed to detect relatively small deviations from the distribution expected of a homogeneous substance (Schumaker, 1963; Oncley, 1969; Jeffrey et al., 1975). However, it is of greater practical concern to have a method that can be used for preparations like whole LDL or HDL, where the polydispersity is usually more obvious. Simple observation of the schlieren pattern may be enough to show that a sample is inhomogeneous, as in Fig. 1.3. If the distribution curve is unimodal and has only two inflections, the sample is polydisperse. If the curve has more than two inflections, the preparation is heterogeneous. However, there are two pitfalls to be avoided, particularly when the degree of dispersion is small. Firstly, the baseline of the pattern is usually markedly curved, which can give an illusion of symmetry to schlieren curves that are actually skewed. Secondly, if the density of the solvent is too great, the over-all variation in effective particle mass that is due to the heterogeneity of the lipoprotein becomes small in relation to the mean effective mass. As a result, the schlieren curve will be narrower and appear to be more symmetrical than it would if the density of the solvent was lower. A superficial observer may then be misled into thinking that the dispersion of his preparation is less than is really the case.

The classical analysis of highly asymmetrical lipoprotein distributions is that developed by Gofman and his collaborators (DeLalla and Gofman, 1954; Ewing et al., 1965; Lindgren et al., 1972). In this procedure, the distribution of the lipoprotein in the centrifuge cell is recorded at set intervals of time during a conventional sedimentation velocity experiment that is performed under standard conditions. The resulting profile is then 'sliced' into several arbitrary sections and the areas of these are converted to their equivalent masses of lipoprotein. The boundaries of these sections are defined by the flotation rates that correspond to the points in the profile at which the boundaries are drawn. The profile can therefore be expressed as a series of intervals of flotation rate e.g. 0–12, 12–20 etc., between which stated masses of lipoprotein will be found. The most recent, computerised versions of this procedure allow the profile to be sliced into as many as 100 sections which give a detailed, quantitative measure of the dispersion of the lipoprotein preparation and, by summation, an estimate of the total weight of lipoprotein. It is interesting to note however, that this method of extracting data from the centrifuge record has not changed in essence since its origin some thirty years ago. The use of digital computers has allowed an increase in the number of measurements that can be made across the profile, and analogue-to-digital converters can now save much labour by recording the coordinates of these points directly, in 'computer-readable' form. However, these developments have probably tended to deter all but the most determined of potential users. Moreover, in its original form, the method is inflexible since it requires each set of observations to be made at precisely the same times, under exactly the same conditions. In an attempt to break this inflexibility and to allow observations to be made at any time during a centrifugation at any speed, Bosanquet et al. (1980) have written a modified computer programme. However, this is apparently based on the same theoretical model that was used by Ewing et al. (1965).

Although this technique is probably still the best available for the routine analysis of ultracentrifuge records of lipoprotein distribution, it suffers from the serious practical disadvantage that it requires so much time and labour to set it up, that it is useless to anyone who only needs to do a few analyses. Unfortunately, little has been done to develop simpler methods that would be useful under these circumstances. Lindgren has described a way of resolving the schlieren curve into several components by means of a commercial 'curve analyser' but this not only requires additional, costly equipment, but also has the disadvantage that the components may be taken to have a real existence, which is not necessarily true. DeLalla et al. (1967) have also described a somewhat simplified and more easily comprehensible version of the earlier procedure of DeLalla and Gofman (1954), but this is still too complex for the occasional user. Fortunately, it can be simplified even more by restricting the conditions of the experiment. Firstly, it must be possible to observe the whole of the lipoprotein preparation in a single schlieren photograph. Although this confines the observations to HDL, LDL or to sub-fractions of VLDL, it is precisely in these relatively well-defined fractions that the high resolving power of the centrifuge is most valuable. Secondly, the consequences of the dependence of sedimentation rate on concentration must be minimised by working at the lowest feasible level, i.e. from 5 to 7.5 mg of lipoprotein/ml in the centrifuge cell.

The *procedure* is as follows: dialyse the sample against a salt solution of such a density that the lipoproteins will float (or sediment) at a convenient rate. Centrifuge this sample, in parallel with its solvent, in a double sector cell in the conventional way, at a known temperature. (If this temperature is chosen to be 26 °C, the comparison of the results with those of earlier workers may be simplified.) Maintain the drive current at a steady value during the acceleration and note the time at which the rotor reaches full speed, which should be set at about 50 000 rev/min. Allow the boundary to move well clear of the base of the cell (or the meniscus), and then take a photograph at a known time after the rotor reached full speed.

In this record is a symmetrical curve of approximately Gaussian shape, the solute may be homogeneous, or nearly so, although this cannot be assumed without closer study. However, most lipoprotein preparations will manifest some degree of inhomogeneity, either by a skewing of the schlieren curve, or by the presence of inflections in the limbs of the curve, or even by the presence of more than one peak. Inflections or multiple peaks show that there is more than one "population" of lipoprotein particles present, although these themselves may not necessarily be homogeneous. Unfortunately, it is difficult to describe such a curve, or its divergence from symmetry, in a simple yet precisely quantitative manner. The following procedure, which may suffice for descriptive purposes, is based on the analysis developed by DeLalla and Gofman (1954).

Make a ten-fold enlarged drawing of the photograph on graph paper, taking care to mark the positions of both reference traces and the base of the cell (or the meniscus if the sample is sedimenting). Mark the points at which the limbs of the schlieren trace cut the base-line, and those of any other singular points such as peaks, troughs or inflections in the limbs. Calculate the distances of these points from the centre of the rotor by making use of the known positions of the reference traces in the usual way, and convert them into the corresponding flotation (or sedimentation) rates by means of the equation:

$$S = \frac{\ln X_1 - \ln X_0}{\omega^2 t}$$

where X_1 is the distance of the singular point from the centre of rotation in cm.

X_0 is the distance of the base of cell from the centre of rotation in cm.

ω is $2\pi f$, where f is the rotor speed in rev/sec.

$t = 0.33t_1 + t_2$ where t_1 is the time in seconds for the rotor to accelerate to full speed, and t_2 is the time from then until the photograph is taken.

With the chief features of the profile thus expressed in terms of a physical parameter, its geometry can be specified after the manner of a series of map references. However, a more useful index of the asymmetry of the curve can be obtained by measuring the areas that are defined by the singularities. This may be particularly helpful when there are no inflections in the limbs of the curve. Draw in the maximum ordinate on the ten-fold enlargement of the schlieren curve and measure the two areas thus defined by the ordinate, the curve and its baseline. Any convenient method can be used to make this measurement, e.g. a planimeter. The asymmetry of the peak can than be represented by the ratio $A = A_L/A_H$, where A_L and A_H are the areas on the low density and the high density side of the maximum ordinate respectively. These areas are proportional to the concentration of lipoprotein in the relevant part of the centrifuge cell and, if there is a preponderance of the denser particles for example, A will be less than one. Note however, that $A = 1$ does not indicate that the preparation is homogeneous, merely that the distribution is probably symmetrical. If there are inflections in the limbs of the peak, these may be used to define smaller parts of the total area in a similar way. For example, consider an LDL which is distributed between S_f 0 and S_f 17, with a peak at S_f 7.5 and an inflection at S_f 4. The proportions of the total area that are defined by these boundary points might be: S_f 0–4 = 15 %, S_f 4–7.5 = 45 %, S_f 7.5–17 = 40 %. Note that this is essentially the same procedure as that used to handle the asymmetry of the HDL distribution. In this case, there is frequently an inflection at a flotation rate of about 3.5 svedbergs (*d* 1.20 g/ml) and this boundary has become the conventional demarcation between HDL_2 and HDL_3.

4.1.2. Sedimentation rate, particle weight and particle diameter

Although these properties are attractive as numerable characteristics of a substance, there is little point in estimating values for a grossly inhomogeneous substance, for which the results will be open to criticism on theoretical grounds. These properties should therefore only be estimated for lipoproteins that show no more than a slight skewing of the boundary during velocity centrifugation, i.e. well defined subfractions of the lipoprotein profile.

It is customary to centrifuge lipoproteins in solvent of high density, in which they float, i.e. their sedimentation rate is negative. However, it is not obligatory to use these conditions except in the case of VLDL. Indeed, the use of high concentrations of salt is not only open to theoretical objections, but introduces the possibility that the behaviour of the lipoprotein particles may differ from that in physiological salt solutions. This question of artifactual behaviour has not been exhaustively studied, but such evidence as is available suggests that the hydrodynamic properties of LDL are not grossly altered in solvents of high ionic strength. For example, the particle weights that were obtained by Adams and Schumaker (1969b) for LDL in NaBr solutions of density about 1.4 g/ml agree satisfactorily with those found by centrifugation at a density of about 1.063 g/ml. It may however, be prudent to avoid very high ionic strengths for the centrifugation of HDL, which appear to be less stable than the low-density lipoproteins.

The rate of sedimentation of lipoproteins, like that of most macromolecules, is dependent on concentration and account must be taken of this effect if the sedimentation coefficient is to be used as a characteristic, or as a means of estimating the weight or diameter of the particle. This is usually done by means of the relation:

$$S = S^{\circ}\ (1 - KC)$$

where S° is the sedimentation (flotation) rate at infinite dilution, C is the concentration of lipoprotein in the ultracentrifuge cell, and K is a constant. Unfortunately, there is no single reliable value of K. The value of 0.89×10^{-2} ml/mg is now used by the Donner laboratory (Ewing et al., 1965; Lindgren et al., 1972; Lindgren, 1974) in the place of an earlier estimate of 1.6×10^{-2} ml/mg, and the value of 1.04×10^{-2} ml/mg is quoted by Adams and Schumaker (1969b). However, the experimental basis for these determinations has not been published and it is uncertain whether they are applicable to lipoproteins of all sizes, at all concentrations, in solvents of all densities. The divergence of these estimates, taken with the evidence of other workers (Lee and Alaupovic, 1974; Mauldin and Fisher, 1970; Fisher, Granade and Mauldin, 1971; Hammond and Fisher, 1971), suggests that K may not be a well-defined constant of universal application. The most reliable practice is therefore to measure the flotation rate at two or three different, low concentrations (i.e. less than 10 mg/ml) and to extrapolate the results to infinite dilution. If no such correction is made, the concentration at which the observation was made must be quoted.

It is the practice, when centrifuging other macromolecules, to recalculate the observed results to standard experimental conditions, namely sedimentation in water at 20 °C. However, this is inconvenient in the case of lipoproteins since some of these have a density close to that of water and the corrected scale of sedimentation rate would have both positive and negative values. More importantly, it is often impossible to do the correction accurately for lipoproteins (a) because the density of the lipoprotein particle is unknown, and (b) the necessary information on the way this property changes with temperature is lacking (Section 2.1). Some workers have assumed that the latter coefficient has the same value as that of an unconjugated protein but, from such evidence as is available (Toro-Goyco, 1958; Mills, 1977), this may be as much as 20% too large. For these reasons, it has become the practice to quote the flotation rate of a lipoprotein under the conditions of observation. In this context, the conditions originally proposed by Gofman and his collaborators have, by wide usage, come to be the effective convention for centrifuging lipoproteins. In this system, the observations are made in salt solutions of density 1.063 g/ml or 1.20 g/ml at a temperature of 26 °C, and this can be recommended for the measurement of flotation rates, or for the *approximate* estimation of particle weights. For the latter purpose, the system used by Adams and Schumaker has the advantage that it minimises the errors that arise when the density of the lipoprotein is not accurately known (cf. below). For the most precise studies on lipoprotein preparations of minimal heterogeneity, for which all the physico-chemical parameters are to be measured, use 0.15 M NaCl as the solvent.

In principle, it is the second moment of the distribution curve that should be used for the determination of the sedimentation rate, rather than the maximum ordinate (Goldberg, 1953), and this procedure should be used for critical work on lipoprotein fractions in which the dispersion is small. When the peak is markedly asymmetric however, it is arguable that the sample is too heterogeneous for meaningful measurements to be made on it, and that it will be better defined by a combination of its peak flotation rate and an index of its polydispersity, than by the flotation rate as estimated from the second moment. Moreover, there is considerable doubt that asymmetry of the peak is entirely due to the presence of particles of different sizes. The association of particles, the exchange of components between them, and the loss of protein by disproportionation during the centrifugation could all contribute to the production of a skewed schlieren curve. Unfortunately, little work has been done on these aspects of the physical chemistry of lipoproteins and their importance cannot be assessed.

4.1.2.1. Measurement of flotation (sedimentation) rate. We have already described a method for extracting the flotation rate from data obtained during the investigation of heterogeneity (Section 4.1.1). This technique is simple but suffers from the disadvantage that it depends on a single observation. The conventional procedure, in which several photographs are taken at known intervals during the centrifugation, gives more accurate results (Chervenka, 1969; Lindgren, 1974).

Procedure. Dialyse the sample against NaCl solution of $d_{26} = 1.063$ g/ml, or NaBr of $d_{26} = 1.20$ g/ml. Then centrifuge in an epoxy resin centrepiece at 26 °C as in Section 4.1.1. Take at least four photographs, at intervals of 10 min, starting when the peak has moved clear of the base of the cell. Determine the positions of the peak relative to the reference trace by measuring the photographs with a travelling microscope, or by measurement of a 5- or 10-fold enlargement with an accurate scale. (If the latter method is adopted, it will be necessary to determine the magnification of the enlarger. For this purpose, the discs that are ruled with 1 mm squares for cell counts make useful standards. Alternatively, a small coin, the diameter of which can be accurately measured with a micrometer, can be used.) Also determine the magnification due to the camera lens by measuring the distance between the reference traces on the photograph. Use this value (which is a constant of the centrifuge) to calculate the real distance of the peak from the centre of the rotor (X cm). A plot of $\ln(X)$ vs. time (seconds after the first exposure) should be a straight line, the slope of which is the apparent flotation rate. If this procedure is repeated at several different concentrations of the sample, the apparent flotation rate can be extrapolated to zero concentration as explained above.

The same procedure can be used to estimate the sedimentation rate of the lipoprotein in 0.15 M NaCl solution.

4.1.2.2. Measurement of particle weight and diameter The estimation of particle weights by the sedimentation velocity method makes use of the equation

$$S = \frac{M\,(1 - \bar{v}\rho)}{Nf} \qquad (1)$$

where S is the sedimentation coefficient of the particle, M is its mass and $\bar{v}$ its partial specific volume. The symbol ρ represents the density of the solvent, N is Avogadro's number and f is the translational frictional coefficient for the particle. It can be seen from this, that two additional pieces of information are needed to allow the particle weight to be deduced. In the classical procedure, the parameter f is eliminated from this expression by substitution from the equation

$$f = \frac{RT}{ND}$$

where D is the diffusion constant of the particle and T the temperature of the experiment, and R is the gas constant. Although the diffusion coefficient can be measured in the untracentrifuge (Chervenka, 1969), the time-consuming nature of the experiment appears to be a powerful deterrent to the majority of potential users, except where the object is an accurate study of a sample of low dispersion. This has led to the use of alternative methods in which the frictional coefficient is eliminated from eqn. (1) by substitution from the Stokes equation

$$f = 6\pi r\eta$$

whence

$$S = \frac{M(1 - \bar{v}\rho)}{6\pi\eta r N} \tag{2}$$

where η is the viscosity of the solvent and r is the radius of the lipoprotein particle. From this equation we can eliminate either M or r, by making use of the fact that, for a spherical particle

$$r^3 = \frac{3M}{4\pi Nd}$$

where d is the density of the particle. Note that there is some ambiguity connected with the use of hydrated density rather than partial specific volume but, for this approximate procedure, the former is acceptable. In this way we obtain the equations

$$S = \frac{4r^2(d-\rho)}{18\eta} \tag{3}$$

and

$$S = \frac{M^{2/3}(d-\rho)}{8.3387 \times 10^{16}\eta d^{2/3}} \tag{4}$$

whence

$$M = d\left[\frac{8.3387 \times 10^{16}\eta S}{d-\rho}\right]^{3/2} \tag{5}$$

Although the introduction of the viscosity term into these equations simplifies the practice of estimating the particle weight, it must be borne in mind that a number of assumptions are implicit in this derivation, notably that the lipoprotein is a rigid, uncharged sphere that is large enough to justify the use of the Stokes equation, and that the value of M derived from the Stokes radius is the same as that in eqn. (1). In an effort to reduce the effect of some of these ambiguities, Adams and Schumaker (1969b) adopted the proposal put forward by Oncley (1941) that the frictional coefficient should be resolved into three terms that represent the components due to the size, shape and hydration of the particle. However, this analysis introduces two additional variables, the axial ratio and the hydration of the particle, both of which are among the least well established properties of lipoproteins. This refinement therefore seems to have little to recommend it at present.

It must also be remembered that the Stokes equation applies to a particle at infinite dilution. The value of S that is used in eqns. (3) and (4) should therefore strictly be corrected to the same condition.

In addition to the uncertainties that are inherent in the derivation of eqn. (5), a practical difficulty arises from the fact that the difference between the density of the lipoprotein particle and that of the solvent in which it is centrifuged is usually small when compared with the conditions used for centrifuging a protein. A small error in partial specific volume (or hydrated density) of the particle will therefore introduce a much larger error in the estimate of its relative mass, e.g. a discrepancy of 1% in $\bar{v}$ can lead to an error of as much as 25% in the estimate of the particle weight of LDL. An accurate value of $\bar{v}$ can be determined by one of the methods that are discussed in Section 4.2, but these procedures are exacting and, with the exception of the expensive vibrating tube densimeter, are tedious. They are therefore not well suited to routine estimations. Adams and Schumaker (1969b) therefore proposed that the centrifugation should be performed in an NaBr solution of density about 1.4 g/ml, when an error of 0.005 g/ml in the density of LDL leads to an error of about 3% in the estimate of its particle weight. Even with HDL, the error would not exceed 4%. Under these circumstances, it may be acceptable to use the approximate estimates of hydrated density that can be obtained from density gradient centrifugation, or from the flotation rate of the lipoprotein in solvents of density 1.063 or 1.20 g/ml. If the latter method is used, the hydrated density must be estimated by

TABLE 4.2
The approximate hydrated density of lipoproteins of known flotation rate.
S_f = flotation rate in NaCl solution of density 1.063 g/ml at 26 °C.
F = flotation rate in NaBr solution of density 1.20 g/ml at 26 °C.
The values quoted are interpolated from published measurements (cf. Table 7.2). Note the sparsity of data for high density lipoproteins.

S_f	d (g/ml)	F	d (g/ml)
0	1.063	0.35	1.145
2	1.051	3.0	1.094
4	1.049		
6	1.037		
8	1.029		
10	1.022		
15	1.010		
20	1.001		
40	0.981		
60	0.969		
80	0.960		
100	0.954		
150	0.942		
200	0.938		
400	0.932		

the interpolation of reported data such as those summarised in Table 4.2.

The simple Adams–Schumaker technique may thus be the most attractive when an approximate estimation of particle weight will suffice. For example, when the polydispersity of the sample is relatively great. However, the result will generally be less reliable than a single estimate of flotation rate at a known lipoprotein concentration. When the sample is relatively homogeneous, it would be appropriate to refine the determination by correcting for concentration dependence and by estimating the partial specific volume (hydrated density) of the lipoprotein.

Procedure. If the particle weight is to be estimated from a flotation (or sedimentation) rate that has been corrected to zero concentration, solutions of at least three different concentrations of the lipoprotein should be prepared. Dialyse all these in the same sealed container for at least 24 hours, against several changes of the chosen solvent solution (e.g. NaCl d = 1.063 g/ml for low density lipoproteins; NaBr d = 1.20 g/ml for HDL). If no correction is to be made, only one lipoprotein solution need be prepared. When equilibrium has been established, determine the density (Section 2.1.1.3) and viscosity of the dialysing solution.

Establish the concentration of each lipoprotein solution by the dry weight method (Section 6.1) and determine the flotation rate in each one (Section 4.1.2.1). Plot the reciprocal of the flotation rate against the concentration (in mg/ml) and extrapolate to obtain the flotation rate at infinite dilution. For an accurate determination of the hydrated density of the lipoprotein, it will be necessary at this stage, to use one of the methods of Section 4.2. If the precision of the work in hand does not justify this, or if the flotation rate is not corrected to infinite dilution, an approximate estimate of the hydrated density can be taken from Table 4.2. With these data, the weight-average particle weight of the lipoprotein can be estimated from eqn. (5). Note however, that flotation rates must be put into this equation in the form of negative sedimentation rates, i.e. the value of S is negative. The average particle diameter can be estimated in a similar way from eqn. (3).

4.2. *Partial specific volume (hydrated density)*

The partial specific volume ($\bar{v}$) of a lipoprotein is not only needed for the estimation of particle weight (Section 4.1.2.2), but is a valuable characteristic of the particle in its own right. However, to make a useful determination of $\bar{v}$ requires that the density of a lipoprotein of accurately known concentration should be measured with great accuracy. For example, the density of a 1% solution should be determined to $\pm 1 \times 10^{-6}$ g/ml and its concentration to $\pm 0.05\%$ (Charlwood, 1957). Weighings must therefore be performed on a microbalance, and the temperature at which the measurement is made must be maintained to within at least ±0.01 °C. These exacting conditions are acceptable when the object is an accurate estimation of the particle weight of a lipoprotein fraction of low dispersion, but they make $\bar{v}$ unattractive as an everyday means of characterisation.

The partial specific volume is evaluated by means of the relation:

$$\bar{v} = \frac{1}{d}\left(1 - \frac{\Delta d}{c}\right)$$

where d is the density of the solvent, Δd is the difference between the density of the solvent and that of the lipoprotein solution, and c is the concentration of lipoprotein in grams per millilitre of solution.

The density of the solutions can be determined either by pycnometry (Schachman, 1957), or with the vibrating tube density meter, both of which methods are described in Section 2.1.1.3. Despite its high precision and economy of material, the density-gradient column of Linderstrom-Lang and Lanz (1938) has apparently not been used for studies on lipoproteins, presumably because it is uncertain whether they will be denatured by the organic solvents from which the gradient is made. Note that, to obtain the required accuracy, allowance must be made for the atmospheric buoyancy when weighing the pycnometer.

Because lipoproteins are insoluble in distilled water, the partial specific volume must be determined in a dilute salt solution and the concentration of the lipoprotein must be obtained from dry weight determinations on the solution and its dialysate (Section 6.1). Note that it is desirable to dialyse at a pH as close to the isoelectric point as is practicable, in order to minimise the Donnan effect. Moreover, the concentration of salt should be kept below 10 mM, if possible, because the amount in the solution may then be assumed to be same as that in an equal volume of the dialysate. Unless a large amount of solution is available, the volume taken for the dry weight estimation may best be determined by weight, since the density is known. Alternatively, it can be carefully measured with a calibrated, precision micrometer syringe.

By contrast with the determination of $\bar{v}$, the hydrated density of the lipoprotein (d_h) can be estimated with relative ease from buoyancy measurements. Although d_h is not identical to $\bar{v}$, it is a close approximation that is almost equally characteristic of the lipoprotein and has often been used in particle weight determinations. It can be measured in two ways, neither of which requires a knowledge of the concentration of the lipoprotein.

(a) By centrifuging the lipoprotein on a density gradient in the preparative ultracentrifuge, and then measuring the density of the gradient at the point of equilibrium. It is important to centrifuge for long enough to ensure that equilibration is complete. The method has the merit that it can be carried out as a part of the preparative procedure that is described in Section 2.1.2, with the reservation that high concentrations of lipoprotein may distort the shape of the gradient. In any event, the method is not capable of high accuracy.

(b) The analytical centrifuge can be used to determine the density of the solvent in which the sedimentation rate is zero. This is more tedious than (a) but is somewhat more accurate. When used with unconjugated proteins, a long extrapolation to zero sedimentation is necessary because the concentration of salt that would be needed to make them float is very high. With lipoproteins however, it is easily possible to obtain data from both flotation and sedimentation experiments and to interpolate them.

Procedure. Prepare at least four salt solutions of densities such that the lipoprotein will float in some and will sediment in the others. At this stage, the density of the solutions need only be approximately known. Dialyse the lipoprotein (1 ml of a 1% solution) for 24 hours against the solution of lowest density, in a vessel that is kept tightly closed to prevent evaporation. Then measure the sedimentation rate at a known temperature, as described in Section 4.1.2.1. Retain the main volume of the solvent in the stoppered dialysing flask. After the centrifugation, re-disperse the lipoprotein by rotating the cell end-over-end for several minutes at a rate of not more than about 10 rev/min. Remove the sample from the cell, dialyse it against the next most dense solvent solution and determine the sedimentation rate as before. Proceed in this way until the lipoprotein has been examined in all the solvents.

At this stage, determine the density (*d*) and viscosity (η) of each solvent (Section 2.1.1.3), at the temperature at which the centrifugations have been performed. Plot the density of the solvent (*d*) against the product ηS, where S is the sedimentation (flotation) rate in svedberg units. The points should lie on a straight line which intersects the density axis at a point that represents the hydrated density of the lipoprotein.

4.3. *The analysis of polydispersity in the preparative ultracentrifuge*

In this technique, the sample is divided into sub-fractions that are then analysed for at least one physical or chemical property of the lipoprotein. Since it is possible to determine several different properties on each fraction, the measure of polydispersity that is obtained can be more discriminating than that given by the analytical ultracentrifuge.

Serial centrifugation is important in this context because it is the only procedure that can be used to fractionate lipoproteins of density less than 1.006 g/ml. Unfortunately, the need to perform several consecutive centrifugations makes it much more time-consuming and therefore more expensive than the alternative density gradient method, which is the one almost universally used for the analysis of the denser lipoproteins. If the method is to be effective, the sample must form a shallow layer at the bottom of the tube, and the infranatant material must not be disturbed when each floating fraction is recovered. Moreover, if the method is to be quantitative, the whole of the infranatant must pass on at each successive stage of the centrifugation. From this it can be seen that a preparative procedure like that of Gustafson et al. (1965) is not well adapted to the purpose, in contrast to the technique developed by Lindgren, in which the layer of lipoprotein at the bottom of the tube is kept as short as possible, and a density gradient is layered above it (Lindgren, 1975; Hatch et al., 1967; Lindgren et al., 1972a,b). The tubes are then centrifuged for carefully calculated times, in a swing-out rotor. After each run, the supernatant lipoprotein layer is aspirated off and the infranatant centrifuged again, without disturbance. With appropriate modifications to the density gradient, this procedure can be used to subfractionate any of the three main classes of lipoproteins and allows the minimum cross-contamination of the fractions to be calculated. It is claimed that this procedure will reduce both the contamination of the products with plasma proteins and the cross-contamination with other lipoproteins. In turn, this can reduce, or eliminate the need to wash the fractions by further centrifugation, thus diminishing the chances of degradation. However, despite its evident value, particularly for the fractionation of VLDL, the technique has been little used, perhaps because it requires access to computing facilities. The increasing availability of cheap desk-top micro-computers, and the relative simplicity of the programmes required, may lead to its wider acceptance.

As it was originally described, this technique made use of the now obsolete Beckman Sw 25.3 rotor which, for the analysis of chylomicrons, was fitted with 6 ml tube adapters. The currently available SW 27.1 uses the same tube and has almost the same radius, and could therefore be substituted with only the minimum of readjustment of the centrifuging times. The following description applies to the fractionation of lipoproteins of density less than 1.006 g/ml, and we shall assume the use of the 102 mm × 16 mm tube. To ensure that the lower boundary of the sample is flat, a hemispherical plastic insert is first put into the tube, with a drop or two of NaCl solution of density 1.065 g/ml to seal it. Three ml of the sample to be analysed (adjusted to a solvent density of 1.065 g/ml) is added, and a discontinuous salt gradient layered above it by the careful superposition of the following sodium chloride solutions, viz. 1 ml each of density 1.0464 and 1.0336 g/ml and then 3 ml each of density 1.0271, 1.0197, 1.0117 and 1.0064 g/ml.

The use of a gradient in this system provides a stability that may be very important when large particles of low density are being analysed. Unfortunately, it greatly complicates the task of calculating the time for which the tube must be centrifuged to recover a particle of given flotation rate. Lindgren does this by regarding the final gradient as made of 12 superimposed homogeneous layers which are numbered from the bottom of the tube. The distance of the bottom of the *j*th layer to the centre of the rotor is designated as r_j, and the density and viscosity of the solvent in this layer are d_j and η_j respectively, at the temperature of the run. The flotation rate in the *j*th layer (F_j), of a particle of given S_f value (S) is then given by

$$F_j = \frac{S(d_j - \sigma)\,1.0173}{(1.063 - \sigma)\,\eta_j}$$

where σ is the hydrated density of the particle in question.

The time t_j that it takes for this particle to pass through the *j*th layer is then given by:

$$t_j = \frac{\ln r_{j+1} - \ln r_j}{\omega^2 F_j}$$

where ω is the angular velocity of the rotor.

The total time taken for the particle to migrate from the bottom of the first layer to the bottom of the top layer (i.e. a nominal 100% recovery of these particles) is then given by:

$$\omega^2 T = \sum_{j=1}^{j=11} \omega^2 t_j$$

This value of T must then be corrected to take account of the periods of acceleration and deceleration in the usual way.

Under the conditions originally described by Lindgren (1975), a centrifugation amounting to 0.739×10^6 g·min was sufficient to concentrate all particles of $S_f > 3200$ in the top 0.5 ml layer. This was removed and the tube re-centrifuged for a further 1.32×10^6 g·min to concentrate the particles of $3200 > S_f > 1100$ into the next 0.5 ml. A third centrifugation for an additional 3.384×10^6 g·min likewise recovered the particles of $1100 > S_f > 400$.

It will be seen from this abbreviated description that the technique requires much preliminary work to set it up. For example, it is necessary to determine all the values of r_j and also the shape of the gradient (which is needed for the estimation of d_j and η_j). The programme of calculations would also be discouragingly laborious if performed by hand, but is fortunately easy to run on a small, suitably programmed, digital computer. It should also be noted that the calculations make use of the hydrated density of the lipoprotein, which is rarely an accurately known property. Moreover, in the case of the very-low-density lipoproteins at least, the gradient is not centrifuged to equilibrium and care must be taken that the gradient does not undergo an appreciable change in shape during the course of the whole series of centrifugations. Altogether, it is evident that this procedure is feasible only in the most well-equipped laboratories and, in any event, is unattractive when only a few analyses are in prospect.

In this context, it should be observed that Zilversmit has devised a similar but simpler technique which can be used to analyse lipoproteins down to a diameter of a few tens of nanometres (Zilversmit, 1963, 1969; Zilversmit et al., 1966). This method is not a serial centrifugation but, like the Lindgren process, uses non-equilibrium conditions on a density gradient. The following summary assumes that the centrifuge tube is 76 mm × 25 mm, into which a linear sucrose gradient 5.5 cm long is introduced (Section 2.1.2.2). This ranges from about 1.18 g/ml (40% sucrose in 0.9% NaCl) at the bottom, to about 1.13 g/ml (30% sucrose in 0.9% NaCl) at the top. The sample is then introduced below the gradient in about 0.75 ml of 50% sucrose and beneath it, a basement layer of 60% sucrose that fills the curved bottom of the tube. The sample is thus made to form a layer about 1.5 mm thick, that is a known distance from the centre of rotation.

During the centrifugation, the velocity of a particle at a distance r from the centre of rotation is given by the relationship:

$$\frac{\mathrm{d}r}{\mathrm{d}t} = \frac{D^2\omega^2 r\,(d_r - \sigma)}{18\,\eta_r}$$

where D and σ are the diameter and hydrated density of the particle respectively; and d_r and η_r are the density and viscosity of the solvent at radius r. For the particular gradient described, it happens that the ratio

$$\frac{r(d_r - \sigma)}{18\eta_r}$$

is almost independent of r. Consequently, at a set rotor speed, the velocity of a given particle up the gradient is approximately constant and proportional to the square of its diameter. The latter can therefore be estimated from the distance the particle has moved up the tube after a set period of centrifugation at a set speed. According to Zilversmit (1969), 60 min (including acceleration and deceleration) at 5000 rev/min in the Beckman SW 25.1 rotor will deposit a particle 131 nm in diameter at a point 0.5 cm from the start, while particles 422 nm in diameter will have moved 5.0 cm. If the rotor speed is 25 000 rev/min, the corresponding particle sizes are 26 nm and 84 nm. It is therefore possible to analyse a wide range of lipoproteins into fractions that are defined by limits of diameter, or by flotation rate.

Because it only involves one centrifugation, this technique is less time-consuming han the Lindgren procedure but this advantage is bought at the cost of lower resolution. It is also worth noting that, because of the high density of the solvent, the Zilversmit method is comparatively insensitive to uncertainties in the value of σ. However, the high concentration of sucrose in the final fractions may be a nuisance if they are to be subjected to other methods of analysis.

Both the foregoing procedures are useful for the sub-fractionation of chylomicron preparations, but great care must be taken to avoid overloading the gradient. If the concentration is too great, convective streamers are likely to form and destroy the gradient. In general, the sample should be the smallest possible and in no circumstances should it exceed 10 mg of lipid/ml.

4.3.1. Equilibrium centrifugation

This method requires the sample to be centrifuged to equilibrium on a density gradient which is then divided into small fractions. The density of each fraction is determined by comparison with a blank gradient (Section 2.1.2.5) and at least one physical or chemical property of the lipoprotein that it contains is measured. In this way the dispersion of the property with respect to density can be determined. For good resolution, the gradient should be as shallow as possible, consistent with the maintenance of stability. However, it will then be able to accommodate only a small load and the lipoproteins may be difficult to locate and analyse accurately. To overcome this problem, some workers have attempted to increase the sensitivity of optical detection by staining the lipoproteins before centrifuging them. But this procedure is open to the objection that it is not the native lipoprotein that is studied and that the observed distribution of optical density is not that of an intrinsic property of the lipoprotein. This objection can also be levelled at methods that enhance the sensitivity by labelling the lipoprotein with radioactive isotopes. For example, it is known that the introduction of radioactive iodine into the protein moiety carries the risk of denaturing some lipoprotein (cf. Weech et al., 1978). There is thus a real chance that the distribution of radioactivity on the gradient does not fully represent that of the native lipoprotein. Moreover, the distribution of activity cannot be equated with the distribution of lipoprotein mass because the protein content, and therefore the specific activity of the lipoprotein, is not the same at all points down the gradient. Unfortunately, it cannot be directly equated with the distribution of apo-lipoprotein either, since some label always enters the lipid moiety. Apart from these objections, this method of labelling suffers from the draw-back that it is time consuming and therefore not well suited to frequent routine analyses. In the method proposed by Foreman et al. (1977), the lipoproteins are labelled by exchange with [^{14}C]cholesterol (or [^{14}C]lecithin) before centrifugation. Their distribution on the gradient is then determined either by conventional counting, or by radioimmuno-assay. This method may be somewhat less laborious and less likely to denature the lipoproteins than the iodination procedure, but the activity of the labelled product cannot be related to any property of the native lipoprotein. Like the iodination technique, this method is therefore better suited to the detection of heterogeneity (e.g. the distinction and estimation of LDL and HDL in small samples) than to estimations of polydispersity, for which the determination of an intrinsic property will require a comparatively large sample. Despite these technical limitations, its high sensitivity makes radioactive labelling the preferred method for the analysis of changes in lipoprotein distribution that are the result of lipoprotein metabolism, and it is widely used for this purpose (see for example Weech and Mills, 1978).

The location of the lipoproteins on the gradient can also be determined by refractometry (Sections 2.1.2.5; 6.1.1), but the method is uncertain at low concentrations because it is difficult to match the experimental and the blank gradients exactly. In principle, this method has the advantage that it will give a quantitative estimate of the total lipoprotein present in the fraction but, in practice, it is subject to the following limitations. Firstly, the refractive increment of the lipoprotein is a function of the concentration of salt in the solvent and will therefore vary down the gradient in a way that is at present ill-defined. Secondly, the refractive increment is not an accurately known property of the lipoproteins in any solvent.

The most widely used way of locating unlabelled lipoproteins is by the estimation of one of their more easily measured components, e.g. cholesterol, phosphorus or protein (Section 6.3). This method avoids the difficulties that are peculiar to the labelling procedure but is limited by the relative insensitivity of some of the analytical methods. In this context, the high sensitivity of the micro-scale method of thin-layer chromatography that is described in Section 6.7 is particularly valuable. By the use of this equipment, a complete analysis of each lipoprotein fraction can be obtained, thus giving not only the mass distribution of the lipoproteins, but also the dispersion of the five main chemical components (Fig. 3.2). This procedure, coupled with an immunochemical analysis of the apolipoprotein distribution in the fractions, is probably the most powerful at present available for the analysis of the polydispersity of a lipoprotein.

Whatever method of location and analysis is used, it must be remarked that the value of the experiment will be greatly diminished if the fractions from the gradient are designated only in an arbitrary manner e.g. by fraction number, or by distance along the gradient. Only if the density of the fractions is determined can the full potential of the analysis be realised. However, it is important, when measuring the density of the fractions (Section 2.1.2.5), to take account of the temperature at which the gradient was run (Sections 2.1, 2.1.2).

The procedure for the preparation, loading and unloading of the gradient is essentially that described in Section 2.1.2. The analysis of the lipoprotein fractions can be performed either by the conventional assay of radioactivity, or by the chemical procedures described in Chapter 6.

4.4. Gel filtration

Unfortunately, as we have already pointed out (Section 3.2.1), the difficulty of obtaining appropriate standards makes gel filtration chromatography better suited to the qualitative analysis of the heterogeneity of lipoproteins than to their quantitative characterisation. However, in view of its simplicity and its potential when used in conjunction with methods of chemical or immunochemical analysis, we shall briefly describe the technique below.

Procedure: The columns used should be about 100 cm × 1 cm diameter and have screens at both ends to ensure that the surfaces of the gel are flat. Either 6% or 4% agarose gels can be used (e.g. Bio-Gel A-5m or A-15m, or Sepharose 6B or 4B). The smaller pore size gives the best resolution of LDL and HDL, whereas the more dilute gel gives a better resolution of VLDL and LDL which is offset by an inferior separation of LDL and HDL.

The composition of the eluting solvent is apparently not critical: 0.15 M NaCl that is adjusted to pH 7 and contains the usual preservatives (Chapter 2) is suitable. Wash the chosen gel and equilibrate it with the buffer in accordance with the manufacturer's recommendations. Then pack the column with great care to ensure that it contains no discontinuities or irregularities to disturb the flow of the eluent (Fischer, 1980). Wash the column with at least three volumes of the buffer before putting it to use. There is no evidence that the temperature of the column has any marked influence on the results obtained, but *fluctuations* in temperature may be disadvantageous. It may therefore be prudent to run the column in a cold room, or at least in an air-conditioned room.

To examine the heterogeneity of a preparation, apply 10–20 mg of *freshly prepared* lipoprotein to the column in a volume of 0.5–1.5 ml, and elute at a rate of 5–10 ml/hour per square cm of column area. Collect fractions of about 5 ml volume and determine the profile of the eluted peak by measuring the absorbance at 280 nm. As in the ultracentrifuge, the presence of multiple peaks, or an asymmetrical peak will show that the specimen is significantly heterogeneous with respect to particle size.

Margolis (1967) has shown that, provided suitable standards are available to calibrate the column, the method of Ackers and Steere (1967) can be used to estimate particle size from the elution volume for the lipoprotein. This paper by Ackers and Steere gives a brief discussion of the theory and practice of the method, which depends on the measurement of the 'sieve coefficient' which is defined by the equation:

$$V = V_o + \sigma V_i \tag{1}$$

where V_o is the void volume of the column; V_i is the additional volume needed to elute particles so small that they have access to essentially all the volume of solvent within the gel, and V is the volume at which a lipoprotein of radius r is eluted. Moreover, it is assumed that the likelihood that the particle will encounter a pore large enough to admit it is given by the probability integral, whence it can be deduced that

$$\sigma = 1 - \frac{2}{\sqrt{\pi}} \int_o^x e^{-z^2}\, dz \tag{2}$$

where $x = (r - r_o)/b$. Here, r_o is the molecular radius that corresponds to the peak of the Gaussian elution curve and b is its standard deviation. The value of the probability integral is tabulated in many manuals of probability and statistics.

Each column of gel must be calibrated by determining the elution volumes of two particles of known radii r_1 and r_2. Insert these values of V_i into eqn. (1) to obtain the equivalent values of σ_1, and σ_2. Then look up these values in the tables of the probability integral to obtain the corresponding values of x_1 and x_2 by means of eqn. (2). Finally, calculate the constants b and r_o from the relations

$$b = (r_1 - r_2) \ / \ (x_1 - x_2)$$

and

$$r_o = 0.5 \ [r_1 + r_2 - b(x_1 + x_2)]$$

Once the calibration is completed, the Stokes radius of the lipoprotein can be estimated as follows. First determine the relevant value of σ from the elution volume by means of eqn. (1). Next, find the corresponding value of x from the probability tables and use it to calculate the radius of the particle from the equation $r = r_o + bx$. An upper limit for the particle weight can then be estimated from

$$M = \frac{4\pi N r^3}{3\,\bar{v}}$$

where N is Avogadro's number and $\bar{v}$ is the partial specific volume of the particle (Section 4.2). This will be an estimate of the hydrated particle weight which, in the case of LDL, can be corrected to the anhydrous condition using the known degree of hydration, namely 0.05 to 0.1 g/g of lipoprotein (Toro-Goyco, 1958; Essex et al., 1977).

4.4.1. The estimation of polydispersity by gel filtration chromatography

This procedure is essentially analogous to density gradient centrifugation (Section 4.3) and has the same advantage, namely that the fractions can be characterized by several different chemical or physical methods. Because chromatography columns can be made to carry large loads, these analyses can be more comprehensive and more easily performed than is often the case with the comparatively small fractions that are isolated from a gradient. However, a far better resolution (by density) can be obtained on the gradient than on the column (by particle size). Moreover, the practising student of lipoproteins is more likely to be constrained by lack of material than to be free to avail himself of the potentially large capacity of the chromatography column. Another disadvantage of the chromatographic analysis lies in the difficulty of making a quantitative estimate of the particle size of each lipoprotein fraction. Consequently, the dispersion is more often than not referred to some arbitrary variable such as elution volume or fraction number. Nonetheless, valuable information on the dispersion of the chemical properties of the lipoprotein can be obtained in this way, as is exemplified in Fig. 3.1.

Prepare a chromatography column of a size appropriate to the amount of lipoprotein to be analysed and run it as described in the previous section. Collect the eluate in fractions which can be analysed by the methods described in Chs. 4–6.

4.5. Electron microscopy

As we have pointed out in Chapter 3, the characterisation of lipoproteins by electron microscopy is open to the objection that the particles observed are denatured and probably give little information about the real lipoprotein. This may not apply to the freeze-etching technique that has been developed by Gulik-Krzywicki et al. (1979), but experience with this method is at present limited. Nonetheless, electron microscopy can sometimes be of value for the detection of certain abnormal lipoproteins that form discs rather than spheres, e.g. lipoprotein-X (Hamilton et al., 1971). It can also be used to demonstrate the absence of particles outside chosen size limits.

However, the shape and size of the spherical particles can only be judged from free-standing examples that have been freshly isolated. In crowded fields that have been negatively stained, the particles frequently form what appears to be a tesselation of polygons and, as the specimens age, there is also a tendency to form random agglomerates. This is particularly noticeable with the VLDL, which can only be successfully stained for microscopy when they are freshly isolated from the plasma. If this precaution is not observed, the field will contain clumps of misshapen particles and, in extreme cases, amorphous deposits. The LDL and HDL lose their integrity more slowly, and will give satisfactory preparations for up to two or three days after their isolation. But it must be remembered that some of this period will be consumed in dialysing the sample in preparation for its transfer to the support grid. Loss of integrity in the case of LDL is often signalled by the appearance of chains of rather flattened particles which are followed, as the sample deteriorates, by larger clumps of these deformed particles. This behaviour is presumably related to the aggregation that is seen with the naked eye in samples that have been stored for a long time.

The study of lipoproteins by electron microscopy was surveyed by Forte and Nicols in 1972 and the following summary is largely based on the techniques used by these authors.

Equipment and materials. We shall assume that the basic equipment and expertise for transmission electron microscopy are available. Details of the procedure will be found in Forte, Nichols and Glaeser (1968). The following solutions are required:

Ammonium acetate buffer: 9.31 g of ammonium acetate
0.23 g of ammonium carbonate
0.5 g of disodium EDTA
water to 1 litre; adjust to pH 7.4

Sodium phosphotungstate solution:
1.0 g of sodium phosphotungstate
water 50 ml; adjust to pH 7.4

Use double distilled water to prepare this reagent and filter it through a 0.8 μ membrane filter immediately before use.

Sample grids. 300 mesh copper grids, coated with Formvar and supported with carbon. According to Kunitake et al. (1978), it is advantageous to allow the prepared grids to age for at least 6 months at room temperature before use. This is said to improve the uniformity with which the lipoprotein particles spread on the surface. However, grids that are stored in this way may become fragile and tend to disintegrate in use.

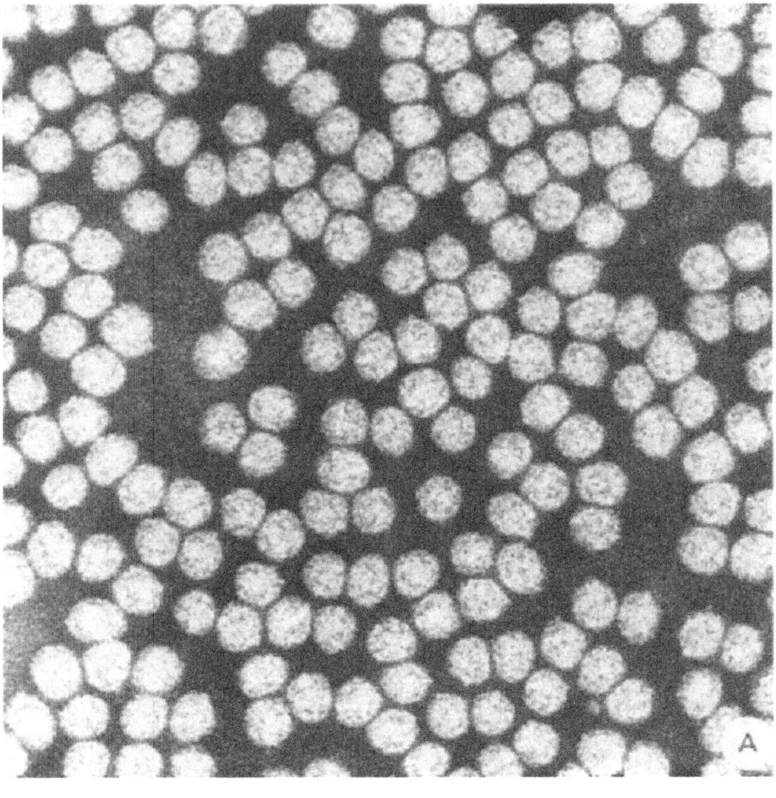

Fig. 4.1A. Freshly prepared LDL visualised by negative-staining electron microscopy (courtesy of Dr. M.J. Chapman).

Procedure. Isolate the lipoproteins in the usual way, and dialyse them against de-aerated ammonium acetate buffer at 4 °C for 24 hours, preferably under N_2. In our experience, the final concentration of lipoprotein that is recommended by Forte et al. (1968) is too high by a factor of about 10. Better results are obtained if the concentration of the dialysed lipoprotein solution is adjusted to about 0.4 mg lipoprotein/ml. The simplest way of transferring the sample to the grid is to mix 25 µl of the lipoprotein solution with 25 µl of sodium

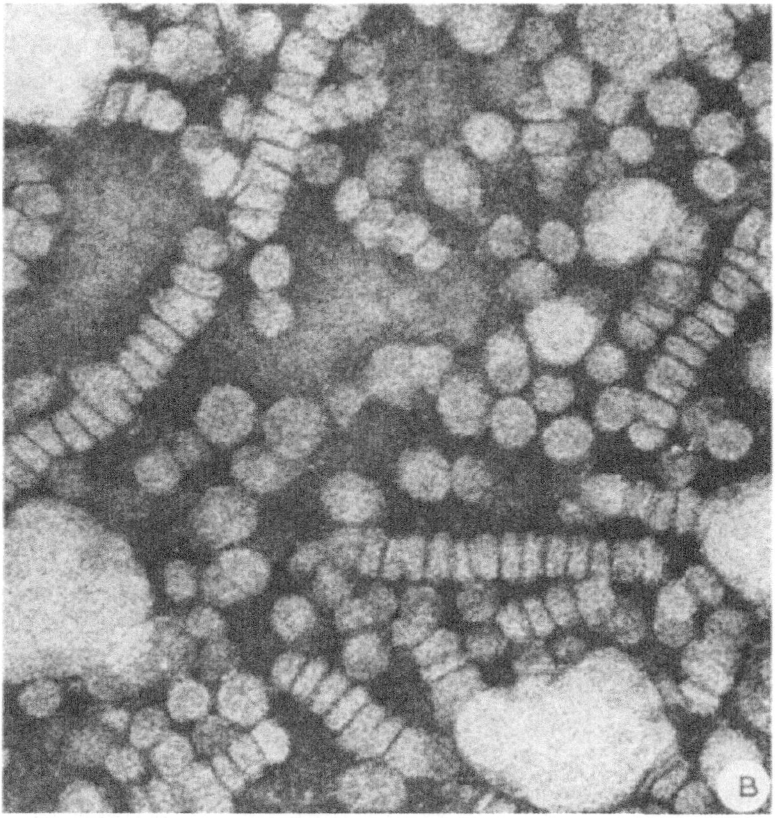

Fig. 4.1B. Aged LDL, showing the clumping that is characteristic of particles whose integrity is impaired (courtesy of Dr. M.J. Chapman).

phosphotungstate solution in a small tube. Use a capillary pipette to deposit a small drop of the mixture onto the grid and allow it to stand for about 60 sec. Then remove the excess fluid by touching the edge of the grid with the tip of a triangular slip of filter paper. Allow the grid to dry at room temperature and examine it immediately in the microscope.

The procedure used by Hamilton et al. (1971) is slightly more complex, but is thought by some to give more uniform preparations of isolated particles. In this technique the stain is added to the lipoprotein on the grid, as follows: add a drop of lipoprotein solution to the grid, stand for 15 sec and then remove the excess fluid with filter paper. Immediately add a drop of phosphotungstate solution, stand for 15 sec and remove the excess as before. When the grid is dry, examine it in the conventional way.

To determine the apparent particle size distribution in a lipoprotein sample, take photographs of fields in which the particles are mostly free-standing. The edge-to-edge diameter of at least 200 isolated particles should be determined, on the negative, using a comparator capable of measuring to at least 0.05 mm. The microscope can be calibrated by any of the conventional procedures, e.g. by observations on a shadowed carbon replica of a diffraction grating.

CHAPTER 5

Electrophoretic analysis of intact lipoproteins

In this chapter we shall be concerned with electrodynamic methods for the separation and characterisation of lipoproteins, both in their native plasma and as isolated preparations.

The electrical properties of the lipoproteins are like those of simple proteins. They have an isoelectric point at about pH 5.5, above which they are negatively charged. These properties have been made the basis of several different methods of analysis by zone electrophoresis. The first of these to be developed uses paper as a support and can resolve the plasma lipoproteins into four main populations of particles. The two main classes, the α- and β-lipoproteins, have mobilities approximately the same as those of the plasma α- and β-globulins. A third class, the pre-β-lipoproteins, has a slightly higher mobility than the β-lipoproteins, but is barely resolved from them. The very large chylomicron particles do not migrate, and form a band at the origin. The sharpness and resolution of these bands of particles was later improved by the introduction of cellulose acetate and agarose gel as support media and, of these, agarose is now considered to give the best results.

In all these procedures, the analysis is performed on a thin strip of the support and the lipoproteins are located by staining with a fat-soluble dye. These simple methods are particularly well suited to the clinical laboratory, where the demand is for a quick analysis of plasma that will decide whether it contains an abnormal lipoprotein or an abnormal distribution of lipoproteins. However, their poor resolution and the insensitivity of the staining process, not to mention the difficulty of making reliable measurements of mobility, make

TABLE 5.1
Index of electrophoretic methods

PLASMA OR
LIPOPROTEIN PREPARATION

- Paper (Section 5.2)
- Cellulose acetate (Section 5.3)
- Agarose (Section 5.4)
- Polyacrylamide gel
  - Uniform-pore gel (Section 5.5.3.1)
  - Discontinuous gel (Section 5.5.3.2)
  - Gradient gel (Section 5.5.3.3)
- Immunoelectrophoresis
  - Simple (Section 5.6)
  - Two-dimensional (Section 5.7)

them much less useful for characterising isolated lipoprotein fractions. For this purpose immuno-electrophoresis is better as a means of detecting contamination with plasma proteins or, for example, the contamination of β-lipoprotein with α-lipoprotein. This technique is also valuable for the study of the distribution of different apo-lipoproteins among the lipoprotein particles. However, its ability to resolve the lipoproteins is reduced, by comparison with that of simple zone electrophoresis, by the diffusion that precedes the immuno-precipitation. A much superior resolution can be obtained on polyacrylamide gel, particularly on gels of graded pore size. This technique will resolve both the α- and the β-lipoproteins into three or four components and thus provides a measure of their heterogeneity. It is potentially the most useful for characterising isolated lipoproteins, despite its essentially non-quantitative nature. Nonetheless, each method of electrophoresis has its own distinct sphere of utility.

The practice of electrophoresis as used for the analysis of lipoproteins does not differ in its essentials from that used for other plasma proteins. It is usually possible to use the same equipment. However, because the concentration of the lipoproteins is relatively low, it is usually necessary to use more plasma than would be needed for a conventional analysis of the proteins. In the sections that follow, we shall confine ourselves largely to a description of the technical features that are peculiar to, or are particularly important for the electrophoresis of lipoproteins. Descriptions of the general principles of zone electrophoresis and of the different types of equipment available may be found in numerous textbooks, e.g. Smith (1976).

5.1. Preparation of samples

The preparation and storage of plasma is discussed in Appendix 1. In the context of electrophoresis, it is particularly important that the plasma should be fresh and that it is never frozen. The period for which the plasma may be stored before analysis may depend on the purpose of the analysis. For the greatest resolution, it should be no more than a few hours. However, it is possible to identify the hyperlipoproteinaemias using plasma that has been stored at 4 °C for 2 or 3 days. It is advantageous to add EDTA to a final concentration of 2.5–5.0 mM. This reduces the oxidation of the lipoproteins and improves the sharpness of the bands. The EDTA may either be used as an anti-coagulant, or may be added to serum after it is recovered from the clot.

It is questionable whether plasma or serum is the better starting material. On balance, the fact that the analysis can be started more promptly when plasma is used may outweigh the disadvantage that fibrinogen is present.

So far as isolated lipoproteins are concerned, it should be noted that they usually migrate rather faster on electrophoresis than they do when run in their native plasma. The reason for this change in electrical properties is not certain but may result from oxidation or other damage during the isolation process. It is therefore important to take the precautions described in Ch. 2.

5.2. Electrophoresis on paper

According to Hatch and Lees (1968) the best results are obtained with the Durrum type of equipment (Williams et al., 1955) in which the paper strip is suspended in the form of an inverted V. However, excellent results can also be obtained with modern designs of apparatus in which the strips are stretched horizontally. These may be more troublesome to set up than the Durrum type, but have the advantage that their volume is smaller and equilibrium is more easily achieved. No matter what apparatus is used, it must be on a level surface, away from draughts and in a position of approximately uniform, constant temperature. In the Durrum apparatus it may help to maintain a uniformly saturated atmosphere if filter paper moistened with buffer is fixed inside the ends of the cover.

In the following sections we shall assume that commercially available equipment will be used, and that it will be set up and operated in accordance with the maker's instructions (cf. Smith, 1976). All the equipment must be kept scrupulously clean and free from deposits or growths. Do not allow salt deposits to accumulate; in some equipment it is possible for this to lead to a short circuit of the power supply.

Electrolyte solution. Although the pH of the electrolyte buffer is almost invariably quoted as 8.6, the compositions recommended often vary quite widely. The following recipe is typical, but it may be found that a small change in the quantity of barbituric acid will improve the resolution obtained with particular equipment or conditions. Some workers also use the electrolyte at 50% of the following concentration.

Buffer solution:	Sodium diethyl barbiturate	20.6 g
	Diethylbarbituric acid	3.68 g
	Disodium EDTA	0.372 g

Dissolve the sodium barbiturate in 500 ml of cold distilled water, and the diethylbarbituric acid in about 150 ml of warm water. Mix the two solutions, add the EDTA and make up to 1.0 litre. If necessary, adjust the pH to 8.6 with HCl or NaOH. *Immediately before use*, add 10 g of bovine serum albumin (Fraction V) to each litre of buffer.

Albumin is added to the buffer to improve the separation and sharpness of the bands. However, samples of albumin from different suppliers may not be equally effective. Hatch and Lees (1968) say that human albumin gives the best results. In their opinion, bovine Fraction V is unsatisfactory as it is normally supplied but will give better results if it is treated with charcoal as described by Chen (1967). By contrast, other writers have recommended bovine albumin but it is not always clear whether they have used the charcoal treatment. If experience suggests that a particular brand of albumin may require this treatment, it can be carried out as follows:

Dissolve 10 g of albumin in 100 ml of distilled water at 23 °C. Mix in 5.0 g of activated charcoal (e.g. Norit) and lower the pH to 3.0 by adding 0.2 N HCl. Stir for 1 hour at 0 °C and then remove the charcoal by centrifugation at 2 °C. Finally, raise the pH to 7.0 with 0.2 N NaOH. When the albumin is added in this form, an appropriate adjustment should be made to the volume of the barbiturate buffer, to allow for dilution.

The resolution of freshly made electrolyte is often poor but improves after a few runs. This situation can be remedied by using old electrolyte to moisten the strips before putting them into the tank. The electrolyte may be used for several runs provided that the polarity of the electrodes is reversed each time.

Preparation of the paper. Whatman No. 1 filter paper is widely used and gives good results but any filter of equivalent quality can be used. Larger amounts of lipoprotein can be applied to Whatman 3MM paper but the resolution may then be reduced.

Cut strips of paper to the size recommended by the makers of the electrophoresis tank. Mark the point at which the sample is to be applied, lightly, with a pencil. This will be in the centre of the strip if the Durrum apparatus is used, but about 25% of the distance from the cathodic end if the strip is run horizontally. Write the sample identification at one end of the paper. If the tank is of Durrum design, suspend the strips in place and moisten the paper with electrolyte from a pipette. Replace the lid and allow the system to equilibrate for at least 1 hour before applying the samples. If the tank is of the horizontal design, dip the strips into electrolyte and blot off the excess with clean filter paper. Then stretch them over the supports in the tank; it is essential that they do not sag in the middle. Close the apparatus and allow it to equilibrate.

Whatever form of tank is used, it is important that there is good contact with the electrolyte. If intermediate wicks are used, they must be clean and of adequate thickness and there must be no air bubbles between the wick and the paper strip. Ensure too, that the level of the buffer in the two electrode compartments is the same by connecting them briefly with a syphon.

Application of samples. Plasma is applied to the line marked on the strips, without removing them from the tank, at the rate of 8–10 µl/cm. If the presence of a 'Type III hyperlipoproteinaemia' is suspected, it will be necessary to isolate the fraction of density less than 1.006 g/ml by ultracentrifugation (Section 2.1.1) and to analyse both this and the complementary fraction by electrophoresis, in parallel with the original plasma, on the same paper strip. Apply 10 μl of the infranatant and 20 μl of the supernatant per cm. In general, isolated lipoprotein fractions can be analysed satisfactorily provided that they are dissolved in 0.15 M NaCl or in the electrolyte buffer. The solution applied to the strip should contain about 5 mg/ml of lipoprotein and should be used at the rate of 8–10 μl/cm.

With care, the sample can be applied directly from a micro-pipette (e.g. a Micro-cap; Drummond Scientific Co.) but it is important to avoid scuffing the surface of the paper. This difficulty can be avoided by using one of the commercially available applicators, which have the added advantage that they not only require less skill but also apply the solution more uniformly. The essential part of the applicator is a grooved plastic strip, or a pair of parallel wires that hold the sample by surface tension. Use a micro-pipette to load the applicator with the required volume. Then transfer the sample to the surface of the damp paper, allowing time for the completion of the process. In tanks of the horizontal design, the pressure of the applicator may displace the paper. This can be avoided by stretching a nylon thread across the tank below the point of application. In any event, check that the paper is still taut after the sample has been applied. Clean the applicator before using it again. (*Note:* Some makes of tank have slots cut in the cover through which the applicator can be inserted. This makes it unnecessary to remove the cover and disturb the atmospheric equilibrium. The slots are normally covered strips of adhesive tape. If there are several of these strips, only a short length of the slot need be open at any one time.)

Electrophoresis. Start the run as soon as possible after the samples have been applied. Any regulated supply of D.C. current can be used provided it has an adequate power rating. Remember to reverse the polarity of the electrodes at each run. The best results are obtained with a constant voltage power supply, though a constant current system can be used if the voltage is adjusted during the early part of the run to compensate for the increase in resistance that occurs. The optimum conditions for each set of equipment will depend on its configuration and on the local conditions, and must be determined by experiment. Do not pass too high a current, as the strips will overheat and this will cause the bands to migrate unevenly. As a guide, the voltage should be adjusted to give a current of 0.1–0.15 mA/cm of strip, which should be continued for 16–18 hours.

Staining. At the end of the run, switch off the power and remove the strips without delay. Dry the papers in an oven at 80–90 °C for about 20 min. Do not overheat, as this may reduce the uptake of stain by the lipoproteins (Hatch and Lees, 1968). Then suspend the strips in a bath of stain (Appendix 4); do not allow them to stick to each other or they will stain unevenly. The optimum duration of the staining process is highly dependent on local laboratory practice and must be determined by trial. Even for the same dye, different authors recommend periods that may range from 1 to 8 hours. As a guide, we suggest the following procedures (note that the staining time can be reduced if the bath is agitated):

Oil Red O. Remove the strips from the bath (which should be maintained at 37 °C) after about 1.5 hours, wash them in running tap water for 1 min, then blot them and hang them to dry at room temperature.

Sudan Black. Stain for about 30 min at room temperature. Remove the strips from the bath, wash briefly in 50% ethanol and then in 40% ethanol until the background is almost white.

5.3. Electrophoresis on cellulose acetate membranes

In this system, the bands of lipoprotein are sharper than those obtained on paper and, because they are adequately resolved after a much shorter migration, the analysis can be completed more quickly. However, the resolution is still not quite so good as that in agarose gel, and the membranes are more difficult to de-stain than either paper or agarose. There is consequently little incentive to use this system for lipoprotein analysis except in the clinical laboratory, where its speed makes it more attractive than paper electrophoresis and the higher cost of agarose gels may be more than enough to offset their marginally better performance. It is interesting to note, in this context, that the cellulose acetate method was considered to be unsuitable for the routine screening studies of the Lipid Research' Clinics Program of the American National Institutes of Health.

The equipment used for electrophoresis of lipoproteins on cellulose acetate is of the conventional 'horizontal' type and is widely available commercially. The technique used is similar to that for electrophoresis on paper (Section 5.2) and we shall confine ourselves to a description of the points of difference.

Electrolyte solution. Use the solution given in Section 5.2 at half-strength, i.e. 0.05 M; albumin may be omitted.

Preparation of the membrane. Strips of cellulose acetate intended for the analysis of a single sample are typically about 15 cm long by 2.5 cm wide. Alternatively, several samples can be applied to a sheet 15 cm × 7.5 cm. Mark the positions at which the samples are to be applied as in Section 5.2. If a large membrane sheet is used, leave a space of 4 mm between each sample and at least 1 cm to the edge of the sheet. Then float the membranes on fresh electrolyte for at least 5 min. Do not plunge them into the solution as this may trap air within the pores. When they are thoroughly wetted, remove the membranes with forceps and blot off the excess buffer with clean filter paper. Position the moistened strips in the tank, making sure that they do not sag in the middle, and equilibrate for at least 1 hour.

Application of samples. Apply 2 µl of plasma (or its equivalent in isolated lipoprotein) across a streak 5 mm long, as described in Section 5.2.

Electrophoresis. Do not exceed a current of 0.5 mA/cm of membrane. The analysis should be complete after 30–60 min, when the migration will be about 2.5–3.0 cm. Remove the strips promptly after switching off the current but be careful that the electrolyte does not swill across them while doing so.

Staining. The methods used for staining paper strips do not always work so well with cellulose acetate. This seems to be due to differences in the manufacture of the various brands of acetate, some forms being less troublesome to destain than others. Chin and Blankenhorn (1968) used Sepraphore III membranes (Gelman Sciences Inc.) which they stained with Oil Red O, and were apparently able to de-stain by washing in water. On the other hand, Fletcher and Styliou (1970) used the same membranes but found it desirable to bleach the background as follows:

Transfer the damp strips from the electrophoresis tank to a bath of Oil Red O at 37 °C (Appendix 4), taking care that they do not stick together. After 18–24 h, remove the strips and wash them with water. The background can then be decolourised by judicious treatment with a solution that contains 0.025% sodium hypochlorite in 5% acetic acid (household bleach will do). The duration of this process will depend on the intensity of the residual stain, and may take several minutes. Finally, wash the strips thoroughly in 5% acetic acid followed by water. Blot off excess liquid and dry between sheets of filter paper under a weight to keep the membrane flat.

Colfs and Verheyden (1967) used Cellogel membranes (Chemetron) as the supporting medium and located the lipoproteins by staining with Sudan Black. To reduce the background staining to a tolerable level, they deacetylated the membranes after the electrophoresis by dipping them into 33% KOH for 20 sec before transferring them to the staining bath. However, the timing of this de-acetylation is very critical and Berends et al. (1972) found it better to use a mixture of 5% NaOH solution and 96% ethanol in the proportions 7:6 by volume as a combined fixing and de-acetylating bath. After soaking in this reagent for 5–10 min, the strips are then stained for 2.5–3 h in a saturated solution of Oil Red O in 60% ethanol.

An alternative method of location that avoids the difficulties associated with the dyes is the ozone–Schiff procedure that was proposed by Kohn (1961) (Appendix 4).

5.4. Electrophoresis in agarose gel

This method, which was originally proposed by Noble (1968), is faster than electrophoresis on paper and has relatively good resolution, but it requires the daily preparation of gels. This disadvantage can be overcome by the use of commercially available ready-made gels, on plastic slides that can be put directly into the electrophoresis tank (e.g. Biogram A: Bio-Rad Laboratories Inc. or Universal Electrophoresis Film: Corning Medical). These give good results and are more easily stored than home-made gels, but are expensive.

Those who must perforce prepare their own gels need to be aware that some brands of agarose give better results than others and that it may be prudent to buy a good stock of a batch that works well. Moreover, highly purified agarose does not give such good separations as preparations that are rather less pure. It is then necessary to adopt the subterfuge used by Noble (1968) and add a small amount of agar to the agarose. This is the procedure that we shall describe. However, the ratio of agar to agarose may need slight adjustments if different brands are used.

The mechanical strength of the gel may also vary from brand to brand and can play a part in deciding the concentration of agarose to be used. Values ranging from 0.5 to 1.2% have been quoted by different sources.

The practice of adding albumin to the gel is also variable: some use it, others do not. In our experience, it is more important to find the

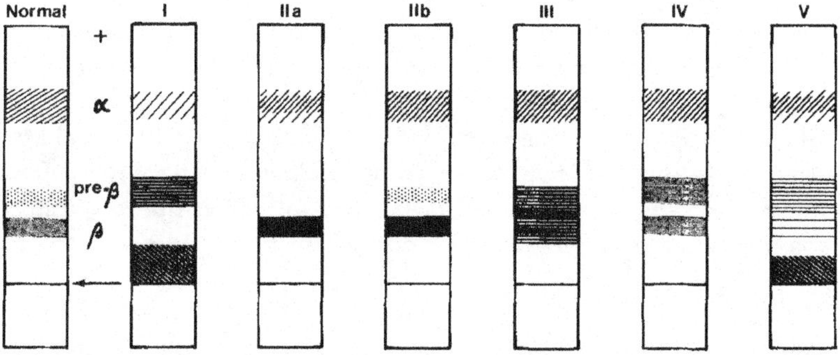

Fig. 5.1. A schematic representation of the distribution of plasma lipoproteins as revealed by electrophoresis in agarose gel. The diagrams show the approximate relative intensities of the stained bands in the hyperlipoproteinaemias of the Fredrickson classification.

appropriate mixture of agarose-agar, the addition of albumin will then affect the analysis of *plasma or serum* to only a small extent. For the analysis of *isolated lipoproteins* however, albumin should be present.

Materials. Although it is possible to prepare gels on glass plates e.g. microscope slides, these are expensive and cumbersome to store. It has therefore become common to use plastic film that can be cut into strips or sheets like paper. The material used is Cronar, clear base, 0.004 inches thick, which is manufactured by E.I. du Pont de Nemours and Co. Inc.

Reliable preparations of agarose can be obtained from several makers, including the following: Pharmacia Fine Chemicals AB, Bio-Rad Laboratories Inc., Industrie Biologique Française S.A. and Serva Feinbiochimica.

Electrolyte solution. Use the buffer described for paper electrophoresis (Section 5.2) diluted with an equal volume of water. Do *NOT* add albumin.

Preparation of plates. If Cronar film is to be used, cut pieces of the required size and flatten them, if need be, by warming on a glass plate at 80–85 °C. To simplify the handling of the film during the coating process it is worth-while to stick it to a glass plate with a few drops of water (Noble, 1968). The plate, which may be warmed to retard the setting of the gel, must be placed in a level surface.

If glass plates are used as a base for the gel, they must be scrupulously clean and must also be set on a level surface. Microscope slides can either be coated individually, or six at a time in plastic frames that are made for the purpose (Gelman Sciences Inc.). Both plastic film and glass slides can be labelled with a felt-tip using a spirit based ink. Alternatively, a more permanent mark can be made with a diamond pencil.

The following procedure is essentially that of Noble (1968), which has given good results in our hands. Weigh 0.4 g of agarose (Bio-Rad Standard, Low m_r) into a flask and add 40 ml of electrolyte solution. Into a second smaller flask or tube, weigh 0.1 g of agar (Difco Noble Agar) and add 10 ml of electrolyte. Warm both flasks in a water bath at 80–90 °C with occasional agitation until the solids are dissolved. Add the agar solution to the agarose and cool the mixture to 45 °C. If albumin is to be included, add 1 ml of a 25% solution of bovine serum albumin (Fraction V) in 0.15 M NaCl. Keep the mixture at 45 °C. (N.B. it is permissible to re-melt agar solutions but not agarose).

The layer of gel on the slide should be as thin as is practicable i.e. 0.75–1.0 mm, which is equivalent to spreading 0.75–1.0 ml of agarose solution on 10 sq.cm of slide. In any event, do not exceed a thickness of 1.5 mm.

If glass slides are used, first smear them with 1 of 2 drops of half-strength agarose-agar and allow them to dry, preferably overnight. This step will improve the adhesion of the full-strength gel and is not necessary when Cronar film is used. To apply the gel, warm a pipette by drawing hot water into it several times and then pipette the appropriate amount of agarose-agar on to the slide or sheet. Spread the solution over the surface with the tip of the pipette to within 1.0 cm of the ends. After a moment or two, but before the gel sets, place short lengths (5–15 mm) of thin steel rod (1.0–1.5 mm diameter) across the gel at the points where the sample is to be applied, e.g. about 20 mm from the end of a microscope slide. Pieces cut from thin nails will do well, if they are coated with a thin film of silicone grease before use and are kept scrupulously clean and dry afterwards. After the gel has set, remove the rods with a magnet to leave a slit-shaped well. If several wells are to be made in a large slab of gel, it is convenient to use a plastic comb with prongs 1.0–1.5 mm thick and of appropriate width, that are about 4 mm apart. The teeth of this comb are pressed into the layer of molten gel and removed when it has set. The gels should be allowed to mature, in a humid chamber at room temperature, for at least three hours before use. Discard after 24 hours at room temperature, or after 5 days if the gels are stored in a closed vessel at 4 °C.

Application of samples. Instal the slabs of gel in the electrophoresis tank *before* adding the samples. Wicks of thick filter paper can be used to make contact between the gel and the electrolyte, but wicks that have a much longer life can be made from dress-maker's non-woven interfacing (e.g. Vilene). Make sure that there are no bubbles between the wick and the agarose, and that there is close contact across the whole width of the gel. When all the gels and wicks are in place, close the tank and allow the system to equilibrate, without passing a current, while the samples are prepared.

It is usual to fill the sample wells with a mixture of the lipoprotein solution with agarose, but this is not essential (Johansson, 1972). A rectangular slot 10 mm × 1.0 mm in a gel 1.0 mm thick requires 10 μl of sample, but a similar well made with a 1.0 mm diameter rod may hold only 6–7 μl. It is important that the wells are not filled to overflowing, or that bubbles are formed. This can be avoided if calibrated micro-pipettes are used, as follows:

Take a capillary pipette (e.g. Microcap; Durrum Scientific Co.) of a size larger than is necessary i.e. a 20 μl pipette if the well is to hold 10 μl of sample. Mark it at the point that defines the desired volume when measured from the *top* of the capillary, i.e. 10 μl in this example. Prepare plenty of these capillaries in advance and keep them in a tube that is mounted in the water bath at 45 °C.

To prepare the sample gel, use a piston type repeating pipettor with disposable plastic tips to transfer 20 μl of the solution to be analysed to a small tube in the water bath at 45 °C. Add 20 μl of agarose solution and mix. Fill one of the marked capillaries with this mixture and promptly express it into the sample well until the meniscus reaches the mark. (It is convenient to reserve one slide for a control sample to which a trace of bromphenol blue is added. This allows the progress of the run to be followed).

Electrophoresis. This is performed in essentially the same manner, and with the same precautions as for horizontally run paper strips (Section 5.2). If the recommended concentration of electrolyte is used, the current that can be passed will be determined mainly by the thickness of the gel. If it is too great, the gel will tend to melt and irregular bands will be formed. A current of 2–3 mA/cm is usually satisfactory and should be continued for 45–90 min according to the length of the gel.

At the end of the run, switch off the current and remove the gels from the tank. Fix the lipoproteins by immersing the gels in a bath of 2% trichloracetic acid for 30 min, and then wash for 30 min in distilled water. Cut strips of Whatman 3MM filter paper (or its equivalent) the same size as the gels, moisten them and press one gently on to each slide, making sure not to trap any air bubbles. Dry overnight at room temperature, or for about 20 min in an oven at 80–85 °C. It is undesirable for Cronar sheets to come into contact with the hot shelf of the oven and the N.I.H. Lipid Research Clinics manual (1975) recommends that they are placed on a pad of paper towelling.

Staining. If it is necessary to stain a very large number of gels on plastic film, it may be worth using the rotating dye-rack described by Noble (1968). With this equipment, the time taken to stain to an acceptable intensity can be significantly reduced. For example, Sudan Black may take up to one hour and Oil Red O may take up to 2 hours (Appendix 4). In most laboratories however, it is feasible to immerse the slides in a static dye-bath for a longer time e.g. up to 6 hours in the case of Oil Red O. As when staining paper strips, care must be taken to ensure that the gel slides to not make contact, or some will be masked from the action of the dye. To avoid this, it may be desirable to use a rack like those used for the staining of microscope slides in histology.

When the staining is judged to be complete, remove the slides from the bath and wash them in running water (if Oil Red O was used) or in 50% ethanol (if Sudan Black was used). Rub the surface of the gel gently with a finger to remove dye that is deposited on the surface: it is as well to wear a disposable glove when doing this. The back of the slide can be cleaned with a tissue moistened with ethanol. The background should be perfectly clear and colourless.

5.5. *Electrophoresis in polyacrylamide gel*

The method of 'disc' electrophoresis in polyacrylamide gel was used to analyse lipoproteins soon after its description by Davis (1964), at first with indifferent results. In later developments however, better use was made of the fact that the permeability of the gel can be adjusted to add a controlled degree of molecular sieving to the process of separation. This allows each of the main lipoprotein classes to be separated into several components. Under favourable conditions, a dozen or so lipoprotein fractions may be detected in plasma by this method and, although their electrophoretic mobility cannot be measured accurately, the technique is still one of the best for the qualitative physico-chemical characterisation of a lipoprotein preparation. Unfortunately, it is not easy to recover the intact lipoproteins from the gel after the electrophoresis and it is therefore difficult to identify the sub-fractions by their chemical properties. However, it is possible to cut out small sections of the gel that contain a single lipoprotein band and to analyse its apo-lipoprotein content by immuno-diffusion (Section 6.5).

The equipment for electrophoresis on polyacrylamide gel is described in detail in many sources, e.g. Davis (1974) and Smith (1976). Moreover, there are many manufactures of the relevant equipment, both for electrophoresis in tubes or in slabs of gel, which is often supplied with detailed operating instructions. Whatever make is chosen, it is desirable that there should be provision for cooling the gels during the analysis.

As with other electrophoresis techniques, the practicalities of the polyacrylamide method vary in detail in different laboratories, as can be exemplified by the practice of de-aerating the monomer solutions. In some laboratories this is apparently not done, whereas others de-gas exhaustively in vacuo. The presence of oxygen in the solutions inhibits the polymerisation reaction and its judicious removal not only contributes to the reliability of the gels, but can also be used to give a measure of control over the rate of the reaction. It is not our purpose to describe the technique in detail, those to whom the method is new should consult one of the many comprehensive texts on this subject (Gordon, 1975; Maurer, 1971; Shore et al., 1980; Edelstein and Scanu, 1980). The following sections will therefore be restricted to finer points which are applicable to the electrophoresis of both lipoproteins and apo-lipoproteins (Section 6.4.2).

5.5.1. Preparation of acrylamide gels

Acrylamide is usually caused to form a gel which is cross-linked with *N,N'*-methylene bisacrylamide. However, other cross-linking agents can be used if a gel of different properties is required e.g. *N,N'*-bis-acrylylcystamine or *N,N'*-diallyltartardiamide will form gels that can be re-dissolved. Note that the total concentration of monomer used to make a gel is reported as a percentage, i.e. g/100 ml of the monomer solution. The concentration of the cross-linking agent is reported as a percentage of the total monomer present.

Hazards in the use of acrylamide monomer. It may not be sufficiently well known that acrylamide and bisacrylamide are highly toxic substances. The following warning is taken from the Bulletin of the Safety Office of the National Institutes of Health, January 1971.

'Acrylamide monomer is a neurotoxin and should be handled accordingly. Care should be taken to avoid skin contact, and mouth pipetting of solutions should be avoided. If any is spilled on the skin, immediately wash the area in running water and then soap and water. The acrylamides produce unusual neurotoxic effects. The disease syndrome is characterised by incoordination, ataxia (irregularity of muscular action), and weakness in the extremities (particularly the legs). Initially the complaints are drowsiness, fatigue and tingling in the fingers. The principal complaint is stumbling with a sense of unsteadiness. The mechanism is not clearly understood since no histological effects are seen in the nervous system. These effects can be produced by all routes of entry: oral, skin contact and inhalation'.

The gels used for analytical electrophoresis (or for isoelectric focussing) can be cast in three forms:

(a) Gel rods of 4–6 mm diameter by up to about 100 mm long which are cast in individual tubes. These are widely used for analytical electrophoresis. They are easily made and the equipment will usually allow from 12 to 18 samples to be examined concurrently.

(b) Rectangular slabs of gel, several mm thick, which are cast between two glass plates and mounted vertically for the analysis of several samples simultaneously. This configuration facilitates the comparison of the samples and the density of the stained gels.

(c) Thin layers of gel cast on glass plates or plastic film. These are economical of material and make the comparison of different samples easy, but there is little experience of their use for the analysis of lipoproteins or apo-lipoproteins.

All these gel preparations are available commercially but these products are not usually of low enough concentration for the analysis of intact lipoproteins.

The laboratory method of casting gels in glass tubes is uniform, varying only in the number and composition of the superimposed gel layers. It is preferable to use the precision bore glass tubes that are sold for this purpose, since the resulting uniformity of voltage gradient and power dissipation in the tubes helps toward achieving reproducibility of resolution. Nonetheless, laboratory tubing can, if necessary, be cut to the appropriate lengths. To prevent these home-made tubes being confused with the precision bore type, they may be cut to a different length.

The tubes must always be scrupulously clean and free from grease. Before each use they should be soaked in chromic acid, rinsed exhaustively with distilled water and then steeped for 10 min in 0.1% Triton X-100 (v/v). They should then be drained and dried at 110 °C. The manufacturers of equipment for electrophoresis often supply small plastic caps with which to close the ends of the tubes before they are filled. However, a double thickness of Parafilm well sealed to the glass, may be preferable since there need be no risk of damaging the gel by suction when it is removed. Fix the required number of these tubes in a rack, with their closed ends downwards, as near to the vertical as possible. Then transfer the necessary amount of the reagent mixture which is to form the separating gel into each tube as quickly as possible, while ensuring that no air bubbles become trapped. This operation is easily and reproducibly carried out with an air-displacement pipette. When the reagent has been added to every tube, each one must be gently overlayered with water to a depth of 3–5 mm to suppress the meniscus. This is most easily done by touching single drops of water, from a syringe with a 25 gauge needle, against the side of the tube 1–2 cm above the monomer mixture. In this way the water can be made to flow down the glass and spread over the surface of the solution.

Polymerisation begins as soon as the persulphate catalyst is added and will start to become manifest some 30–60 min later as a developing translucence of the solution. About 2 hours should be allowed for the completion of the reaction, during which time the tubes must not be disturbed. The supernatant water layer can then be removed from the gels by inverting the tubes and absorbing the fluid with filter paper. Be sure to absorb all the water, since any drops that remain will dilute the stacking gel. The polymerised gels may be stored at 4 °C for several days if the tubes are sealed at the top with Parafilm, but note that a premature removal to the refrigerator may lead to non-uniform gels which contain bubbles.

Preparation of stacking gel. This must be prepared in the most reproducible manner possible if the highest resolution of the protein bands is to be maintained. Moreover, it must be prepared immediately before the electrophoresis and under no circumstances should tubes be stored with the stacking gel already cast.

Set the tubes in the vertical position and pipette into each one 0.3 ml of the reagent mixture that is to form the stacking gel, running it carefully down the sides of the tubes as before. Again layer each tube with 3–5 mm of water. If a persulphate catalyst is used, the gels should be ready for use in 30 min. However, if riboflavin is used the polymerisation is induced photochemically by irradiation with a tubular fluorescent lamp. This should be fixed so that a rack of tubes can be set directly beneath it, at a distance of a few inches. The light should be able to pass directly down the tubes without travelling through the glass walls. Polymerisation will take about 20 min and will be seen as an opalescence in the reagent solution.

Preparation of samples. Electrophoresis or isoelectric focussing can be used to examine either the intact lipoproteins or solutions of their apo-lipoproteins, of which the latter often contain protein-denaturing agents. Specific details of the methods of sample preparation that are particular to an electrophoresis system will be found in the appropriate section. In general however, the sample should contain sufficient protein to give stainable bands after the analysis i.e. 10–200 μg depending on the sample and the purpose of the experiment. If the apo-lipoproteins are dry, they can be dissolved in the upper electrode buffer, to which denaturants such as urea or sodium dodecyl sulphate can be added as necessary. The protein sample also contains about 5 μl of tracking dye (usually bromphenol blue) to indicate the position of the electrophoretic front and a drop of glycerol or 20% sucrose to increase its density. The volume of this mixture should be as small as is practicable, and should preferably not exceed 200 μl. If a larger sample is unavoidable, it is wise to make the volume of the stacking gel twice the volume of the sample, and to prepare similar gels for reference samples to facilitate the identification of the components of the mixture. The sample should contain the lowest practicable concentration of salt, the equivalent of NaCl of density 1.019 g/ml being a workable upper limit. Higher concentrations than this will prevent stacking of the proteins and lead to low migration rates and diffuse, trailing protein bands. However, some organic ions that are commonly present e.g. acetate or guanidine have less detrimental effects on the resolution, although they will necessitate a longer time of electrophoresis. Uncharged denaturing agents such as urea and tetramethylurea do not interfere in this way, but β-mercaptoethanol may cause small changes in the mobility even of proteins that do not contain disulphide residues.

Setting up the equipment. Remove the layer of water from above the stacking gels by inversion over filter paper and then carefully remove the rubber caps from the bottom of the tubes. If this is done negligently, the gels can easily be disturbed by suction. If the tubes were closed with Parafilm, this should be peeled away, with the aid of a scalpel if necessary.

In order to achieve the maximum reproducibility, the tracking dye should be allowed to migrate the same distance in each gel in every different electrophoretic run. This distance e.g. 80 mm in a 100 mm tube, can be marked on the outside of the tube with an indelible marker, using two lines drawn on a card as a template. First however, wipe the tube clean or the mark may float off in the electrode buffer, a difficulty that is not uncommon when sodium dodecyl sulphate is present. Then locate the interface between the stacking and the separating gels over one line and mark the position of the other on the tube. Finally, push the tubes, from below, into their rubber grommets in the upper electrode chamber.

Transfer the prepared sample solutions onto the stacking gels with a fine pasteur pipette. Rinse the tube that contained the sample and the pipette with a small volume of the upper electrode buffer and layer it over the sample. The upper electrode buffer must been be carefully superimposed on this until the electrophoresis tube is full. It now only remains to fill the lower electrode chamber with buffer, put the upper chamber in place and fill it with buffer before starting the electrophoresis. In this last operation, be mindful that eddies in the buffer may cause the samples to swirl into the layers above them. This mischance can be avoided by pouring the buffer slowly and steadily through a large plastic funnel, the stem of which is cut off square and is held below the surface of the solution.

After the electrophoresis, the gels must be extracted from their supporting tubes. Provided that the tubes were treated with Triton X100, this can easily be done with the aid of 21 gauge hypodermic needle that is attached to a syringe filled with water. Insert the needle between the gel and the glass and then carefully move it sideways around the surface of the tube while simultaneously injecting water. The detached gel can then be extruded by gently squeezing a rubber teat fitted over the end of the tube. Gels of more than 7% concentration will slide out easily, but those of less than 5% concentration are not so rigid and must be handled carefully to prevent them from breaking or from distorting.

5.5.2. Staining procedures

Probably no other aspect of lipoprotein analysis is more bedevilled by lack of procedural uniformity than the staining of the lipoproteins and their application to the acrylamide gel. We shall therefore briefly discuss this topic before describing how the electrophoresis is carried out.

Although it is possible to stain the lipoproteins after the analysis (Pratt and Dangerfield, 1969), as in paper electrophoresis, it is more usual to stain them beforehand with Sudan Black. This procedure dates from at least as early as 1955 and has been re-published in a bewildering number of modifications. In all of these it is the removal of the excess dye that is likely to prove the most tiresome problem. The stain consists of Sudan Black in an organic solvent and, when this is added to the plasma or lipoprotein solution, much of the dye is precipitated. If these particles are not removed, the lipoproteins will migrate through the gel in irregular, streaky bands.

In the earliest procedures, the dye was dissolved in a solvent such as ethanol and this example was followed by Frings et al. (1971). However, fears that the solvent might cause damage to the lipoproteins led McDonald and Ribeiro (1959) to recommend that propylene glycol or ethylene glycol should be used instead. This method, or the modification developed by Ressler et al. (1961), is now probably the most widely used of those available for pre-staining lipoproteins. Nonetheless, these two techniques differ considerably in their procedural details and there have been many other variants published since. McDonald and Ribeiro mixed plasma with a saturated solution of Sudan Black in glycol in the proportion 5:1 and allowed it to stand for 45 min at room temperature. Ressler et al. used the same stain solution but reduced the proportion of plasma to 2:1 and stored the mixture for at least one day in the refrigerator before analysis. Others have raised the proportion of stain to even higher levels and have incubated the mixture for various times between 1 and 24 hours before removing the excess dye. It is not clear that any of these procedures is really superior to the others provided that sufficient time is allowed for the relatively slow transfer of the stain to the lipoprotein.

The conventional way of removing precipitated particles of dye is to subject the stained plasma to a brisk centrifugation before adding it to the acrylamide gel. However, Frings et al. (1971) simply added both dye and plasma to an activated acrylamide solution that was polymerised on top of the stacking gel. Staining therefore took place in the pores of the gel, where particles of excess dye were retained when the electrophoresis was begun. Subsequently, Naito et al. (1973) showed that this procedure could also be used with Sudan Black dissolved in ethylene glycol. The use of a sample gel in this way has the merit that it not only retains the excess dye, but also makes it impossible to disturb the sample layer by the clumsy addition of electrolyte. However, Narayan (1975) has pointed out that the presence of substances such as glycol or Sudan Black can affect the polymerisation of the gel and he believes that more reproducible results are obtained when the density of the sample is raised by mixing it with a concentrated sucrose solution. In this context, it may be noted that the use of a linearly polymerised acrylamide as recommended by Pratt and Dangerfield (1969) appears to offer little advantage.

In general, although there are several different ways of staining the lipoproteins with fat soluble dyes and of introducing them onto the gel, there seems to be little to choose between them if they are carried out with proper care and attention to detail. The procedures described in Appendix 4 are based on the technique of McDonald and Ribeiro (1959), for which there is evidence that it does not produce artefacts (Narayan, 1975). But it may be necessary for each laboratory to make adjustments in detail to compensate for variations between batches of stain and in local factors such as the average ambient temperature.

It is worth reiterating however, that in contrast to the debate about staining procedures, there is one point in respect of which there can be no controversy: plasma samples must always be fresh.

5.5.3. Acrylamide gels for lipoprotein analysis

Three different types of separating gel have been used for this purpose, namely:

5.5.3.1. Gel of uniform pore size. This is essentially a straightforward application of the method of Davis (1964) with the concentration of the separating gel adjusted to take account of the large size of the lipoprotein particles. From their careful study of the technique, Naito et al. (1973) decided that 3.6% gel gave the best results but that concentrations between 3.0 and 4.0% would give an acceptable resolution. The procedure used by this group is very like the earlier one of Frings et al. (1971), on which our description is based.

The resolution that can be obtained with this system is limited. Both Frings et al. (1971) and Naito et al. (1973) were able to separate the plasma lipoproteins into the three main classes, and some resolution of the β-lipoproteins was occasionally seen. For the analysis of the lipoproteins in native plasma therefore, this method has little more to offer than electrophoresis in agarose gel and is more tedious to do. This limitation arises from the fact that a gel that is permeable enough to allow large lipoproteins to penetrate, will not effectively sieve small ones. The greatest potential of this technique therefore seems to lie in the analysis of isolated lipoprotein fractions, for which the concentration of the separating gel can be adjusted to give the optimum resolution. Even so, although the α-lipoproteins (HDL) can be resolved into several components on gels of about 7% acrylamide, it is not possible to make a gel that is fully permeable to VLDL.

Reagents (see Table 5.2). Acrylamide and bisacrylamide should be kept in a cool, dry place and used within six months of opening. It is advisable to use the products that are specially purified for electrophoresis (Sigma Chemical Co.; Bio-Rad Laboratories Inc.; Pharmacia Fine Chemicals AB; BDH Chemicals Ltd.; Eastman Kodak Co.).

Procedure. These directions assume the use of 7 mm × 75 mm gel tubes, which must be scrupulously clean and free from grease. Bring

TABLE 5.2

Electrolyte solution: 6.0 g of Tris
28.8 g of glycine
Water to 1.0 litre. The pH should be 8.3

Gel reagents	Solution						
	A	B	C	D	E	F	G
Tris	36.6 g	5.98 g					
1.0 N HCl	48 ml	48 ml					
Acrylamide			15.0 g	10.0 g			
Bisacrylamide			0.4 g	2.5 g			
TEMED[a]	0.23 ml	0.46 ml					
Riboflavin					8.0 mg		
Ammonium persulphate							0.14 g
Sucrose						40 g	
Water to (ml)	200	100	100	100	100	100	100
pH	8.9	6.7					

Solution G should be freshly prepared for each experiment, but the others can be stored for up to three months at 4 °C, in the dark.

[a] TEMED, *N,N,N′,N′*-tetramethylethylenediamine.

the gel reagents to room temperature. Close the bottom of each gel tube (Section 5.5.1) and fix the appropriate number in a stand, making certain that they are vertical. Then add the gel reagents as follows:

Separating gel. Mix solutions A, C and G in the proportions 1:1:2 by volume in a tube or small flask that can be evacuated to a pressure of 10–15 mm Hg. Maintain the vacuum for 30–60 sec. Pipette 1.0 ml of the mixture into each tube and float a little water on the surface (Section 5.5.1). Do not allow droplets of the acrylamide solution to adhere to the upper part of the tube. Leave the tubes to stand undisturbed until polymerisation is complete and then remove the excess liquid (Section 5.5.1).

Stacking gel. Mix solutions B, D, E and F in the ratio 1:2:1:4 by volume. Add 0.1 ml of this mixture to each electrophoresis tube and overlayer with water. Then expose to the light of a fluorescent tube until polymerisation is complete (Section 5.5.1).

Application of the sample. Add to the stained plasma or lipoprotein solution (Appendix 4), an equal volume of solution F. Transfer 50 μl of this mixture to the top of the gel and carefully layer electrolyte on the surface until the tube is full (Section 5.5.1).

Alternatively, mix the stain prepared according to Naito et al. (Appendix 4) with solutions B, D and E in the respective proportions 2:1:2:1 by volume. Add 0.2 ml of this mixture and 20 μl of plasma or lipoprotein solution to each tube and mix by inversion. Overlay with electrolyte buffer and expose to light until the gel has formed. According to Naito et al. (1973), the Sudan Black will precipitate from the reagent mixture if the glassware is not scrupulously clean. If this happens, the gel must be prepared again. These authors also believe that the separating gels should be used immediately after their preparation if the best resolution is to be obtained.

Finally, set up the apparatus (Section 5.5.1) and pass a current of 3 mA/tube for 35–45 min. The α-lipoproteins will migrate to the bottom of the tube and the β-lipoproteins will move about 25% of this distance. The larger pre-β-lipoproteins will barely penetrate the separating gel, and chylomicrons will remain in the sample gel.

5.5.3.2. Discontinuous separating gels. This method attempts to gain an improved resolution by the use of a separating gel that is made in two layers. The lower part has a relatively high concentration of acrylamide and is intended to resolve the HDL, while the top portion has a more permeable gel that will resolve the LDL. Despite its comparative simplicity, the method seems to have given rather variable results. Feliste et al. (1973) claim that the α-lipoproteins can be resolved into six fractions, the β-lipoproteins into three and the pre-β-lipoproteins into three or four, though the latter were mainly in the stacking gel which was of 3% concentration. On the other hand, Wollenweber and Kahlke (1970), who used what appears to have been an almost identical system, were able to resolve only two α-lipoproteins and could not resolve the other classes at all. However, it must be said that neither of these groups described their procedures very clearly and their discordant results may have been due to undisclosed differences in technique. However, Wada et al. (1973) who used a similar system in the form of a gel slab, could also resolve only two or three bands of α-lipoprotein. This may be explained by their use of a 5.6% gel instead of the 7% that was used by Feliste et al. (1973). But is seems less likely that the inability of Wada et al. to resolve the β-lipoproteins was due to their use of 3.75% instead of 3.5% acrylamide for the upper part of the separating gel.

The method that follows is that of Feliste et al. (1973).

Reagents: as in Section 5.5.3.1.

Electrolyte solution. Use the buffer given in Section 5.5.3.1 but at twice that concentration.

Gel reagents. All the solutions A to G are as described in Section 5.5.3.1, except for solution C which should have the following composition:

28.0 g of acrylamide
0.735 g of bisacrylamide
Water to 100 ml

Procedure. These directions should be read in conjunction with those given in Section 5.5.1 and 5.5.3.1. We shall assume that tubes 7 mm × 75 mm will be used.

Separating gel. Prepare the first layer of gel as follows: Mix solutions A, C and G with water in the proportions 1:2:4:1 respectively, by volume. Add 1.2 ml of this mixture to each tube and carefully pipette 0.1–0.2 ml of water onto the surface. Set the tubes in a shaded place to polymerise. When the reaction is complete, remove the excess liquid.

For the next layer of separating gel, mix solutions A, C and G with water in the proportions 3:2:8:3 by volume. Add 0.6 ml of this to each tube, layer with water and allow to polymerise.

Stacking gel. Mix solutions B, D, E and F in the ratio 2:3:1:4 by volume. Put 0.3 ml of the mixture onto the separating gel and carefully overlay it with water. Polymerise by exposing to light and remove the excess liquid.

Charging and running the gel: proceed as in Section 5.5.3.1.

5.5.3.3. Polyacrylamide gradient gels. In this method, which is due to Pratt and Dangerfield (1969), the concentration of acrylamide increases continuously down the length of the separating gel. The gradient can run from 2.5% to 8% and offers the best available resolution of those lipoproteins that are able to enter the gel. However, it is more difficult to prepare these gels reproducibly than, for example, the two-part discontinuous type. The method used by Pratt and Dangerfield (1969) is clearly incapable of giving strictly reproducible, uniform gradients. The best technique is probably that of Melish and Waterhouse (1972), who used large glass tubes (12 cm × 1.3 cm diam.) that could accommodate the entire contents of their gradient former. The special problems of making gradients in small tubes will be dealt with below.

The final resolution that can be obtained is a function both of the gradient and of the length of the gel, for example, Melish and Waterhouse (1972) were able to resolve more components on a gradient 10.5 cm long than Pratt and Dangerfield (1969) could detect on a similar gradient only 5 cm long. Nonetheless, any gradient that is used for the analysis of the entire range of plasma lipoproteins will be a compromise that may not give the best resolution of a single class such as the α-lipoproteins. It may therefore be necessary to adjust the gradient to obtain the optimum resolution of particles of special interest, perhaps by the use of non-linear rather than the more usual linear gradients.

Equipment. The apparatus used for conventional electrophoresis in polyacrylamide gels can also be used for gradient gels. The additional equipment needed to produce the gradients is as follows. Any gradient former of an appropriate size can be used, for example the type illustrated in Fig. 2.3. If the gels are to be cast in small tubes (e.g. 75 mm × 7 mm), a device like that shown in Fig. 5.2 will be needed (Castle et al., 1975). Such 'towers' are commercially available but can easily be made from sheet acrylic plastic. It is even easier to make the apparatus from a large diameter acrylic tube, but this design suffers from the disadvantage that the middle of the chamber becomes overheated during the polymerisation, causing the gel in the central

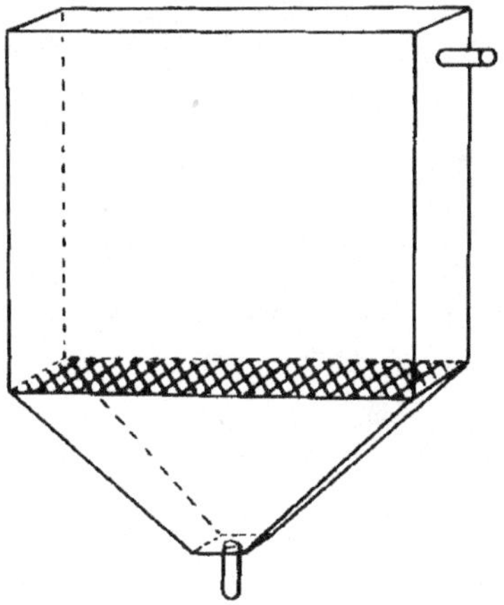

Fig. 5.2. Device for filling tubes for gradient gel electrophoresis. Constructed of acrylic sheet, with a grid at the bottom to support the tubes, the width of the box should be such as to accept two electrophoresis tubes side-by-side. The distance from the grid to the overflow tube must be greater than the length of the tubes. The inlet tube at the bottom is connected by fine-bore plastic tubing to the gradient maker illustrated in Fig. 2.3.

tubes to have different characteristics from that in the outer tubes. It is better therefore, to make the chamber rectangular, to hold not more than two ranks of tubes. If the cylindrical form is used, it is advisable to blank off the centre with a cylinder of plastic or a water-filled test-tube. No matter what the shape of the chamber, it is important that there should be an over-flow level with the top of the gel tubes, or that the height of the chamber should exceed the length of the gel tubes by enough to ensure that the volume above the tubes is at least equal to that of the funnel below them.

Procedure. Connect the gradient maker to the polymerising chamber by means of a fine-bore tube (about 1.2 mm diam.) that can be closed with a clamp. The gradient maker must be at a slightly higher level than the chamber and must be equiped with an adequately vigorous stirrer. Fill the chamber with clean gel tubes.

TABLE 5.3

Reagents 1.0 M H_3PO_4 in addition to those listed in Section 5.5.3.1

Cathodic electrolyte:	11.25 g of glycine
	0.4 g of NaOH
	Water to 1.0 litre: final pH 8.6
Anodic electrolyte:	18.17 g of Tris
	5 ml of 1.0 N HCl
	Water to 1.0 litre: final pH 8.6

Gel reagents	Solution				
	A	B	C	D	E
Tris		35 g			
Acrylamide	19 g				
Bisacrylamide	1 g				
TEMED					1 ml
Ammonium persulphate				1 g	
Sucrose			50 g		
Water to (ml)	100	100	100	10	10

Solutions D and E should be prepared immediately before use. Do not make more than one week's supply of A at a time.

Prepare the following two mixtures immediately before use:

'Heavy' solution	'Light' solution
7.5 ml of A	2.5 ml of A
2.5 ml of B	0.75 ml of 1.0 N H_3PO_4
2.5 ml of C	0.05 ml of D
0.05 ml of D	0.05 ml of E
0.05 ml of E	
1.75 ml of 1.0 N HCl	

Dilute both solutions to 25 ml with water and de-aerate by suction at 15–20 mm Hg for 1 min. These mixtures will make a gradient that is adequate for general use, i.e. it will resolve the β-lipoproteins into about 6 bands, but the α-lipoproteins run fast and give only three bands. For better resolution of the α-lipoproteins (at the expense of that of the larger particles) use the following mixtures:

'Heavy' solution. Substitute 10 ml of A, 0.08 ml of D and 0.08 ml of E for the quantities given in the first recipe.

'Light' solution. Substitute 5 ml of A, 0.08 ml of D and 0.08 ml of E for the amounts quoted above.

N.B. it is important that the polymerisation does not proceed too rapidly; the connecting tube between the gradient maker and the chamber must not become blocked, and the gel tubes must be completely filled before gelling begins.

Procedure. It is necessary to know in advance the volume of liquid needed to fill the polymerising chamber from the top of the 'funnel' to the top of the gel tubes, when the latter are in place. Let this be V ml.

To prepare the gels, start by filling the 'funnel' at the bottom of the chamber and the tube that connects it to the gradient maker, with 'light' solution. Clamp the tube and note the volume of the solution used, which we shall assume to be X ml. Fill the bore of the cock between the reservoirs of the gradient maker with 'light' solution and close it. Ensure that there are no bubbles in either the cock or the exit tube. Next, put equal weights of 'light' and 'heavy' solutions into the appropriate reservoirs ('light' nearest the exit tube). The total volume of these two solutions should be V ml but note that the volumes of each cannot be equal since the hydrostatic pressure in the two reservoirs must be the same, although the densities of the solutions are different. Open the cock and switch on the stirrer. Open the exit tube and allow the contents of the gradient maker to flow into the polymerisation chamber: clamp the tube again before air can enter it. It should take about 20 min to reach this stage, when the gel tubes should be full to the brim. Add a further 3 ml of 'heavy' solution to the gradient maker and allow it to enter the chamber. Finally, fill the exit tube and the 'funnel' with 50% sucrose solution (i.e. X ml). Allow the apparatus to stand undisturbed until polymerisation is complete, which should be about 60 min after the tubes are filled. Remove the tubes from the chamber and mature the gels for 48 hours at 4 °C before use.

The methods of staining and loading the lipoproteins onto the gel, and of performing the electrophoresis are as described in Section 5.5.3.1.

5.6. Immunoelectrophoresis

The full potential of this useful technique cannot be exploited unless mono-specific antisera to all the apo-lipoproteins are available. However, even with the limited number of commercial antisera, it can yield useful information on the cross-contamination of preparations of isolated lipoproteins and on their contamination with other plasma proteins. It is of little use for the routine analysis of plasma, although it may have some value for the detection of lipoprotein-X as is discussed in Section 5.6.1.

The procedure is essentially standard (Grabar and Williams, 1955). It is described in most manuals of immunological technique (e.g. Clausen, 1974) or of electrophoresis (e.g. Smith, 1976), and is specifically dealt with by Weeke (1973). Detailed working instructions are also given with the Gelman equipment (Gelman Sciences Inc.).

The equipment and materials are essentially those required for electrophoresis in agarose gel (Section 5.4) and for immunodiffusion (Section 6.5).

Procedure. If the samples to be examined are isolated lipoprotein fractions, dialyse them against the electrolyte buffer for 24 hours before the gels are prepared.

Prepare gels of agarose on microscope slides, exactly as described in Section 5.4, and punch two 3-mm wells across the slide about 20 mm from the cathodal end (Fig. 5.3a). Some laboratories make the gels in electrolyte buffer which is 50% of the concentration used for simple agarose electrophoresis, since this has the effect of making the migration of the lipoproteins more rapid. Under these conditions, care must be taken that the slides do not overheat during the run.

In order to minimise the handling of the loaded gels and to to ensure prompt starting of the run after loading, it is advisable to instal the slides in the electrophoresis tank before filling the sample wells. Under

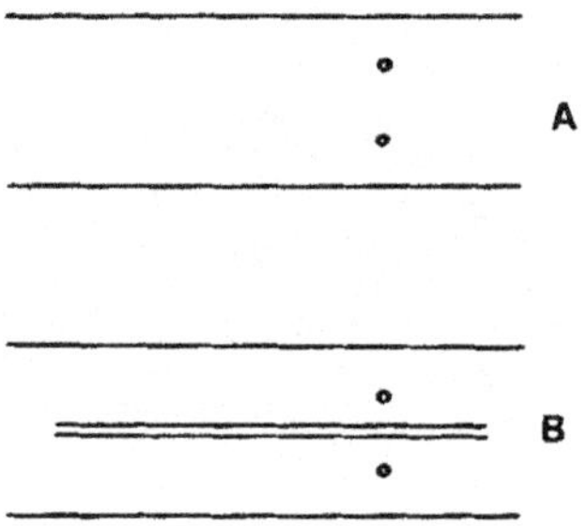

Fig. 5.3. The disposition of the wells used for immuno-electrophoresis in agarose gel.

these circumstances, the wells can be filled directly with the liquid samples. If the gels are loaded on the bench, use a mixture of equal volumes of sample and agarose as described in Section 5.4. When all the slides are prepared, close the tank and pass a current of 2–3 mA/cm width of gel for 60–90 min. At the end of the run, switch off the power, remove the slides and cut a trough for antiserum between each pair of wells (Fig. 5.3b). This should be 1 mm or 2 mm wide and should be filled with from 50 to 150 µl of antiserum according to its titre and the nature of the sample being examined. Place the slides in a closed humid chamber and allow diffusion to take place at 4 °C for at least three days, with periodic inspections. The slides can then be processed as for immunodiffusion (Section 6.5).

5.6.1. Electrophoresis of lipoprotein-X (LP-X)

This lipoprotein can easily be distinguished from those usually present in the plasma because it moves towards the cathode during electrophoresis in agar gel (Fig. 5.4). This analysis is potentially valuable because the presence of LP-X has been said to be diagnostic for obstructive liver disease (Seidel et al., 1969, 1970). In practice, it has given erratic results because the nature of the agar affects both the extent of the cathodic migration and the sensitivity of the detection method. The published reports have not sufficiently emphasised that

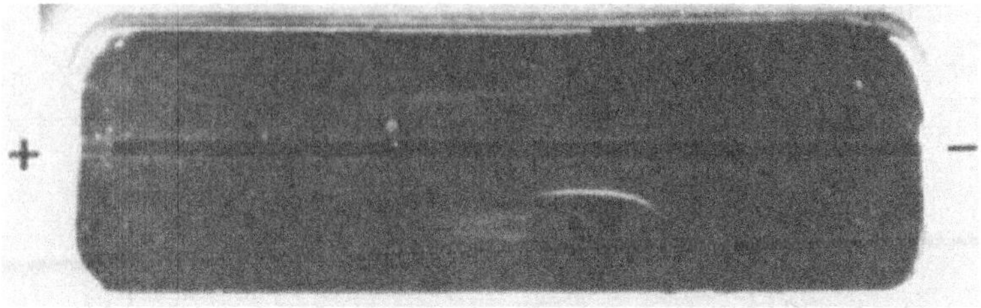

Fig. 5.4. The electrophoretic migration of LP-X in agar gel. Upper well, normal plasma. Lower well, plasma from a patient with obstructive jaundice. Trough, antiserum to LP-X.

different batches of agar do not all perform equally well, although Seidel et al. (1970) do point out that the characteristic migration of LP-X does not occur in purified agar and recommend the use of Difco Bacto Agar. In later work on the quantitative estimation of LP-X, Magnani and Alaupovic (1972) used Difco Special Agar Noble and remarked that it was necessary to select batches that were satisfactory. It is significant that the laboratories that have found the highest correlation between cholestasis and the presence of LP-X have almost invariably used a commercial electrophoresis kit ('Rapidophor all in for LP-X'; Immuno AG). Those who have prepared their own gels have generally obtained less impressive results. In our experience, it is extremely difficult to find, among the products of laboratory suppliers, a batch of agar that will reproduce the results that are claimed by the chief protagonists of this method. It is possible however, to modify the properties of the agar so that the results approach those described in the more optimistic reports. Nonetheless, the fact that the results of the test are dependent on undefined technical parameters that must be carefully tuned to give the 'correct' result must seriously reduce the level of confidence in their reliability. It is unfortunate that the method has not been the subject of published systematic examination and development.

Materials. Plasma must be fresh and on no account be frozen, as this can bring about gross changes in the properties of the LP-X.

Agar 'suitable for LP-X determination' can be obtained from Hoechst AG, but we shall describe a way of preparing a modified agar in the laboratory. This has two ingredients, a conventional agar such as Difco Noble Agar and an unpurified agar that is used as a modifier. In England, a suitable material for the latter purpose is to be found in health-food shops, which supply it as a substitute for gelatine in vegetarian cuisine. It is produced by Cadbury-Schweppes Ltd., and forms a cloudy gel of low strength.

The following reagents are also needed to detect the presence of LP-X after electrophoresis. Two procedures are available:

(a) Immuno-precipitation with anti-serum to LP-X. This is the more specific method, since it detects the presence of the characteristic apo-lipoprotein C in an anomalous position on the cathode side of the origin. Anti-LP-X serum can be obtained commercially (Immuno AG; Hoechst AG) but can be made, if need be, by isolating LP-X from the blood of a patient with cholestatic jaundice (Section 2.5.2) and injecting this into rabbits in the conventional way (Clausen, 1974).

(b) Precipitation with polyanions. This is somewhat less specific than (a) since there may be other proteins that migrate towards the cathode and also react with these reagents. In practice, this does not seem to be a significant defect. The reagents are closely akin to those used for the preparative precipitation of lipoproteins (Section 2.2.3) and are prepared as follows:

(I) 0.02 g sodium heparin (160–170 units/mg)
0.20 g $MgCl_2 \cdot 6H_2O$
0.10 g NaCl
Water to 10 ml

or (II) 0.06 g dextran sulphate
0.12 g $CaCl_2$
Water to 10 ml

Note that heparin for injection may be used for this reagent but that dextran sulphate must be of high molecular weight (ca. 500 000).

Procedure. Prepare 1% solutions of Difco Noble Agar and of the unpurified agar in electrolyte buffer (Section 5.4). These stock solutions can be cooled and re-melted again as required. Mix 9 parts of

Noble agar with 1 part of the unpurified agar and use this solution to prepare gels on 7.5 cm × 2.5 cm microscope slides as described in Section 5.4. Note that the proportions of this mixture may need adjustment if different preparations of agar are used, and that albumin should not be added. (If unmodified Bacto-Agar is used, it may be advantageous to prepare a 1.5% gel and to pour it to twice the normal thickness.)

Punch two pairs of 3 mm wells on each slide as shown in Fig. 5.5a (with care, three wells can be made as in Fig. 5.5b). Put 8 µl of sample into each well, set the slides in the electrophoresis chamber and pass a current of 2 mA per slide for 60–90 min. Note that the run must not be continued for too long or the anodically migrating components of one sample may mix with the cathodically moving substances from its neighbour.

If the sample contains a high concentration of LP-X, it can often be detected as a crescent shaped, transparent area in the agar. This may be as much as 10 mm from the well. Smaller amounts of LP-X cannot be seen in this way but must be precipitated with antiserum or with polyanion. The reaction is then usually nearer to the origin. If antiserum is used as the reagent, punch a new well 7.5 mm to the

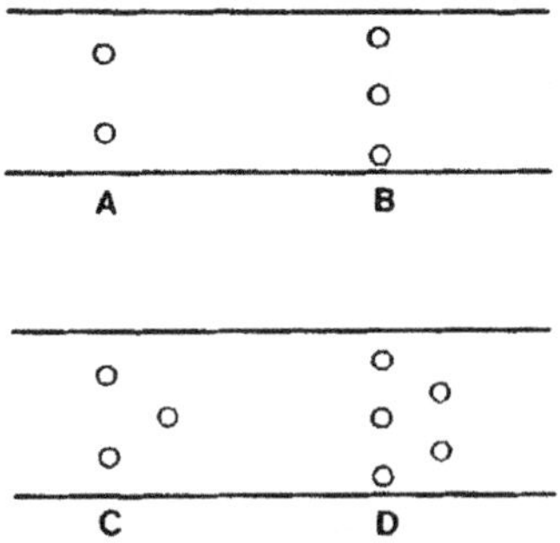

Fig. 5.5. Alternative arrangement of the wells for immuno-electrophoresis of LP-X in agar gel. Sample wells 3 mm in diam. can be set in pairs (A) or in threes (B). Two such arrays can be made 4.5 cm apart on a 7.5 cm × 2.5 cm gel slide. After the electrophoresis, wells for the antiserum can be punched 7.5 to 10 mm on the cathode side of the sample wells, as in C and D.

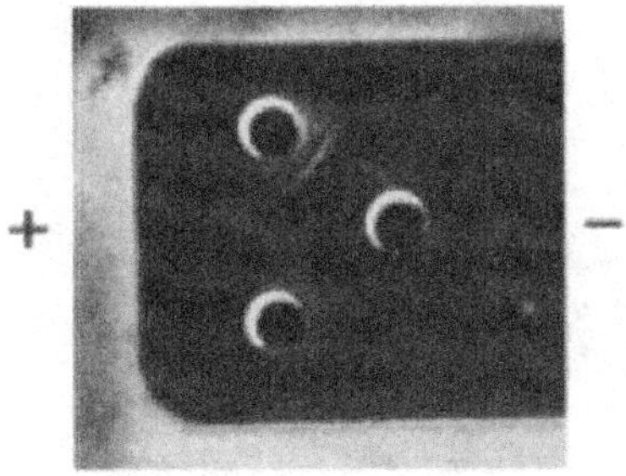

Fig. 5.6. Electrophoresis of LP-X in agar gel. Upper well, plasma from a patient with obstructive jaundice. Lower well, normal plasma. Right-hand well, antiserum to LP-X.

cathode side of each pair of sample wells, as shown in Fig. 5.5c. Fill this with the antiserum and stand the slides in a humid chamber at room temperature. Examine them carefully for the formation of an immunoprecipitate like the typical positive reaction shown in Fig. 5.6.

A modification of this procedure that is even more economical of antiserum was described by Seidel et al. (1973). In this case, do not punch wells for the antiserum, but apply 10 μl directly to the surface of the gel about 5 mm to the cathode side of the origin. Incubate the slide in a humid chamber at 37 °C for 5 hours to develop the precipitin reaction. Its originators claim that this method has the special merit that it avoids the possibility of an excess of antibody when testing weakly positive samples.

A somewhat similar procedure is used with the polyanion precipitants. For example, a few drops of the heparin–$MgCl_2$ reagent can be put on to the slide and spread across it with the side of a pasteur pipette. The reaction will be complete in a few minutes. As can be seen in the example of Fig. 5.7, a diffuse precipitate sometimes forms on the cathode side of the well but this will wash out in 1% NaCl and does not indicate the presence of LP-X.

The exact function of the impurities in the agar that seem to be necessary for the working of this analysis is not clear. They indubitably lead to an increased electroendosmosis which contributes to the cathodic migration of the LP-X. This consequence of raising the

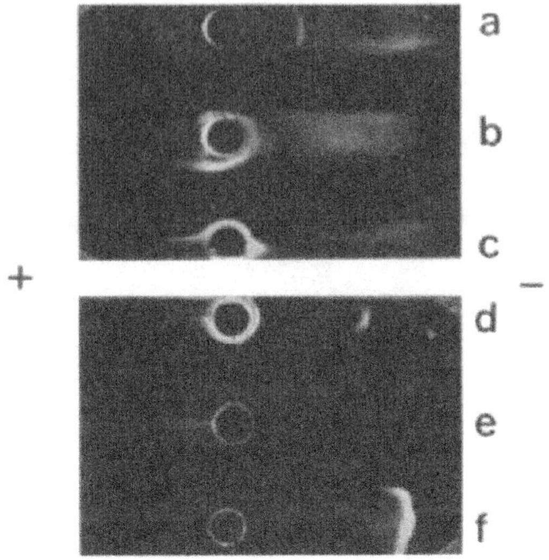

Fig. 5.7. The detection of LP-X with the heparin-manganese reagent: electrophoresis in agar gel. Wells a, d and f contained LP-X positive sera. Wells b, c and e: normal sera.

proportion of 'impurity' can be seen in Fig. 5.8, which also shows that the intensity of the precipitin reaction is greater in the doctored gels. This enhancement may allow LP-X to be detected even in a sample that does not give a positive reaction in normal agar. However, if the doctoring is carried too far, confusing non-specific precipitates may form, particularly when the polyanion reagents are used. It is therefore advisable to titrate each new batch of normal agar to find the proportion of 'impure' agar that must be added to give the best results.

According to Neubeck and Seidel (1975), native LP-X in plasma and the isolated lipoprotein both migrate in the same direction on electrophoresis in agar. However, this is not always true, as Fig. 5.9 shows. The reason for this discordancy is not known but may be at least partly related to the quality of the agar. However, if the isolated LP-X is mixed with lipoprotein-free plasma, its 'normal' electropho-

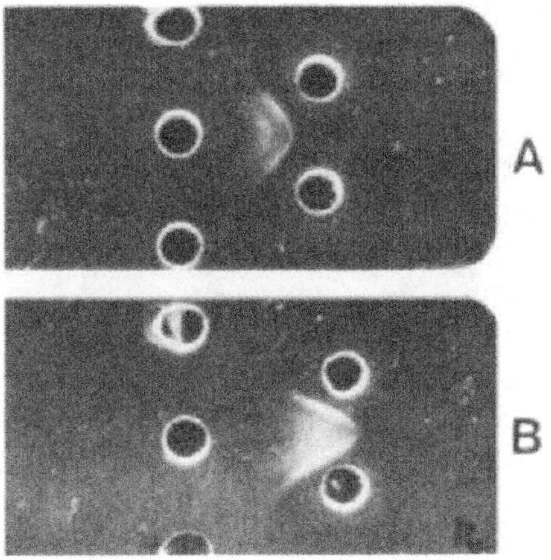

Fig. 5.8. Slide A, gel composition 90% Noble agar + 10% crude agar. Slide B, gel composition 80% Noble agar + 20% crude agar. The same sera were used on both slides. Left-hand wells: top, weakly positive LP-X serum; centre, strongly positive LP-X serum; bottom, LP-X negative serum. All right-hand wells contained antiserum to LP-X.

retic properties are restored, presumably because the LP-X has become re-associated with another constituent of the plasma. This is, in some respects, like the effect of albumin on the bile lipoprotein that was described by Manzato et al. (1976), although this lipoprotein

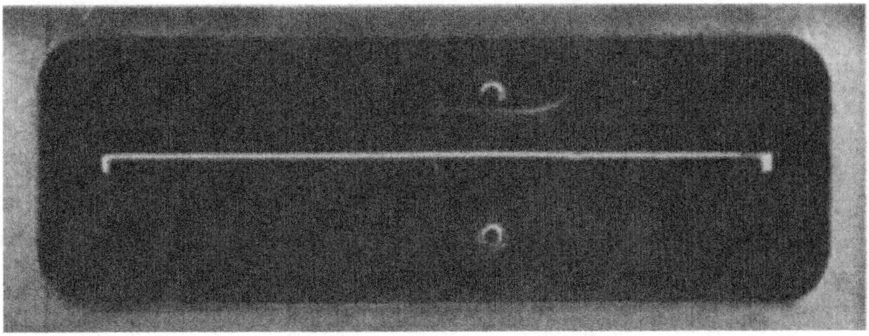

Fig. 5.9. Upper well, LP-X positive serum. Lower well, isolated LP-X. Trough, antiserum to LP-X.

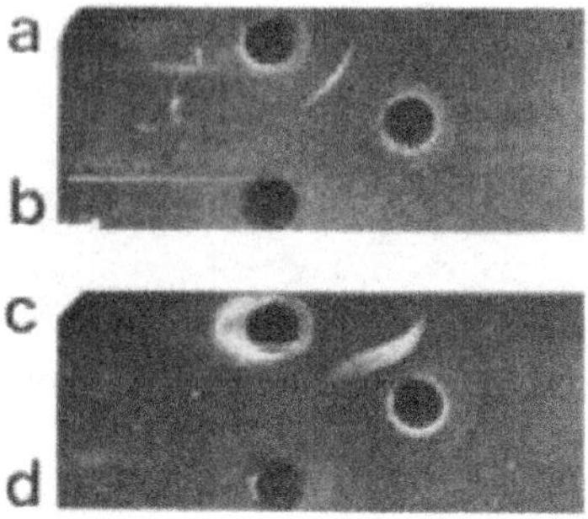

Fig. 5.10. Well (a), LP-X. Well (c), LP-X + bovine serum albumin. Wells (b) and (d), normal serum. All right-hand wells contained antiserum to LP-X.

differs from LP-X in several important points. Moreover, both the cathodic migration and the precipitin reaction of LP-X can be increased if bovine serum albumin is added to the sample before the electrophoresis (Fig. 5.10). It is prudent therefore, to check the behaviour of the electrophoretic system before analysing samples of isolated LP-X and to add bovine serum albumin if necessary (mix 5 μl of the sample with 5 μl of 10% albumin and transfer to the sample well).

5.7. Two-dimensional immuno-electrophoresis

This is an extremely powerful technique which has a potential for lipoprotein analysis that has not yet been fully exploited. Unfortunately, although the method has been successfully used to quantify proteins, it has proved to be difficult to estimate lipoproteins in this way. In any event, it is easier to do a quantitative immunoassay by the one-dimensional 'rocket' technique that is described in Section 8.9.3. Nonetheless, the two-dimensional method is valuable because it can distinguish lipoproteins according to the nature of their protein moieties and can show how these are distributed among particles of different electrophoretic behaviour. The kind of results that may be obtained are exemplified in Fig. 5.11, which shows (a) a serum in

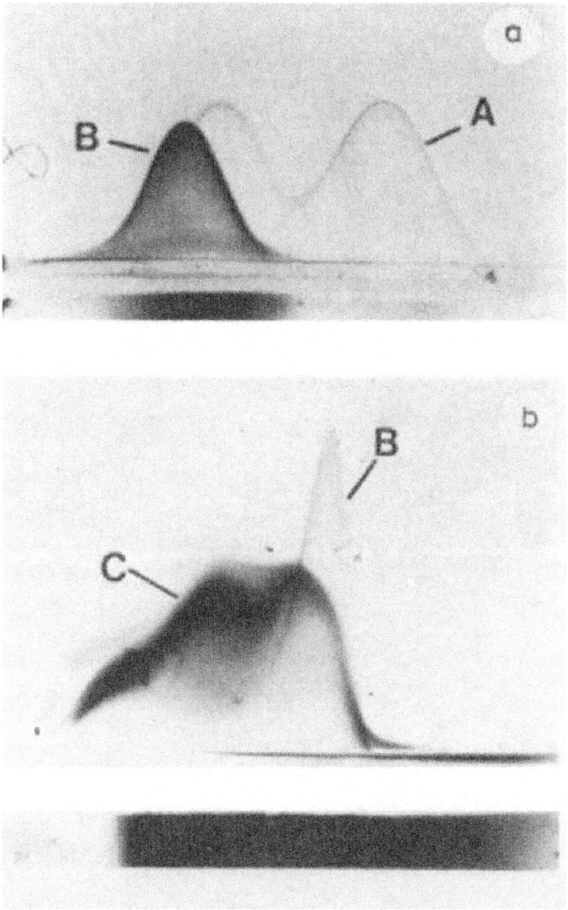

Fig. 5.11. Two different sera analysed by two-dimensional immuno-electrophoresis. (a) First dimension, migration left to right; second dimension, migration into a mixture of antisera to apoliproteins A and B. (b) First dimension, migration right to left; second dimension, migration into a mixture of antisera to apolipoproteins B and C.

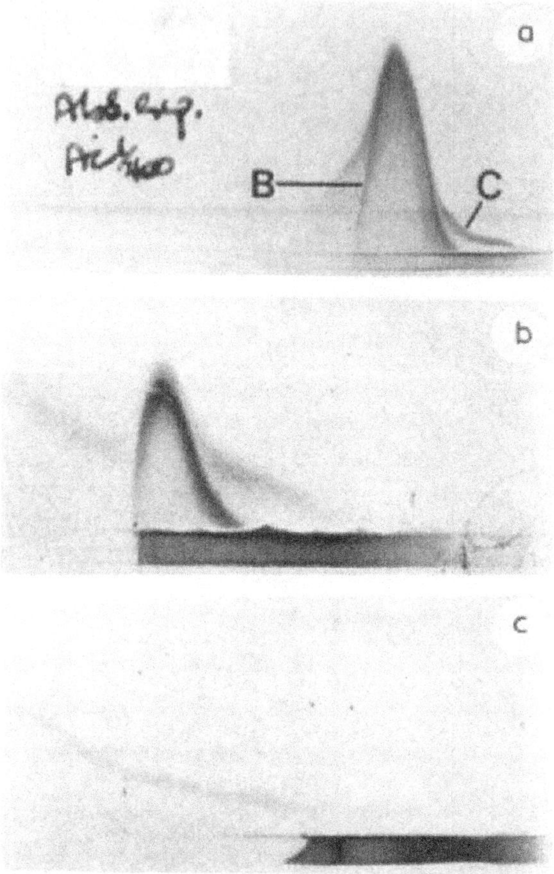

Fig. 5.12. Three stages in the immuno-electrophoretic analysis of the lipoproteins in a hypertriglyceridaemic serum: migration in the first dimension is from right to left. In all three examples, the second electrophoresis was into a mixture of antisera to apo-lipoproteins B and C. In (a), there was no antiserum in the first gel. In (b) and (c) the first gel contained antiserum to apo-lipoprotein B (c>b).

which there is substantial level of apo-lipoprotein A in the β-lipoprotein and, (b) the distribution of apo-lipoprotein B and apo-lipoprotein C in a lipaemic serum. By modifying the procedure slightly, it is possible to carry the analysis of the lipaemic serum a stage further, as shown in the example of Fig. 5.12. Here, the gel used for the first electrophoresis contains antiserum to apo-lipoprotein B, which prevents the migration of lipoproteins in which the B protein is a significant constituent. The gel for the second electrophoresis contains antisera to both the B and the C apo-lipoproteins and it can be seen that the latter are not affected by the presence of anti-apolipoprotein B in the first stage. Once the migration of the B-containing lipoproteins has been completely suppressed, the distribution of the C-containing particles can be seen. Note however that the precipitin line formed by reaction with anti-apolipoprotein C does not cross that formed by anti-apolipoprotein B. Under the conditions that normally exist in the gel, it therefore appears that these two species of lipopro-

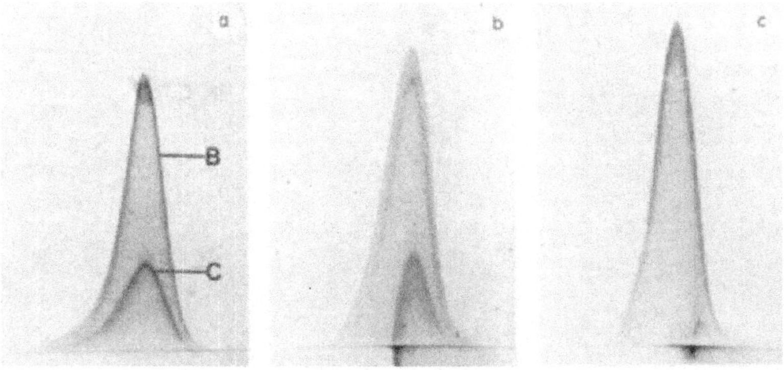

Fig. 5.13. Analysis of LDL by two-dimensional immuno-electrophoresis. Equal amounts of the same LDL were applied to each of three gels. The gels used for the first stage electrophoresis contained anti-serum to apo-lipoprotein C in increasing concentration (a = 0). Migration in the second dimension is, in each case, into the same mixture of antisera to apo-lipoproteins B and C. The C-containing particles can be completely precipitated in the first stage without affecting the subsequent migration of the B-containing particles.

tein particle migrate as a loose complex that is only split up when the B-containing lipoproteins are precipitated.

This co-migration of 'B' and 'C' particles apparently does not occur with the smaller lipoproteins. In LDL for example (Fig. 5.13), the two classes seem to react independently against the relevant antisera. However, whereas the C-containing lipoproteins can be completely precipated without affecting the migration of the B-containing particles (Fig. 5.13c), absorption with anti-apolipoprotein B also removes the C-containing lipoproteins (Fig. 5.14). The reason for this is not understood.

A detailed analysis of the identifiable apo-lipoproteins that are present in a sample and the way that they are distributed between the α-, β- and pre- β-lipoproteins requires the use of monospecific antisera to all the apo-lipoproteins. As in the case of simple immuno-elec-

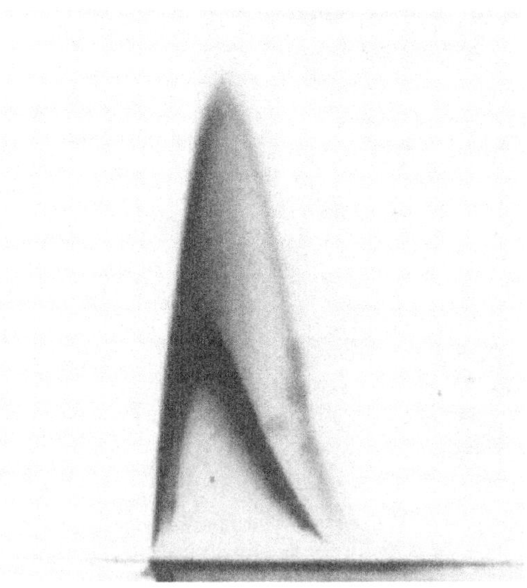

Fig. 5.14. This analysis of LDL differs from that shown in Fig. 5.13 only insofar as antiserum to apo-lipoprotein B was substituted for antiserum to apo-lipoprotein C in the first stage gel.

trophoresis, this imposes a strict limit to the full use of the two-dimensional technique in many laboratories.

When preparing gels for this method, do not use mixtures of agar and agarose that contain more than 20% agar. This is particularly important in the first stage electrophoresis, since too much agar can distort the distribution pattern and lead to the formation of precipitin lines that cross even though the antiserum is mono-specific (Fig. 5.15). The inclusion of agar in the second stage gel does not alter the shape of the pattern, or change the number of 'components' that are detected, but it causes the peaks to become smaller and to have a crisper outline. Consequently, although it may be desirable to use pure agarose for the first stage, it is probably an advantage to include a small proportion of agar in the gel for the second stage. The exact amount must be determined by experiment.

Fig. 5.16 shows an apparent splitting of the peak of the precipitin line which suggests the presence of two components with partial immunochemical identity. This is not related to the composition of the gel but is an artifact that arises when the plane of precipitation

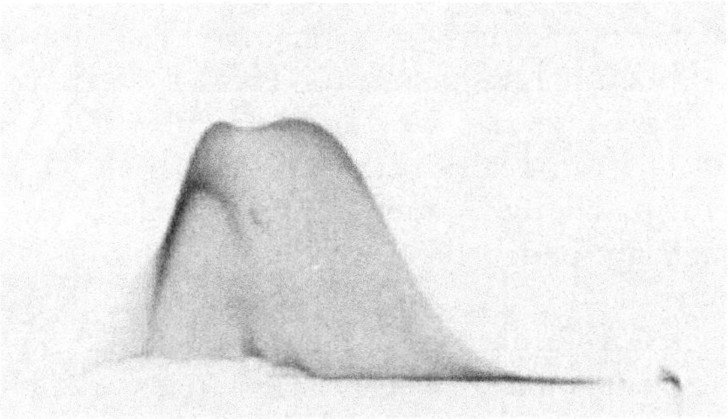

Fig. 5.15. The effect of an excess of agar on the pattern produced by two-dimensional electrophoretic analysis. In this example, the gels used for both dimensions contained 25% agar. The antiserum used in the second dimension was against apo-lipoprotein B. An asymmetric precipitin curve was produced which gives the appearance of having more than one component. This should be compared with the analysis of the same serum which is shown in Fig. 5.16.

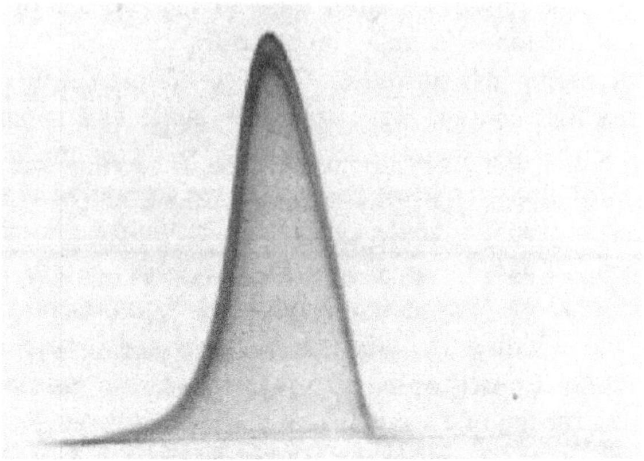

Fig. 5.16. In this analysis the conditions were identical to those used for Fig. 5.15 except for the use of 100% agarose for the first stage gel. Note that the precipitin curve is symmetrical and appears to be homogeneous. The double line around the edge of the curve is not the result of heterogeneity but is because the plane in which the immuno-precipitation took place was not perpendicular to the plane of the slide.

is not vertical. When this is the case, the upper and lower edges of the zone of precipitation do not superimpose when the gel is dried and this gives the false impression that two lines have been formed. This obliquity of the front can be reduced by using thin gels and by ensuring that the run is carried out under equilibrium conditions, i.e. at a low current, to reduce heating, and in a draught-free environment of uniform temperature.

The technique of two-dimensional (or 'crossed') immunoelectrophoresis has been described in detail by Ganrot (1972) and by Weeke (1973), and the following procedure differs from these only in minor respects.

Equipment and materials. These are as used for electrophoresis in agarose gel, on microscope slides (Section 5.4). In addition, square slides are needed for the second electrophoresis. The method we shall describe makes uses of 50 mm × 50 mm slides which are economical in the consumption of antiserum. If these are too small to bridge the gap between the electrolyte reservoirs in the electrophoresis tank, support them on a larger glass or plastic plate.

The antisera used should be of the highest titre available and as nearly monospecific as possible. The two-dimensional technique is an excellent way of detecting heterogeneity in an antiserum and if spurious precipitin lines are present in the lipoprotein analysis, the patterns can be very difficult to interpret. The amount of antiserum that must be added to the agarose for the second stage of the electrophoresis will depend on the purpose of the experiment and on the titre of the antiserum. Whenever a new batch of antiserum is brought into use, an approximate value for the working concentration must be determined before the analyses can be begun. The easiest way of doing this is to run a series of one-dimensional 'rocket' analyses (Section 8.9.3) on a standard serum. Although the titre of antiserum that gives a satisfactory result in this system will be slightly greater than the optimum for the two-dimensional technique, the additional dilution required should be only about two-fold, and can be precisely determined by trial.

Procedure. The technique is, in general, the same as that used for simple electrophoresis in agarose (Section 5.4).

Begin by pre-coating all the slides to be used with 0.5% agarose before the running gel is prepared. The latter is a 1% solution of agarose-agar mixture (4:1 by wt.) in electrolyte buffer to which 0.5% bovine serum albumin is added. Pipette 2.5 ml of this warm mixture on to each 75 mm × 25 mm microscope slide, and make a slot 20 mm × 1.5 mm parallel to, and 20 mm from what is to be the cathodal end. Leave the matured slides to equilibrate in the closed electrophoresis tank, without passing a current, while the samples are prepared. Dilute plasma, or its equivalent in a solution of isolated lipoprotein, with an equal volume of warm agarose-agar solution (Section 5.4), and pipette 25 μl into the slot in the gel slide. When all the slides are loaded, pass a current of 2 mA/slide for 2 hours.

Shortly before this first run is completed, make enough agarose-agar-albumin solution to coat the required number of 50 mm × 50 mm slides with 3.5 ml each. Keep this mixture at 45 °C in a water bath.

Put the appropriate antiserum into a tube and set this also at 45 °C. Immediately before use, add to the agarose solution sufficient of the antiserum to give the final dilution that was pre-determined to be the optimum. This may range from 1:50 to 1:1000 according to the titre of the antiserum and the concentration of the lipoproteins in the sample.

End the first electrophoretic analysis when the albumin has moved about 40 mm. This can be judged by including a marker of bromophenol blue on one slide. Switch off the current and remove one slide from the tank, leaving the remainder in place in the closed chamber. Use a trough cutter (e.g. Gelman Sciences Inc.) to cut a filament 1 mm wide from the middle of the slide. Carefully pick it up with a lightly greased gel knife or microspatula and transfer it to a 50 mm × 50 mm photographic projector cover glass that is on a level surface. Trim off the filament that was on the cathode side of the sample slot and discard it. Position the larger portion parallel to, and 10 mm away from what will be the cathodal edge of the square slide

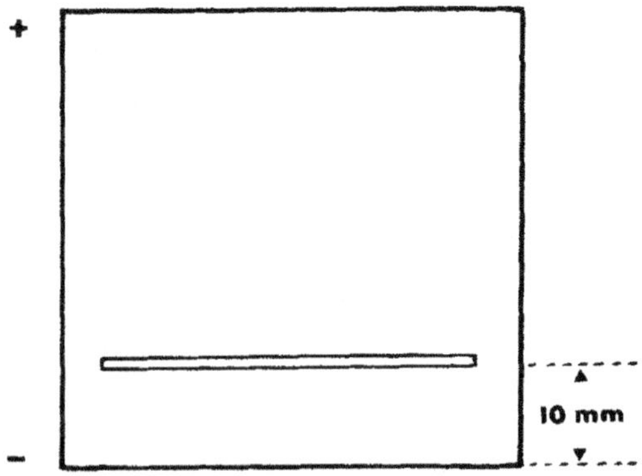

Fig. 5.17. Two-dimensional immuno-electrophoresis as performed on a 5 cm × 5 cm glass slide. A filament 1 mm wide is cut from the first stage gel and positioned along a line parallel to, and 10 mm above the lower edge of the slide. After the surface has been covered with 3.5 ml of molten agarose which contains antiserum, the current is passed in the direction shown.

(Fig. 5.17). Then, without disturbing the filament, cover the slide with 3.5 ml of the agarose-antiserum mixture and allow it to set.

Up to three filaments can be cut from each 75 mm × 25 mm gel, but they must be kept well within the extremities of the sample trough, where the migration of the lipoproteins is likely to be distorted. These three filaments can either be run as replicates, or against different antisera, or against different dilutions of the same antiserum.

When all second-stage gels have been prepared, position the slides in the electrophoresis tank and apply 50 mm wicks soaked in electrolyte to establish electrical continuity with the reservoirs. These wicks must not overlap the gel by more than 8 mm. Close the tank and pass a current of 0.5 mA per slide overnight or, less satisfactorily, 4 mA per slide for 2 hours. Meanwhile, fix what remains of the first electrophoresis gels in 2% trichloracetic acid for at least 2 hours and then wash them in water for 24 hours.

When the second electrophoresis is completed, switch off the current, remove the slides from the tank and soak them for at least 48 hours in a large volume of 1% NaCl solution. Next, cut a 2 mm slot just below the filament that was the origin for the second stage and insert into it a 2 mm strip cut from the corresponding first stage gel. This strip must correspond as nearly as possible with the original filament and can be fixed in place with a drop or two of warm agarose. Finally, process the slides for staining as in Section 6.5.

5.8. Isoelectric focussing

This technique has been little used for the analysis of lipoproteins, despite optimistic reports by Kostner et al. (1969, 1972), who claimed that it was capable of distinguishing eight lipoproteins in plasma. Since at least some of these substances were said to contain only one apolipoprotein (by immunological analysis), it may be thought surprising that the method has not received more active attention. Godolphin and Stinson (1974) suggested that it was a good way of diagnosing Type III hyperlipoproteinaemia but even this limited application does not seem to have found favour. In a later report Emes et al. (1976) described the analysis of plasma lipoproteins by a two-dimensional technique in which electrofocussing along the X-axis was followed by polyacrylamide gel electrophoresis in the Y-direction. However, this does not seem to yield enough additional information to warrant the considerable extra labour involved.

In the method developed by Kostner et al. (1969) the solution of ampholytes also contained 33% ethylene glycol and it is not clear to what extent this may have modified the electrical properties of the lipoproteins. If it can be shown that the fractions produced are not artefacts, this technique may deserve further examination.

CHAPTER 6

The chemical characterisation of lipoproteins

This chapter is concerned with the determination of the fourth of the parameters mentioned at the beginning of Chapter 3. However, since methods for the estimation of lipids and protein are readily available in many textbooks and reviews (Kates, 1972; Dittmer and Wells, 1969; Perkins, 1975), it will be largely confined to those matters of technique that are particularly relevant to the analysis of lipoproteins.

At the outset it should be said that the literature relating to lipoproteins contains many references to useless partial analyses. A quantitative chemical analysis is a classical way of defining a substance and an incomplete analysis would normally be regarded as unacceptable. Why a partial analysis should be thought to be satisfactory in the case of lipoproteins is not clear. While it is admittedly laborious to determine all five of the different molecules that are normally the major components of a lipoprotein, it must be remembered that the time spent in measuring any fewer is time wasted. It is also normal analytical practice to begin with a known weight of sample and to check that this is equaled by the total of the estimated weights of each component. That this is rarely done for a lipoprotein preparation leaves the analysis unsubstantiated by any kind of check on its reliability. Moreover, it assumes that the qualitative composition of the lipoprotein is known, without providing any means of checking that this is true. The second criticism may be relatively unimportant in the case of lipoproteins from man and other mammals but, in those of more primitive animals, partial glycerides, monoalkyldiacylglycerols, hydrocarbons or wax esters may be present in appreciable amounts. The composition of any lipoprotein of unusual provenance should therefore be confirmed by exhaustive qualitative analysis before quantitative studies are begun.

Aside from the avoidable errors and uncertainties that can arise from perfunctory analytical practice, the accuracy of the analysis can be degraded by the chemical heterogeneity of the lipoproteins. This exists at two levels and may be impossible, or at least impracticable to overcome. Not only do the proportions of the major constituents of the lipoprotein vary but, with the exception of unesterified cholesterol, each of these is itself heterogeneous. For example, the relative proportions of the fatty acids that compose the phospholipids, triglycerides and cholesteryl esters may vary considerably from sample to sample. But, because it is often difficult to determine these substances gravimetrically, they are frequently estimated by the determination of a single parameter (e.g. cholesterol) which is converted to the 'weight' of ester on the assumption that a single fatty acid is present. A better estimate can be made if the real distribution of fatty acids in the lipid is also determined. Their mean molecular weight can then be calculated. In man, this usually differs but little from the value for a C18 acid but in other animals, the differences can be significant.

An additional uncertainty is introduced into the estimation of total phospholipid by the common procedure of measuring the lipid phosphorus and multiplying the result by 25. This assumes that lecithin is the only phospholipid present and a better estimate may result if each phospholipid is determined separately. Fortunately, lecithin (for which the factor is close to 25) usually accounts for about 65% of the phospholipid, and sphingomyelin (factor about 23) for approximately 25%. In practice therefore, the use of the factor 25 leads to a variable error that probably amounts to an over-estimate of about 2%.

To summarise, care must be taken, when designing a study of lipoproteins, to ensure that it gets the quality of analytical work that it deserves.

In general, the first step in the analysis of a lipoprotein is the separation of the proteins from the lipids, followed by the analysis of these two fractions according to the scheme summarised in Table 6.1. However, the newly developing technique of quantitative micro-

TABLE 6.1
Schematic illustration of the stages of lipoprotein analysis

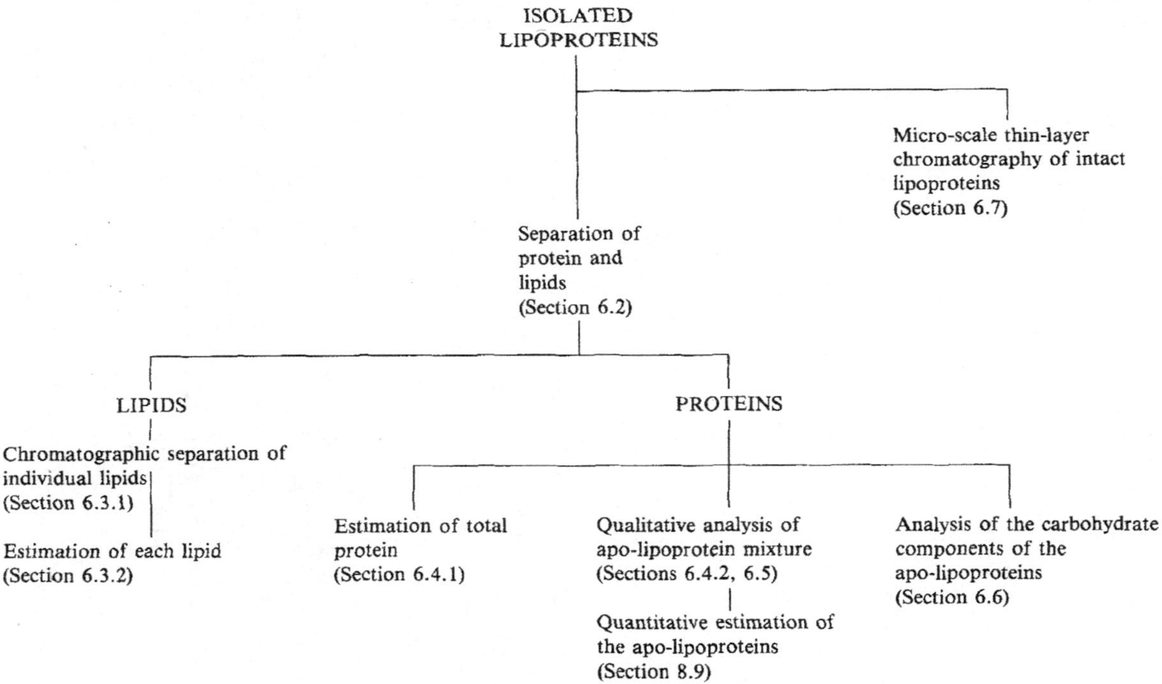

scale thin-layer chromatography allows all the components of the lipoprotein to be determined without the preliminary extraction of the lipids, and this is described in Section 6.7.

6.1. Estimation of total lipoprotein

The only accurate way of estimating the total lipoprotein content of a purified preparation is by weighing. The method is simple and requires no special equipment, but it is time-consuming and requires several milligrams of lipoprotein that are denatured by the drying process. There is also the technical difficulty that lipoproteins are not soluble in pure water. Moreover, the concentration of salt in the solution is often high e.g. up to 20% in a sample prepared by ultracentrifugation. When this is the case, it is not permissible to deduct the dry weight of the solvent from that of an equal volume of solution unless a correction is made for the partial volume of the lipoproteins. The condition for this correction to become negligible depends (a) on the desired accuracy of the result and (b) on the concentration of the lipoproteins but, if it is assumed that the weight of the lipoprotein is to be determined to better than 0.1%, the salt concentration must not exceed 20 mM NaCl or its mass equivalent. In some experiments, it may be possible to reduce the concentration of salt below this limit by dialysis. If the weight ratio of salt to water in the solvent (R) is known, the amount of salt present in the solution may then be estimated as WR, where W is loss of weight on drying. However, because dialysis usually results in a change of volume and is a difficult technique to perform quantitatively, this operation must be done before aliquots are taken for analysis. Even then, there is the likelihood of variable errors due to the leaching of material from the membrane into the interior of the bag.

For more accurate work, it is necessary to adopt a procedure like that used by Hanig and Shainoff (1956). In this, the amount of NaCl in the solution is estimated by titration and subtracted from the total solids obtained by weighing. More recently, Harvie (1973) suggested that it would be simpler to dialyse the lipoprotein solution against water in a pre-weighed dialysis bag, then to dry the bag with its contents and subtract from this the weight of the empty bag. However, material continues to leach from the semi-permeable membrane for a considerable time, making it difficult to maintain accuracy and precision when using this technique.

As with other protein dry-weight determinations, it is sometimes difficult to decide on the best way to dry the specimen. Hanig and Shainoff (1956) lyophilised their samples at $-50\,°C$ and then heated them at $57\,°C$ in vacuo over P_2O_5 for 12 hours. However, the fluffy material that results from freeze-drying is easily disturbed, and Harvie (1973) found it impossible to use this method to dry lipoprotein solutions in open weighing bottles because they became electrically charged. She therefore dried the specimens at $110\,°C$ for 48 hours. But this procedure is contrary to the experience of Armstrong et al. (1947) who found that heating at $110\,°C$ caused very large losses in the weight of lipoprotein preparations. After comparing the behaviour of both proteins and lipoproteins under different conditions, these authors concluded that lipoproteins should be dried at between 50 and $80\,°C$ in vacuo, over P_2O_5. In our laboratory it has also proved to be possible to dry small, thinly spread samples in a vacuum desiccator over P_2O_5 at room temperature. From the evidence available, the method of Armstrong et al. appears to be the most reliable for general use but, whatever procedure is adopted, it is important that drying is continued to constant weight.

One of the least precise aspects of any of these procedures is the measurement of the small volumes that are usually involved. This is preferably done with a micrometer syringe which, for the most accurate work, should be calibrated for the temperature of use.

The weighings should be performed on a properly maintained balance capable of working to at least 0.01 mg. Moreover, the weighing bottles should be handled with tongs and must be in thermal equilibrium at the time of weighing.

6.1.1. Estimation of lipoprotein by refractometry

If a less accurate estimate of the total lipoprotein concentration can be tolerated, it can be obtained by refractometry. This procedure was developed by Lindgren et al. (1964) and requires the measurement of the refractive index of the lipoproteins solution and that of the solvent by means of a precision refractometer. The specification of the instrument and its method of use are described in Section 2.1.2.5. The technique requires only a small sample and is a convenient way of monitoring the concentration of lipoproteins during the process of their isolation, provided that a sample of the solvent (e.g. a dialysate) is available. However, the refractometry of lipoproteins is subject to two potential sources of uncertainty:

(1) the apparent refractive index of a homogeneous particle in solution depends on the size of the particle.

(2) the lipoproteins are not homogeneous particles and, since the refractive index of lipid (1.45–1.50) is less than that of protein (ca. 1.6), their index is composition dependent.

Studies on polymer latices (Heller and Pugh, 1957) suggest that the size effect may be negligible for lipoproteins that are less than about 400 nm in diameter and is therefore probably of little practical consequence. However, since the composition of lipoprotein particles is related to their size, the apparent refractive index of a given sample will depend both on the size distribution of its particles and on the extent to which their mean protein:lipid ratio may differ from the 'normal' for that particular fraction. Lipoprotein concentrations that are determined by refractometry are consequently always subject to an uncertainty through the use of average values of the refractive increment that do not take account of these sources of variance. Unfortunately, the specific refractive increment (Δn) for lipoproteins has not been studied systematically and determinations are both sparse and imprecise. Moreover, these measurements have been made in only a few salt solutions and it may often be found that the refractive increment appropriate to the solvent in use is not known.

Under these circumstances, an approximate value can be obtained by means of the equation:

$$\frac{dn}{dc} = \frac{3n_1\ (m^2-1)}{2D\ (m^2+2)}$$

where D is the density of the anhydrous lipoprotein, n_1 the refractive index of the solvent and m is the ratio of the refractive index of the lipoprotein to n_1. A measured value of the refractive increment can be used to calculate a value for the refractive index of the lipoprotein, but an uncertainty of unknown magnitude will attach both to this and to the value of D.

Lindgren et al. (1966) summarised the data for Δn which were then available and which are reproduced in Table 6.2.

Note that the values for HDL were considered to be insufficiently accurate to be worthy of correction to a wavelength of 589 nm. Subsequently, the value of Δn for an LDL fraction of density 1.019–1.054 g/ml was determined in our laboratory, with results that are slightly higher than those quoted in Table 6.2, i.e.

$\Delta n = 0.173 \pm 0.0054$ at 589 nm in NaCl of density 1.007 g/ml

and $\Delta n = 0.159 \pm 0.0055$ at 589 nm in NaCl of density 1.063 g/ml.

Note that all these determinations were made at 26 °C.

Procedure. Dialyse the lipoprotein to equilibrium with a salt so-

TABLE 6.2

Lipoprotein	Solvent	Δn/g/ml	Wavelength (nm)
VLDL	0.199 molal NaCl	0.158 (meas.)	589
LDL	1.745 molal NaCl	0.154 (meas.)	589
	0.199 molal NaCl	0.166 (calc.)	589
HDL	0.199 molal NaCl + 2.771 molal NaBr	0.149 (meas.)	546
	0.199 molal NaCl	0.173 (calc.)	546

lution of known concentration. Determine the refractive index of both solutions as described in Section 2.1.2.5. The concentration of the lipoprotein in mg/ml can then be calculated from the relation:

$$C = \frac{100\ (n_2 - n_1)}{\Delta n}$$

where n_1 and n_2 are the refractive indexes of solvent and lipoprotein solution respectively.

Practical details of the refractometric determination are given in Section 7.1.

6.2. *Separation of the lipid and protein moieties*

This is dealt with in detail in Appendix 3. Once the lipids have been extracted, they should be analysed as soon as possible. If a delay is unavoidable, the extract should be evaporated to dryness at low temperature under a stream of nitrogen and stored as described in Appendix 3. Equally, lipids that have been isolated during the analysis of the extract should be determined immediately. If, for some unforeseen reason, this is found to be impossible, the fractions should also be dried and stored with the precautions used for the original mixture.

6.3. *Analysis of the lipid moiety*

There is a confusing multiplicity of methods available for the determination of lipids, and particularly of cholesterol. To some extent, the choice must be guided by the sensitivity that is required. Highly sensitive enzymatic methods of estimating cholesterol and triglyceride are now available in kit form from several manufacturers (e.g. Boehringer Mannheim GmbH; BDH Chemicals Ltd.; E. Merck). These are provided with full working instructions that make it unnecessary to describe them in detail here. However, these methods may not be suitable for the analysis of lipoproteins that have been prepared by methods that involve precipitation with divalent cations (Section 2.2.3) unless care has been taken to ensure that these have been removed (Steele et al., 1976).

So far as the traditional methods of analysis are concerned, there may be little to choose between one and another when they are working well. However, some are easier to set up than others and many factors, including the skill and temperament of the analyst, will decide the final choice. For example, it might be unwise to pick on the Liebermann–Burchard reaction for the determination of cholesterol in a laboratory where the ambient temperature is high. The analyst should therefore look for the method that gives the best performance in his hands, under the local conditions, and should keep to it, not forgetting to make regular checks that he is maintaining his performance.

The classical procedure for the chemical analysis of a lipoprotein depends on the fact that the four classes of lipid that are normally present can be determined by specific reactions and therefore need not be completely separated. This procedure may not be applicable if other lipids are present, for example the hydrocarbons or glyceryl ethers that are found in some fish lipoproteins. This deficiency, coupled with the need for relatively large amounts of material has led to the replacement of this technique by methods in which the lipids are completely separated before they are estimated.

6.3.1. Chromatographic separation of lipids

This can be done either by thin-layer chromatography, or on columns. The former has the advantage of good resolution and is well suited to the analysis of small samples. Moreover, it also provides visual evidence for the presence of each component, which is a partial assurance that no unusual lipids are present. On the other hand, the process of estimation is more tedious and possibly less precise than when column chromatography is used, since each component must be recovered from the thin-layer plate and extracted from the adsorbent before it can be determined. If substances such as squalene, glyceryl ethers or partial glycerides are present, it may be difficult to establish conditions under which these are completely separated on a column. It may then be more satisfactory to isolate groups of lipids on a column and to resolve these by thin-layer chromatography.

Detailed descriptions of these methods are given elsewhere (e.g. Dittmer and Wells, 1969; Kates, 1975). However, these systems are sensitive to the source and treatment of the adsorbent, and to the purity and dryness of the solvents used. It is therefore important to confirm each one one with authentic lipids and to make such adjustments as may be necessary to suit local circumstances. In this context, it is worth noting the value of thin-layer chromatography (particularly on microscope slides) as a way of monitoring the efficiency of separations on columns.

6.3.1.1. Chromatography on columns of silicic acid The separation of lipids in this system reached a high state of efficiency as a result of the work of Börgstrom (1952) and of Hirsch and Ahrens (1958). Subsequently, several variations on this theme have been described, a number of which have been reviewed by Kates (1972). The method we shall describe is derived from that of Hirsch and Ahrens (1958).

Materials. Silicic acid; Bio-Sil HA (325 mesh) (Bio-Rad Laboratories Inc.). Similar products are available from other manufacturers but these may have slightly different properties which may require the procedure to be modified. Activate the silicic acid by heating at 110 °C for 18 hours and store it in a desiccator over phosphorus pentoxide. Do not activate a large amount at a time.

Solvents. Petroleum ether (boiling range 40–60 °C), benzene, chloroform and methanol, all of analytical grade. Distill all the solvents before use and dry the benzene over sodium.

Method. Note that, in this system, free fatty acids will elute with the triglycerides and are best removed by a preliminary extraction, as follows:

Dissolve the lipoprotein extract in about 5 ml of petroleum ether and extract four times with 1 ml of 4% aqueous sodium carbonate.

Wash the combined extracts twice with 2 ml of diethyl ether and return these washings to the original solution. Then acidify the sodium carbonate solution with 2 N sulphuric acid and extract the fatty acids with three washes of about 2 ml of petroleum ether. Finally, remove any traces of mineral acid from these combined extracts by washing them three times with 1 ml of water. Evaporate both the extracted lipids and the free fatty acids to dryness at low temperature, under a stream of nitrogen.

To prepare the column, slurry 1 g of activated silicic acid in a solution of 5% benzene in petroleum ether and pour it into a chromatography column about 15 cm long by 0.7 cm diameter. Dissolve 2–3 mg of the lipoprotein extract in 0.5–0.75 ml of 5% benzene-petroleum ether and apply this carefully to the column. Elute the lipids from the silicic acid according to the scheme shown in Table 6.3. A flow rate of about 2 ml/min can be achieved by applying nitrogen to the top of the column at a pressure of 0.15–0.3 kg/cm^2. Evaporate the fractions to dryness without delay, under nitrogen at 50 °C, and estimate the lipids as in Section 6.3.2.

6.3.1.2. Thin-layer chromatography *Separation of neutral lipids.* Any system that gives the required resolution can be used, and several are reviewed by Skipski and Barclay (1969) and by Kates (1975). The system devised by Mangold and Malins (1960), which is briefly described below, is satisfactory for the analysis of the lipids that are found in mammalian lipoproteins. When additional classes are present, as in some of the more primitive animals, extra stages of chromatography may be needed. These are briefly reviewed at the end of this section.

Materials. Silica gel G. Diethyl ether, acetic acid and petroleum ether (boiling range 60–80 °C), all of analytical grade. A solution of 0.2 g of 2′,7′-diclorofluorescein in 100 ml of 95% ethanol. Authentic lipid standards (Sigma Chemical Co.; Supelco Inc.; Applied Science Laboratories Inc.). Glass plates approximately 20 cm × 20 cm.

Method. Weigh out an amount of silica gel equal to 7 g for each plate to be covered, and suspend it in about twice its weight of distilled

TABLE 6.3
Scheme for the elution of lipids from a 1 g silicic acid column

Fraction	Lipid constituents	Eluent	Volume (ml)
I	Hydrocarbons	5% benzene in petroleum ether	100
II	Cholesteryl esters and waxes	40% benzene in petroleum ether	20
III	Triacylglycerols, alkyldiacylglycerols and free fatty acids	80% benzene in petroleum ether	100
IV	Cholesterol	100% benzene	50
V	*Diglycerides* and *monoglycerides*	*Chloroform*	50
VI	Phospholipids	Methanol:chloroform:water (87:10:3 v/v/v)	40

water. Spread the plates with a layer 0.25 mm thick in the usual way. Several plates can be spread at once and kept over calcium chloride in a closed cabinet until required. Before use, activate the silica gel by heating the plates at 110 °C for 30 min. Apply the mixtures of lipids along a line about 2.5 cm from the lower edge of the plate, at the rate of 50–100 µg on a single spot. Keep the size of each spot as small as possible. Develop the plates in a mixture of petroleum ether-diethyl ether-acetic acid (90:10:1 by volume) and then allow them to dry at room temperature for 20–30 min. Locate the spots of resolved lipids by spraying with dichlorofluorescein and viewing in ultraviolet light. This method will not detect phospholipids, which should be located by putting the plate into a closed container filled with the vapour from a few iodine crystals. Mark the positions of the spots and scrape off the adsorbent from each into separate stoppered tubes. Also take an equivalent area from an unused part of the plate as a control for each lipid.

To elute triglycerides, diglycerides or glyceryl ethers, add 10 ml of diethyl ether to the tube and shake for 10 min. Centrifuge and pipette off the ether into a second tube. Repeat this procedure twice with 5 ml of ether. Combine the ether eluates and evaporate them to dryness under nitrogen.

To elute cholesteryl esters, unesterified cholesterol and monoglycerides, add 10 ml of chloroform–methanol (4:1 v/v) and shake in a water bath at 40 °C for 10 min. After centrifuging down the silica, pipette off the eluate and repeat the process twice, with 5 ml of the solvent. Combine the eluates and evaporate to dryness under nitrogen at a temperature below 50 °C.

Separation of phospholipids. This method is described in great detail by Skipski and Barclay (1969), to whose paper reference should be made.

Materials. Silica gel H. 1.0 mM sodium carbonate. Chloroform, methanol and acetic acid, all of analytical grade. Authentic phospholipid standards as 0.02% solution in chloroform–methanol (2:1 v/v). 1.0 N HCl in methanol.

Method. Slurry 40 g of silica gel in about 90 ml of 1.0 mM Na_2CO_3

and spread a 0.5 mm layer on 20 cm × 20 cm glass plates. This will make about 5 plates, which should be allowed to dry at room temperature and can then be stored for several days in a cabinet with precautions to prevent their contamination. Activate the plates by heating for one hour at 110 °C before use.

Evaporate a sample of the phospholipid mixture, equivalent to about 20 μg of phsophorus, to dryness under nitrogen. Dissolve the residue in 50 μl of chloroform-methanol (2:1 v/v) and apply it as a streak to the chromatoplate. Wash the tube twice with 50 μl of solvent and apply these washings to the streak. Also apply appropriate standards to a near-by section of the plate. Develop the chromatogram with chloroform-methanol-acetic acid-water (25:15:4:2 by volume). It may be necessary to adjust the proportions of this solvent slightly to achieve the required separation. When the solvent front is 1–2 cm from the top, remove the plate from the tank and allow it to dry at room temperature for 20 min. Locate the phospholipids by standing the plate in a tank of iodine vapour, and circle them with a needle. Also mark equivalent areas from a blank part of the plate. Then keep the plate at room temperature until the iodine has evaporated, when each marked area can be carefully scraped into a separate stoppered tube. Add 3 ml of 1.0 N HCl in methanol to each tube, mix thoroughly and then shake for 15 min in a water bath at 40 °C. Centrifuge the tubes for 10 min at 500 *g* and aspirate the solvent layers into separate tubes. Repeat this procedure twice more, combining the supernatants from each sample of silica. Evaporate these eluates to dryness under nitrogen at 50 °C and determine the phosphorus in the residue (Section 6.3.2.3).

The separation of unusual lipids. Hydrocarbons, wax esters, glyceryl ethers and partial glycerides are said to be present in trace amounts in the lipoproteins of mammals, and form a high proportion of the plasma lipids in some more primitive animals. To resolve these substances it will be necessary to make a preliminary fractionation by one of the methods already described, and then to subject the appropriate fraction to further analysis by one of the methods which are briefly summarised as follows:

(a) Cholesteryl esters can be separated from wax esters on silica gel G in *n*-hexane–diethyl ether (95:5 v/v) (Nicolaides et al., 1970). But, note that the cholesteryl esters of the very long chain, highly unsaturated fatty acids that are found in fish, are also resolved from the esters of the C16 and C18 acids. The longer chain esters run in the same position as wax esters and confirmatory tests should be made on any material which occupies the position normally characteristic of the waxes (Mills et al., 1977).

(b) Triglycerides can be resolved from alkyl-diacylglycerols (Fraction III, Table 6.3) by the system petroleum ether: diethyl ether: acetic acid (90:10:1 by volume) on silica gel (Malins et al. 1965).

(c) Partial glycerides (Fraction V, Table 6.3) can be separated on silica gel G, using the solvent diethyl ether:benzene:ethanol:acetic acid (40:50:2:0.2 by volume) (Freeman and West, 1966).

6.3.1.3. Gas chromatography This technique offers the attractions of high resolving power coupled with a highly sensitive method of quantification, and the ability to carry out an analysis comparatively quickly. However, these are partially offset by the difficulty of working with substances of low volatility, particularly the phospholipids, the need to use internal standards, and the non-specific nature of the detector. Perhaps the most useful procedure of this kind that is available at present, was developed by Kuksis et al. (1975). In this, the phospholipids are first hydrolysed to partial glycerides and ceramides with phospholipase C. The resulting mixture of neutral lipids is then silylated with hexamethyldisilazane-trimethylchlorosilane-pyridine (cf. Section 6.6) and analysed by gas chromatography on a temperature-programmed column of 3% OV-1 on Gas Chrom Q. Tridecanoylglycerol is used as an internal standard.

Triglycerides and cholesteryl esters of different molecular weights can be resolved on this system, as can the partial glycerides that are derived from phospholipids. Moreover, if the phospholipids are removed from the original mixture at the outset, partial glycerides that are native to the lipoprotein can be estimated. However, this is a technically exacting technique which requires an extensive use of correction factors. The ability to resolve, at least partially, some of the lipid esters may be useful under some circumstances but the unequivocal identification of the chromatographic peaks may also be difficult. Furthermore, it may often be possible to obtain information of similar value from an analysis of the fatty acids that are liberated by saponification of the esters (Section 6.3.2.4).

At this stage in its development, it appears that the gas-chromatographic analysis of lipids may be of value for the study of large numbers of samples, or of very small quantities of material, but it also seems unlikely that many laboratories will be able to make full use of its potential.

6.3.2. Estimation of lipids

Saponification. The following procedure can be used to liberate the fatty acids from acylglycerols, cholesteryl esters or phospholipids prior to their analysis by gas chromatography. The cholesterol set free can be measured, and triglycerides can also be estimated by the titration of the liberated fatty acids. The method is essentially that of Albrink (1959) and is capable of hydrolysing the phospholipids that are present in plasma lipoproteins. Less vigorous conditions will suffice for the saponification of the triglycerides.

Evaporate a sample of the lipid solution to dryness in a stoppered tube, under a stream of nitrogen at 50 °C. As soon as it is dry, add 5 ml of alcoholic KOH (made by adding 6 ml of 33% aqueous KOH to 94 ml of ethanol) and heat the mixture in a water bath at 70–80 °C for 1 h (30 min for triglycerides or cholesteryl esters). Next add 5 ml of water to the tube and heat it for 10 min more in the water bath, before cooling it to toom temperature. To recover cholesterol, extract the alkaline solution twice with 5 ml portions of hexane, which should be combined and evaporated under nitrogen.

To recover the liberated fatty acids, re-heat the tube in the water bath and acidify with 1 ml of 1.8 N HCl (or H_2SO_4). Mix the contents of the tube thoroughly and cool to room temperature. Then extract the precipitated acids by washing twice with 5 ml portions of hexane, which are combined and dried under nitrogen.

6.3.2.1. Cholesterol The isolation and determination of cholesterol was reviewed by Kabara (1962). The method of estimation which follows is substantially that of Bloor (1916), which is based on the Liebermann–Burchard reaction (see Kenny, 1952, for a valuable study of the method). The more recently developed methods that use ferric chloride as the chromogen are favoured by some, and may be superior under certain conditions (Kates, 1975).

Reagents. Chloroform, acetic anhydride and sulphuric acid, all of analytical grade. A standard solution containing 1 mg/ml of pure, dry cholesterol in chloroform is also needed. Store this in the dark at 4 °C, in a tightly stoppered bottle, and dilute it 10-fold to give the working standard as required. Suitable standard solutions are available commercially and these save the considerable trouble that is involved in preparing them in the laboratory. If the latter course is unavoidable, the cholesterol should first be purified via the dibromide (Fieser, 1963) and subsequently stored, closely stoppered, in the deep-freeze. Note that crystalline cholesterol deteriorates on storage and it is therefore unwise to keep a large stock of it. Gross deterioration is easily detected by smell, pure cholesterol being almost odourless. Suspect material may be re-crystallised from ethanol. Do not crystallise from glacial acetic acid, since the product is sometimes prone to particularly rapid deterioration. The stock solution in chloroform is stable for a period of months in the refrigerator, but may slowly become more concentrated through evaporation of the solvent.

Method. Run all samples at least in duplicate.

Cholesteryl esters should be hydrolysed and the liberated sterol extracted as in Section 6.3.2.

Set up at least four cholesterol standards in the range from 25 to 300 μg by pipetting appropriate amounts of the working standard into clean test tubes, likewise take replicate aliquots of each sample to be assayed. Evaporate all the solutions to dryness under nitrogen at 50 °C. Cool the tubes and add 4 ml of chloroform, followed by 1.6

ml of acetic anhydride. Mix thoroughly. The next operation should be timed with a stop-clock: add 0.1 ml of concentrated sulphuric acid to each tube, with an interval of 2 min between each, and mixing after each addition. Observe the rate at which the colour develops in the first tube, by spectrophotometry at 620 nm. The time at which the maximum intensity is reached will depend on the ambient temperature and may be between 5 and 12 min after the addition of the sulphuric acid. Each subsequent tube should then be read at this same time interval. Do not attempt to deal with more than 6–12 tubes at a sitting lest the timing of the readings gets out of phase. If the temperature of the laboratory is high or is liable to fluctuate considerably, the kinetics of the colour development can be stabilised by incubating the tubes in a constant temperature bath at about 20 °C.

Note that the colour produced should be a clear, bright green, with no trace of yellow. 'Muddiness' of the colour shows that the procedure is at fault, for example that the sample contains an excess of salt, or is wet, or that the quality of the sulphuric acid has deteriorated. Under these circumstances, abandon the assay.

A calibration curve should be plotted for each set of determinations. This allows a check to be made on the reproducibility of the procedure and on the integrity of the standard.

Where the cholesterol has been obtained by hydrolysis of a human ester fraction, the weight of the original esters can be estimated by multiplying the amount of cholesterol by 1.67. This factor will need adjustment if the distribution of fatty acids in the esters differs appreciably from that in man.

6.3.2.2. Glycerides These can be estimated by determining either the glycerol or the fatty acids that are liberated by hydrolysis. The determination of glycerol (Van Handel and Zilversmit, 1957; Carlson, 1963; Dittmer and Wells, 1969; Kates, 1972) is the more precise and sensitive method, but the measurement of fatty acids is technically simpler and can also be used to estimate free fatty acids, or monoalkyldiacyl glycerols.

Estimation of glyceride via glycerol (Carlson 1963):

Reagents. 0.7 M sulphuric acid. 0.02 M sodium periodate (store in the cold)
Ethanolic KOH: 1 ml of 2.5% aqueous KOH
19 ml of ethanol
Sodium arsenite: 0.9 g of NaOH
2.0 g of As_2O_3
Water to 100 ml

Chromotropic acid (4,5-dihydroxynaphthalene-2,7-disulphonic acid disodium salt): first dilute 150 ml of concentrated sulphuric acid with 75 ml of distilled water. When this is cold, add to it 0.5 g of chromotropic acid dissolved in 50 ml of water. The reagent may be stored in a dark bottle for up to two weeks.

Glycerol standard: Weigh out 800 mg of analytical grade anhydrous glycerol and make it up to 100 ml with water. Dilute this to give working standards that contain 80, 40 and 20 µg of glycerol/ml. (Glycerol can be dehydrated by heating slowly to 175 °C.) Carlson (1963) recommends the use of tripalmitin as a standard, but this requires careful purification by chromatography first.

Procedure. Pipette into a stoppered tube (in replicate) an amount of the triglyceride to be analysed that is equivalent to about 60 µg of glycerol, and evaporate to dryness under nitrogen at 50 °C. Similarly, pipette out 0.25 ml of each working standard and dry in an oven or a vacuum desiccator. Add 1 ml of ethanolic KOH to all the tubes, stopper loosely and heat for 30 min at 60 °C. Cool to room temperature and add 0.8 ml of 0.7 M sulphuric acid, followed by 4 ml of diethyl ether. Shake well and remove the ether layer. Pipette 0.25 ml of the aqueous phase (which amounts to 1.0 ml) into a tube and add 0.1 ml of 0.02 M sodium periodate. After 10 min add 0.1 ml of 0.2 M sodium arsenite and mix. Stand for at least 5 min, then add 2.8 ml of the chromotropic acid reagent. This, and the subsequent operations should be done in a shaded place. Heat the tubes in a boiling water bath for 30 min and read the colour at 570 nm.

A close approximation to the weight of triglyceride in the original sample can be found by multiplying the total weight of glycerol by 9.39. This factor is derived from the average distribution of fatty acids in human plasma triglyceride, as determined by gas chromatography.

Estimation of glyceride via fatty acids (Albrink 1959):

Reagents. Hexane. 1.8 N HCl. Standardised 0.01 N NaOH. Ethanolic KOH (prepared as required, Section 6.2.2).

Nile Blue indicator: 0.02% Nile Blue in water. Wash this solution four times with hexane and then dilute 1:10 with absolute ethanol. The nitrogen that is used for purging the titration mixture must be freed from carbon dioxide by passing it through a soda lime tube.

Procedure. Saponify the triglyceride as in Section 6.2.2. Transfer 3 ml of the hexane solution of fatty acids to a tube that contains 3 ml of Nile Blue solution. Pass a stream of nitrogen through the mixture for 1 min before, and also during the titration. Add 0.01 N NaOH from a microburette or from a micrometer syringe (e.g. Agla, Wellcome Reagents Ltd.), through a very narrow bore plastic tube (0.25–0.50 mm) that dips below the surface of the fatty acid solution. Be sure that this tube is completely filled with NaOH before the titration is begun. The end point is shown by a change in colour from blue to the first perceptible shade of pink, and it is advisable to use one of the control tubes as a standard of comparison for those that follow.

The weight of triglyceride in the sample, in mg, can be closely approximated by multiplying the total amount of fatty acid present, expressed in milli-equivalents, by the factor 288.

6.3.2.3. Phosphorus This is a modification of the method of Allen (1940).

Reagents. 72% perchloric acid. 8.5% ammonium molybdate in water

Stannous chloride: 0.25 ml of 40% $SnCl_2$ in concentrated HCl; Water to 50 ml

Standard phosphate solution: 0.8776 g of KH_2PO_4 in 100 ml of water. Dilute this 10-fold to give a working standard that contains 20 µg P/ml.

Procedure. The reaction is performed in a 20 ml tube that is calibrated with a mark at 10 ml. Transfer samples of phospholipid solution (equivalent to 5–10 µg P) in replicate to tubes that have been thoroughly cleaned with chromic acid. Evaporate to dryness under nitrogen at 50 °C and add 0.8 ml of 72% perchloric acid. Heat at 180–190 °C for 45 min, when the liquid should be clear and colourless. When the tubes are cold, wash down the sides with 5 ml of water. Prepare a suitable range of standard and control tubes, to which 0.8

ml of perchloric acid and 5 ml of water are added. Finally, add 0.5 ml of ammonium molybdate and 0.5 ml of stannous chloride to each tube. Make up to 10 ml with water and mix by inversion. Allow the colour to develop for 10 min and then read the optical density at 680 nm.

It is customary to estimate the weight of total phospholipid in the sample by multiplying the weight of phosphorus by 25. For individual phospholipids that have been isolated by chromatography, the ratio appropriate to the specific substance should be used.

6.3.2.4. Fatty acid analysis The total amount of fatty acid that is liberated by saponifaction of the lipid esters, or is extracted as free fatty acid from the lipoprotein, can be determined by titration as described above. The individual acids present can be identified, and estimated quantitatively if an internal standard is added, by gas chromatography of their methyl esters. Various methods of esterification are available, but the rapidity and smoothness of the reaction with diazomethane goes far to outweigh the need to prepare and handle this highly toxic substance with care. Although diazomethane can be prepared without undue hazard if the procedure is carried out correctly, it is prudent to take strict precautions during the operation. Set up the equipment in an efficient fume cupboard and wear a visor or use a shatter-proof screen as protection in the event of an explosion. Do not inhale the vapour.

Procedure. Assemble an all-glass distillation apparatus, with a flask of 50 ml capacity connected to an efficient condenser. The receiver should have a volume of about 15 ml and should be cooled at least to 0 °C. Lightly lubricate all joints with silicone grease, and keep the dead space in the apparatus as small as possible.

Weigh 1.1 g of *N*-methyl-toluene-*p*-sulphonyl nitrosamide into the distilling flask and dissolve it in 15 ml of dry ether. Add to this a solution of 0.2 g of KOH in 20 ml of methanol. Mix, and immediately connect the flask to the still and fit the thermometer. Warm the flask with a water bath at 45–50 °C and collect the distillate until it runs colourless (about 10 ml). Keep the dark yellow solution in a tightly stoppered tube in the deep-freeze, where it will be stable for some weeks. Prolonged storage, particularly at higher temperatures, will lead to the production of spurious peaks on the gas chromatogram.

To esterify the fatty acids, dry off the sample in a small tube and add the ethereal solution of diazomethane dropwise, until a slight permanent yellow colour is present. Evaporate the solvent and excess reagent under a stream of nitrogen and transfer an appropriate amount of the methyl esters to the gas chromatograph in the usual way (Kates, 1975). For many purposes a 2 metre column of 20% diethyleneglycol succinate on Chromosorb W (80–100 mesh) run at 160–170 °C will give satisfactory results. The individual components of the mixture can be identified by reference to authentic standards and quantified in the conventional way.

6.3.2.5. Determination of unusual components The unequivocal identification of hydrocarbons may be difficult unless there is sufficient material for several physical and chemical tests. Infrared spectroscopy may be valuable in this context, and useful information can be obtained by gas chromatography on a column of silicone SE 30 (2.5% on Chromosorb AW-DMDC, 60–80 mesh) with a programmed increase in temperature from 170 to 260 °C at the rate of 1.5 °C/min. Total hydrocarbons can be quantitatively estimated gravimetrically on the eluate from a silicic acid column (Section 6.3.1.1).

Monoalkyl diacylglycerols can also be determined gravimetrically (Thompson and Kapoulas, 1969), or by methods analogous to those used for triglycerides (Kates, 1975; Section 6.3.2.2). The ethers liberated by hydrolysis can be identified by gas chromatography of their isopropylidene derivatives (Hanahan et al., 1963; Thompson and Kapoulas, 1969).

6.4. Analysis of the protein moiety

The final stage in the chemical characterisation of a lipoprotein is the identification and estimation of the apo-lipoproteins that are present.

Unfortunately, this is also the most difficult of lipoprotein analyses, (1) because it is technically exacting and (2) because there is no clear definition of exactly what an apo-lipoprotein is. The latter complication was discussed in Chapter 1, where the accepted apo-lipoproteins were listed. The known physico-chemical properties of these proteins, by which they may in principle be characterised, are given in Table 6.4. A second list of proteins that have been reported to be minor constituents of lipoproteins, is given in Table 6.5. The evidence that these are genuine apo-lipoproteins is often flimsy and their physico-chemical characterisation is sometimes poor. Despite their controversial role however, it is undeniable that these substances may sometimes be found in a lipoprotein preparation, to the perplexity of the analyst.

At present, the only practical way of unequivocally identifying the apo-lipoproteins on a micro-scale is by reaction with mono-specific antisera. This has the advantage that both qualitative and quantitative analyses can be done without the need to separate the proteins first. Indeed, this is the simplest way of identifying apo-lipoprotein B, which cannot easily be re-dissolved once it has been freed of lipid. However, this procedural simplicity is counter-balanced by the fact that many of the antisera are not readily available, thanks to the difficulty of isolating the relevant purified apo-lipoproteins. A complete immuno-chemical analysis of the apo-lipoprotein distribution may therefore be impracticable in many laboratories.

The conventional alternative of isolating each protein and then identifying, or estimating them chemically also presents problems which stem mainly from the strong tendency of the apo-lipoproteins to form intermolecular complexes, or to be adsorbed at surfaces. Thus chromatographic methods of analysis are unsatisfactory because they give low recoveries and poor resolution. At the time of writing, there is some evidence that reversed-phase high-performance liquid chromatography (HPLC) may be more satisfactory but the application of this technique to apo-lipoproteins is not yet fully developed and, in any event, it requires the use of expensive equipment. The technique that is most commonly used for the rapid qualitative micro-analysis

TABLE 6.4
Physico-chemical characteristics of the major apo-lipoproteins

Apo-lipo-protein	Synonyms	Mol. Wt. (M_r)	Polymorphs	Isoelectric point (pI)		Number of aminoacid residues	Terminal aminoacids		Aminoacids absent	Carbohydrates present	Reference to aminoacid composition
							C	N			
A-I	Apo-Gln-I R-Thr	28 145	A-I_1 A-I_2 A-I_3 A-I_4	5.62 5.53 5.45 5.36	ref. 2	243	Gln	Asp	Ile, Cys		6, 7
A-II	Apo-Gln-II R-Gln	17 400	—	5.10	ref. 2	77 × 2	Gln	PCA	His, Arg, Trp		8
A-IV	—	46 000	—	5.05	ref. 1	?	?	Gln	Cys		9
B	Apo-Ser	?	?	?		?		?		Man, Gal, $GluNH_2$ Fuc, sialic acid	10
C-I	R-Ser	6 631	—	6.5	ref. 3	57	Ser	Thr	His, Tyr, Cys		11, 12
C-II	R-Glu	8 837	—	5.0 4.78	ref. 3 ref. 4	78	Glu	Thr	His, Cys		13
C-III	R-Ala	8 764	C-III_0 C-III_1 C-III_2	4.93 4.72 4.54	ref. 4	79	Ala	Ser	Ile, Cys	Gal, $GalNH_2$ sialic acid	14
D	A-III 'thin-line peptide'	22 100	—	?		?		?		Gal, Glu, $GluNH_2$ Man, sialic acid	
E	Arginine-rich	35 000– 39 000	E-1 E-2 E-3 E-4 E-5	5.7 5.8 5.9 6.0 6.2	ref. 3	?	Ala	Lys	E-4 lacks Cys	Gal, Glu, Man, $GluNH_2$, $GalNH_2$ sialic acid (ref. 5)	15, 16

Abbreviations: PCA, pyrrolidone carboxylic acid; Gal, galactose; Glu, glucose; Fuc, fucose; $GalNH_2$, galactosamine; $GluNH_2$, glucosamine.
References: (1) Beisiegel, 1979. (2) Nestruck et al., 1980. (3) Marcel et al., 1979. (4) Catapano et al., 1978. (5) Jain and Quarfordt, 1979. (6) Brewer et al., 1978. (7) Baker et al., 1974. (8) Brewer et al., 1972. (9) Uterman and Beisiegel, 1979. (10) Lee and Alaupovic, 1974. (11) Schulman et al., 1975. (12) Jackson et al., 1974. (13) Jackson et al., 1977. (14) Brewer et al., 1974. (15) Weisgraber et al., 1980. (16) Weisgraber et al., 1981.

Apo-lipoprotein	Synonyms	Mol. Wt. (M_r)	Polymorphs	Isoelectric point (pI)		Terminal aminoacids C	Terminal aminoacids N	Aminoacids absent	Reference to aminoacid compn.
F		28 000	—	3.7	ref. 1	?			
H	β_2-glycoprotein-I	54 000	1	5.75	ref. 2	?			
			2	6.0					
			3	6.2					
	Threonine-poor	11 000 44 000	—	6.0 6.5	ref. 3	?			
	D-2	7 000	—	?		Ser	Leu		
	Serum Amyloid-A proteins	11 300–11 900	1	6.1–6.2	ref. 4		Arg	Cys	4, 7
			2	5.6–5.7			Ser		
			4	?		Tyr	Arg		
			5	?		Tyr	Ser		
	'SV-D' proteins	8 000–11 000	2	6.85			Ser		6
			3	6.55					
			4	6.10			Arg		
			5	5.70			Ser		
			6	5.20					
			7	5.0					

With the exception of apo-lipoprotein F and β_2-glycoprotein-I, the proteins listed in this table have only been isolated from the plasma lipoproteins of pathological cases. In addition, a trace of 'proline-rich' protein (M_r 74 000) has been detected in chylomicrons, but is said to be absent from other lipoproteins (Sata et al., 1976). Lp(a) is also said to contain a specific apo-lipoprotein but this appears not to have been characterised.

References: (1) Olofsson et al., 1978. (2) Polz et al., 1981. (3) Shore et al., 1978. (4) Eriksen and Benditt, 1980. (5) Malmendier et al., 1979. (6) Malmendier et al., 1980. (7) Parmelee et al., 1982.

of apo-lipoproteins is that of electrophoresis in polyacrylamide gel. This uses simple apparatus but is less sensitive than the immunochemical method, and is not free from the objection that each protein band is contaminated with smaller amounts of other components of the mixture. Moreover, electrophoresis suffers from the handicap that the proteins are identified only by their rates of migration, which can be an uncertain criterion, particularly in inexperienced hands. However, a considerable degree of flexibility can be achieved by making adjustments to the concentration of the polyacrylamide gel, or by the use of acid, alkaline or detergent-containing buffers. Moreover, the isoelectric focussing technique is valuable for distinguishing between the so-called isoforms or polymorphs of which some apo-lipoproteins are composed. The identification of a particular protein can therefore be made more certain by observing its behaviour in more than one system. Where a limited catalogue of antisera is available, the identity of at least some of the proteins can be confirmed by immunochemical challenge of bands that are dissected from the gel (Section 6.5).

Because of the difficulty of making a complete analysis of the apo-lipoproteins, many laboratories will be forced to adopt a compromise procedure such as the following:

(a) Determine the total protein content of the lipoprotein (Section 6.4.1) and also the proportion that is precipitable with tetramethylurea (Section 6.4.1.2).

(b) Determine the qualitative distribution of the apo-lipoproteins by polyacrylamide gel electrophoresis, with confirmatory immunological analyses with such antisera as are commercially available or are easily made.

(c) Make immunological estimations of those apo-lipoproteins for which antisera are available, or determine (albeit only semi-quantitatively) the relative proportions of the proteins soluble in tetramethylurea, by densitometric scanning of electrophoresis gels.

6.4.1. Quantitative estimation of total apo-lipoproteins

This section will deal with the estimation of the total amount of apo-lipoprotein present in a lipoprotein preparation, and with the estimation of the relative proportions of apo-lipoprotein B and the 'soluble' proteins A, C, D and E.

6.4.1.1. Total protein The 'Lowry' method (Lowry et al., 1951) is by far the most widely used procedure for the estimation of total apo-lipoprotein, although gravimetric analysis, the classical nitrogen determination, or total aminoacid analysis have also been used. However, there is some uncertainty attached to the choice of the standard protein. Human or bovine albumin is commonly used and these seem to have a specific chromogenicity that is indistinguishable from that of the soluble apo-lipoproteins, although Curry et al. (1976a) have some evidence that apo-lipoprotein A-I gives 5% less colour than albumin. By contrast, it seems certain that apo-lipoprotein B is less chromogenic than albumin, although the magnitude of the difference is controversial. According to Margolis and Langdon (1966) 77 μg of bovine serum albumin are equivalent to 100 μg of protein from LDL. However, this would have been a mixture of apo-lipoprotein B with small amounts of other apo-lipoproteins. A slightly higher ratio (82:100) was found by Albers et al. (1975), while values of 0.90 and 0.92 have been quoted by Alaupovic and his collaborators (Curry et al., 1978; Lee and Alaupovic, 1974). Since the latter estimates were made with purified apo-lipoprotein B, they would appear to be the more reliable but it is clear that this needs to be confirmed by further studies. Although albumin may therefore be used as a standard for the estimation of apo-HDL, it is plain that no single protein can serve as a perfectly satisfactory standard for LDL and VLDL, which contain varying proportions of apo-lipoprotein B.

The Lowry procedure can be applied directly to a solution of native lipoprotein but the presence of lipids will cause the reaction mixture to become opalescent. This must either be cleared by extraction with ether or hexane, or prevented by the addition of sodium dodecyl sulphate to the reagents. Alternatively, the lipid can be extracted from the protein as described in Appendix 3 and the residue used for the protein determination. This technique may require a higher level of analytical expertise than the first, and would be difficult to use on a very small sample.

The calibration curves for both albumin and apo-lipoprotein consist of two intersecting straight lines. Since the disjunction may not occur at the same point on the two curves, it is important to ensure that the correct part of the standard curve is used for the determination. This effect is slearly shown by van der Bijl and van Gent (1975), who also discuss the effect of sodium chloride on the development of the colour. It is also worth noting that the subject of interference in the Lowry assay has been given detailed attention in a review by Peterson (1979).

Reagents

Solution A: 2% Na_2CO_3 in 0.1 N NaOH.

B: A freshly prepared mixture of equal volumes of solutions of 1% copper sulphate and 2% sodium potassium tartrate.

C: 1 ml of B mixed with 50 ml of A immediately before use.

D: 10 ml of Folin–Ciocalteu reagent (available commercially) diluted with 13.6 ml of water; this solution is 1 N with respect to the phosphomolybdic–phosphotungstic acid reagent and must be prepared immediately before use.

E: A standard solution of protein (human or bovine serum albumin) that contains 100 μg/ml in 0.1 N NaOH. Ash and moisture determinations should be carried out on the protein used and the standard corrected accordingly.

Method. Dissolve the protein obtained by extraction of the lipoprotein in 0.1 M NaOH. If the protein is wet with ether it usually dissolves without difficulty.

Carry out the estimation in replicate as follows: Pipette into a series of scrupulously clean test tubes, volumes of the solution to be assayed that are equivalent to various amounts of protein up to 100 μg. Add water to each tube to give a total volume of 1.0 ml. Prepare dilutions of the standard protein solution in the same way. Add 4 ml of reagent C to each tube, mix and stand for 10 min. Then add 0.5 ml of D, mix vigorously, stand for 30 min and read the optical density at 660 nm against a reagent blank prepared with 0.1 N NaOH instead of the protein standard.

If the estimation is to be made on the intact lipoprotein, substitute this solution for the re-dissolved protein in the fore-going method.

The turbidity that results from the presence of the lipid can be cleared by washing the final reaction mixture twice with 10 ml of diethyl ether or hexane, after the development of the colour is complete. Treat the standards in the same way. Alternatively, the turbidity can be suppressed by the addition of sodium dodecyl sulphate (Curry et al., 1978; Markwell et al., 1981) as follows.

Reagents (Markwell et al. 1981)

Solution A: 2% Na_2CO_3, 0.16% sodium tartrate and 1% sodium dodecyl sulphate in 0.1 N NaOH.

B: 4% copper sulphate.

C: 1 ml of B mixed with 100 ml of A immediately before use.

D: Dilute Folin–Ciocalteu reagent to half-strength with distilled water immediately before use.

E: 100 µg of bovine serum albumin/ml in distilled water.

Note that sodium tartrate is used in reagent A because potassium dodecyl sulphate is sparingly soluble. In any event, the detergent may precipitate at temperatures below 20 °C but can be redissolved by gentle warming.

Procedure. Prepare dilutions of the standard and of the specimens to be assayed, as above. Add 3 ml of reagent C to each tube, mix and stand at room temperature for at least 10 min. Then add 0.3 ml of D, mix vigorously, stand at room temperature for 45 min and read at 660 nm against a reagent blank. The colour is stable for at least 40 min.

The foregoing methods can be used to determine protein in amounts down to about 5 µg. When the quantity is smaller than this, the assay volume can be scaled down by 10 or 20-fold if micro-cuvettes are available. Alternatively, the optical density can be read at 750 nm (Markwell et al., 1981).

6.4.1.2. Chemical estimation of apo-lipoprotein B The most sensitive and specific methods of estimating apo-lipoprotein B are immuno-chemical (Section 8.9) but there are also simpler chemical methods of making an assay of this protein. One is to dissolve the mixture of apo-lipoproteins in 2 mM sodium decyl sulphate, separate the B protein by chromatography on Sephadex G150 and then to estimate it by the Lowry method (Brown et al., 1969). However, this is a tedious procedure that is not well adapted to the micro-scale. Several laboratories have therefore been led to investigate methods in which the apo-lipoprotein B is precipitated by an organic solvent in which the other apo-lipoproteins are soluble. It is difficult to be certain that these methods are strictly quantitative but, despite this, they can be of value. The most widely used procedure is that due to Kane (1973, 1975), in which apo-lipoprotein B is precipitated, together with the lipids, by the addition of 1,1,3,3-tetramethyl urea (TMU) to a concentration of 4.2 M (i.e. 50% TMU in water). Both the soluble and the insoluble proteins can then be estimated by the Lowry method. The technique requires less than 1 mg of protein but the reagent is both unstable and unpleasant. Moreover, even if 4.2 M TMU is the optimum concentration for the separation of the human apo-lipoproteins (which has been questioned), this may not be true for the apo-lipoproteins of other animals. It has been claimed that some of the disadvantages of TMU can be overcome by the use of iso-propanol (Holmquist et al. 1978) but this has yet to be substantiated.

Reagents. 0.15 M NaCl. 1,1,3,3-Tetramethyl urea (Burdick and Jackson; Sigma; Merck): this should be re-distilled (175.5–176.5 °C) and kept in a closely sealed bottle in the dark. Since this reagent slowly oxidises it is imprudent to re-distil large amounts at a time.

Procedure. Adjust the concentration of the lipoprotein solution to be equivalent to about 0.6 mg of protein/ml, in a buffer of ionic strength not less than 0.05 and pH 6.0–9.0. Warm both this solution and the TMU separately to 37 °C in a water bath. Add 0.25 ml of TMU to 0.25 ml of sample in a small tube, mix thoroughly and keep at 37 °C for 30 min. Then centrifuge at 2000 *g* for 10 min at room temperature. The mixture of lipid and apo-lipoproteins B that is precipitated from VLDL usually forms a floating pellicle on a clear infranatant. If the tube is tilted, the pellicle can be induced to stick to the side of the tube, leaving a clear surface through which to insert a pipette. The precipitate from LDL is more dense and is sometimes difficult to sediment even with a brisker centrifugation. Withdraw 0.1

ml of the liquid phase and add it to 0.4 ml of 0.15 M NaCl. Mix well and take samples of this solution for protein estimation by the Lowry method (Section 6.4.1.1). Because TMU affects the colour reaction, the protein standard must be made up in 10% TMU.

This procedure estimates the amount of soluble apo-lipoprotein, which must be subtracted from the total protein to obtain the quantity of the apo-lipoprotein B. In a modification of the method, Reichl et al. (1977) filtered off the precipitated protein and measured it directly. This is a useful technique where a low-density lipoprotein has been radioactively labelled and the specific activity of the label in the apo-lipoprotein B is required. The lower limit of the method appears to be of the order of 1 μg of protein but this, like the precision, is not well established.

Materials. 0.15 M NaCl. Chloroform. Methanol. Re-distilled TMU. Whatman glass-fibre filter discs (type GF/C), 2.5 cm diam. Millipore Pyrex Microanalysis Filter Holder (XX10-02500) and a matching Buchner flask.

Procedure. To 0.5 ml of the lipoprotein solution, add an equal volume of TMU and stand the mixture at room temperature for 2 hours. Collect the precipitate under gentle suction, on a glass-fibre filter in the filter holder. Wash the reaction tube with 1 ml of 4.2 M TMU followed by 1 ml of 0.15 M NaCl, and add these washings to the filter. Then wash the filter successively with 10 ml of a mixture of TMU with 0.15 M NaCl (1:1, v/v), 5 ml of distilled water and 5 ml of chloroform–methanol (1:1, v/v). In the case of radioactive samples, the washed filter can be transferred to a tube for radioassay, after which the protein can be recovered for estimation by the following modification of the Lowry procedure (Section 6.4.1.1).

Transfer the filter disc to a 5 ml test tube, add 0.2 ml of 2 M NaOH, which must entirely soak the filter, and stand overnight at room temperature. Then add 2 ml of a solution containing 1.0% $CuSO_4$ and 2.0% sodium potassium tartrate. Stand for 10 min at room temperature and then add 0.2 ml of 1.0 M H_2SO_4, followed by 0.2 ml of Folin–Ciocalteu reagent that has been diluted to 0.5 M. Mix thoroughly and centrifuge for 5 min at 2000 *g* to remove fragments of glass fibre. Read the optical density of the supernatant solution at 660 nm, 30 min after adding the reagent. Calibrate the method by performing the entire procedure on samples of LDL of known protein content.

6.4.2. Electrophoretic analysis of apo-lipoproteins

The techniques used are derived from that of Davis (1964), in which the stabilising agent is polyacrylamide gel (Section 5.5). Because of the tendency of the isolated apo-lipoproteins to aggregate, a dissociating agent such as urea, or sodium dodecyl sulphate (SDS) must be present in the gel. However, apo-lipoprotein B is almost insoluble even in 8 M urea and for this protein it is necessary to use gels containing SDS. It is important to note that urea hydrolyses in aqueous solutions to form cyanate (Marier and Rose, 1964), which will carbamylate the amino groups of the proteins and give rise to spurious bands on electrophoresis. It is therefore best, albeit expensive, to prepare the gels with ultra-pure urea, or with analytical grade urea that has been treated with an ion-exchange resin (e.g. Amberlite MB-1) immediately before use. Using this grade of urea, it is possible to make up stock gel reagents in 8 M urea and to keep these solutions for up to 7 days in the dark at 4 °C (Shore et al., 1980). In practice however, good results can be obtained with analytical grade urea provided that it is not added to the gel reagents until immediately before they are used. Likewise, urea should not be added to the protein sample until it is about to be put onto the gel.

The properties of the polyacrylamide gel and thus of the resolution obtained are critically dependent on the quality and the relative proportions of the reagents. To ensure that the analyses are reproducible, both within a run and from day to day, it is important to use the best quality materials. The acrylamide should either be the commercial product that is specially prepared for electrophoresis, or should be re-crystallised from chloroform and ethyl acetate (Maurer, 1971). The preparation of the reagents, the samples and the buffers should all be carried out with great care, and it is also important to control the presence of oxygen or of other substances e.g. mercaptoethanol, that inhibit the polymerisation.

The staining of the apo-lipoproteins after the electrophoresis also needs close attention. The proteins do not all bind the dye equally strongly, for example apo-lipoproteins A-I and A-II will stain well under conditions which leave apo-lipoprotein C-I only weakly coloured (Shore et al., 1980). In addition, some of the proteins are more difficult to fix than others. Consequently, if the pattern of apo-lipoproteins that is revealed by the electrophoretic analysis is to reflect accurately the real composition of the sample, the staining procedure itself must be consistent and reproducible.

6.4.2.1. Techniques for staining gels The method used for staining the proteins will depend on the purpose for which the electrophoresis was carried out.

(a) The most usual purpose is the qualitative analysis of the apo-lipoproteins, for which the stain should be capable of detecting components that account for not more than 2% of the total protein when the gel is adequately loaded. In the past, Amido Black (also called Buffalo Black or Amido Schwartz) was widely used but this has now been largely supplanted by Coomassie Blue R250, which is much more sensitive. Moreover, this dye can be used to stain the proteins in a simple one-step procedure in which fixing and staining take place at the same time. This technique depends on the fact that a colloidal suspension of the dye is formed when an aqueous solution of Coomassie Blue is added to 10% trichloracetic acid, in which the dye is insoluble. If the gel to be stained is immersed in this solution, the dye is taken up by the proteins and, if its initial concentration is not too great, there will be insufficient left to colour the background. Thus there is little need to de-stain the gel. The process may take up to 48 hours at room temperature but can be completed in one hour in a water bath at 60 °C. However, some de-staining is usually needed after this hot process.

Procedure. This technique can be used with acidic or basic gels that contain urea, or with gels that contain sodium dodecyl sulphate.

Stir 0.1 g of Coomassie Blue R250 in 5 ml of water continuously until it is completely dissolved. Then slowly mix this solution into 100 ml of 10% trichloracetic acid, stirring continuously. It is important to add the dye to the acid: the reverse procedure frequently causes the dye to precipitate and the product is useless as a stain for polyacrylamide gels. The colloidal suspension may slowly aggregate and form a precipitate that can be filtered off (Whatman No. 1 paper), but this may leave the stain too dilute for use.

Immediately after each gel has been removed from its electrophoresis tube, immerse it completely in a test-tube filled with the stain. This tube may be inverted occasionally to re-mix the buffer that diffuses out of the gel. On no account should the tubes be 'topped-up' with fresh stain, for dye may then precipitate on the surface of the gel. When the proteins have stained to the desired intensity, transfer the gel to a clean, screw-capped tube that is completely filled with water, in which it may be photographed. If it is thought necessary to preserve the gel for a short time, the tube should be filled with 7% acetic acid.

With this technique there is usually no need to de-stain the gels but, if this should become necessary, they can be soaked in a mixture of methanol:acetic acid:water in the proportions 40:10:50 by volume.

The above process may take a considerable time to complete but, if need be, the staining can be made more rapid by placing the tube containing the gel and the stain in a water-bath at 60 °C for 1 h. After this treatment, de-staining is essential. The methanol:acetic acid:water mixture can be used for acidic or basic gels that contain urea, but it is better to use isopropanol:acetic acid:water (10:10:80 by vol.) for gels that contain sodium dodecyl sulphate.

When the routine de-staining of large numbers of gels is necessary, it may be advantageous to use a continuously stirred bath (Edelstein and Scanu, 1980; Shore et al., 1980). In this easily constructed equipment, the gels are put into rimmed plastic tubes in which holes have been drilled. These tubes are then supported by their rims in holes that are drilled in a plastic plate that fits the top of a large beaker. The latter is filled with the de-staining solution and stands on a magnetic stirrer. If the solution is changed occasionally, this arrangement allows the de-staining process to be appreciably shortened.

(b) It should be noted that the foregoing method generally gives unsatisfactory results when used on isoelectric focussing gels because Coomassie Blue R250 is taken up by the ampholytes and is consequently difficult to de-stain. Shore et al. (1980) also find that the stained apo-lipoproteins are easily washed out of gels that contain ampholytes. To overcome this, they first remove some of the ampholytes by a preliminary fixation and washing. They then carry out both the staining and the de-staining operations in reagents that tend to promote the denaturation of the proteins. However, good results with isoelectric focussing gels, or with paper prints taken from them, can be obtained by the following simpler method.

Procedure. Prepare a concentrated solution of trichloracetic acid by dissolving 1 kg in sufficient water to give a final volume of 1 litre. To make the stain solution, dissolve 0.75 g of Coomassie Blue R250 in 225 ml of methanol by stirring for 10 min. Pour this solution into 465 ml of water while mixing thoroughly. Continue to stir, and add 22.5 g of sulphosalicylic acid. When this has dissolved, slowly add 75 ml of the trichloracetic acid solution. The final volume is about 690 ml. Any insoluble residue should be removed by filtration through a Whatman 111 paper.

Stain the gels in this solution until the proteins have become adequately visible, but do not continue for more than 5 or 6 hours or the de-staining time will be impracticably long. To de-stain, transfer the gels to a mixture of ethanol:acetic acid:water (3:1:8 by vol.) at room temperature. After several hours, the background intensity should reach an acceptably low level and the gels can be transferred to water.

Note that method (c) below can also be used for isoelectric focussing gels.

(c) The object of the staining may be to mark the position of proteins that are to be excised from a duplicate gel that has been run in parallel with the one that is stained. In experiments of this kind the sample load is usually comparatively heavy and a highly sensitive stain may not be necessary. But it must be quick-acting, or there will be loss of resolution in the unstained gel through diffusion of the proteins. A suitable combination of these attributes is offered by the Coomassie Blue G-perchloric acid system (Reisner et al., 1975). This procedure also has the advantage that de-staining is not required, even in the case of isoelectric focussing gels, since ampholytes do not take up this dye. Note however, that it is not suitable for gels that contain sodium dodecyl sulphate.

Procedure. Dissolve 0.04 g of Coomassie Blue G250 in 100 ml of 3.5% (w/v) perchloric acid, with stirring, at room temperature. After 1 h, filter through a Whatman No. 1 paper. The resulting solution is stable at room temperature.

The procedure for staining the gels in this solution is the same as that described for Coomassie Blue R250, but major proteins will become visible as blue bands against a pale orange background after about 5 min. Staining should be adequate after 30 min, but the intensity will increase overnight without any change in background. The gels may be transferred to a solution of 0.005% Coomassie Blue G250 in 3.5% perchloric acid for photography. This technique can also be used for isoelectric focussing gels, and will give good results with paper absorption prints taken from preparative flat-bed gels.

When the stained gel is to be used as a marker, it can be set alongside a duplicate that was run in parallel with it, from which the unstained proteins can be sliced out before appreciable diffusion as taken place. If this cannot be done immediately, the gels can be frozen at −20 °C, to be cut up later.

(d) It is possible to test for the presence of glycoproteins by the chromogenic reaction of their carbohydrate residues with Schiff reagent after oxidation with periodic acid. This test can be used as supporting evidence for the identity of those apo-lipoproteins that contain carbohydrate residues (Table 6.4), although it is not so satisfactory as the identification and estimation of these carbohydrates. Because the carbohydrate amounts to only a small fraction of the apo-lipoprotein, this is a relatively insensitive method of staining and the loading of the gel must be several times greater than would be sufficient if the proteins were to be stained with Coomassie Blue.

Procedure A. For polyacrylamide-urea gels (from Kapitany and Zebrowski, 1973). Schiff's reagent is prepared as in Appendix 4.

Immediately after the electrophoresis is completed, soak the gel in 12.5% trichloracetic acid at room temperature for 1 h to fix the proteins. Transfer the gel to a 0.2% solution of periodic acid in a clean tube and allow oxidation to proceed at 4 °C for 45–60 min.

Pour off the periodic acid from the gel, immediately refill the tube with cold Schiff reagent and store for 45 min at 4 °C.

At this stage, the whole gel will be stained pink and must be de-stained to develop the contrast between the background and the stained proteins. However, the proteins also become less intensely coloured and the gel must be carefully observed during the de-staining in order to record the pattern when the contrast is greatest. To de-stain, soak the gel in several changes of 10% acetic acid at room temperature. The life-time of the stained bands can be prolonged by storing the gel in 10% acetic acid at 4 °C but, if necessary, the proteins can be re-stained by passing the gel through periodic acid and Schiff a second time.

Note that light should be excluded from all these staining and washing procedures.

Procedure B. For gels that contain sodium dodecyl sulphate (Glossmann and Neville, 1971).

Procedure A may be unsatisfactory if the gels contain SDS, which can lead to the production of artifacts. In the following procedure therefore, the preliminary fixation is carried out in a medium that will extract the SDS. It is desirable that these washings are performed in a continuously stirred bath.

Immediately after the electrophoresis, wash the gel for 24 h in at least two changes of 7% acetic acid in 40% methanol (Glossmann and Neville recommend 2.5 litres for 12 gels). Transfer each washed gel to a clean tube that contains 1% periodic acid in 7% acetic acid and stand for 1 h in the refrigerator. After the oxidation, wash out the periodic acid by soaking for 24 h in two changes of 7% acetic acid, transfer the gels to separate tubes that contain Schiff reagent and stand them in the refrigerator for 1 h. Finally, wash each gel 3–4 times with 1% sodium metabisulphite in 0.1 N HCl, in which they can also be stored.

(e) Stained gels can also be used for the semi-quantitative analysis of an apo-lipoprotein mixture by densitometric scanning. This technique is discussed in detail in Section 8.9.1, together with methods for the quantitative analysis of apo-lipoproteins.

As a general appendix to the foregoing methods it may be said that de-staining by electrophoresis is not generally necessary and may be best avoided for apo-lipoproteins. If large volumes of de-staining solutions are used, it may be economical to re-use them several times instead of discarding them after a single use. The dyes that are leached out of the stained gels can be adsorbed out of the de-staining solution with charcoal. However, urea, SDS and buffer salts that are also extracted, will accumulate with each use, while the acetic acid, ethanol or methanol that was originally present will become progressively more dilute. It is these factors that will ultimately determine the number of times the de-staining solution can be reused. The simplest way to adsorb dye out of a de-staining solution is to pour it into a filter funnel that is fitted with a filter paper, into which several spoonfulls of activated charcoal powder have been put. This funnel and charcoal can be kept in use until it no longer yields a colourless filtrate.

6.4.2.2. Recording the electrophoretic patterns Since it is impracticable to store stained gels, it is necessary to make a permanent record of their appearance in some other form. This can be done in three ways:

(1) By making a pencil drawing of the pattern. This is difficult to make convincingly representational and to do accurately.

(2) By photography. The speed and reliability with which the details of the electrophoretic pattern can be recorded by modern automatic cameras, in an easily recognisable form, have made this the method of choice. Procedural details are given in Appendix 5.

(3) By optical scanning densitometry, the results of which are

plotted on a chart recorder. This relatively expensive technique can give an objective estimate of the position of a protein band which can lead to a better measurement of its relative mobility. In this respect, the method has some value as a complement to photographic recording, particularly when the photograph and the densitometric trace can be mounted in register. Scanning densitometry can also be used to make a semi-quantitative estimate of the relative proportions of the apo-lipoproteins in a mixture, or to determine whether an unidentified protein and an authentic apo-lipoprotein will co-electrophorese with the same mobility (Section 6.4.3).

6.4.2.3. The polyacrylamide gel-urea system This is a simple and rapid way of separating the apo-lipoproteins on a micro-scale through their differing electrophoretic mobilities in polyacrylamide gel. Unfortunately, mobility is a complex function of the electrical and morphological properties of a protein which it is not easy to measure reproducibly. In itself therefore, the method is not well suited to the unequivocal identification of an individual apo-lipoprotein. Despite this, it can play a valuable role in the preliminary analysis of the different apo-lipoprotein mixtures that are found, for example, in the fractions collected from chromatography columns. Moreover, when used in conjunction with other electrophoretic and immunological techniques, it forms an important part of the system of qualitative analysis by which apo-lipoproteins may be fully identified (Section 6.4.3). In this context, note that, although the electrophoresis is usually performed at an alkaline pH, valuable information on the identity of some apo-lipoproteins can be obtained from experiments at an acid pH. Directions for both systems are given below.

Although it is possible to carry out the electrophoresis in thin slabs of gel, it is far more common to prepare separate gels in glass tubes as described by Davis (1964). The necessary equipment can be obtained from many manufacturers and can also be used for electrophoresis in gels that contain sodium dodecyl sulphate (some are also designed for use with the isoelectric focussing technique (Section 6.4.2.6). It is highly desirable that the apparatus incorporates effective provision for cooling the gels during the run, since higher current densities can then be used without damaging the gels. Although the principles on which the method operates are as described by Ornstein (1964), there are many minor variations in the practice of the technique to be found in the literature. It is not often clear what advantages these variants have to offer. The procedure we shall describe for electrophoresis in alkaline gels is based on that of Kane (1973), while that for acidic gels is adapted from Reisfeld et al. (1962). We shall assume the use of gel tubes 145 mm long by 6 mm bore, but smaller tubes can be used without significant change to the technique. The construction of the gels is shown diagrammatically in Fig. 6.1.

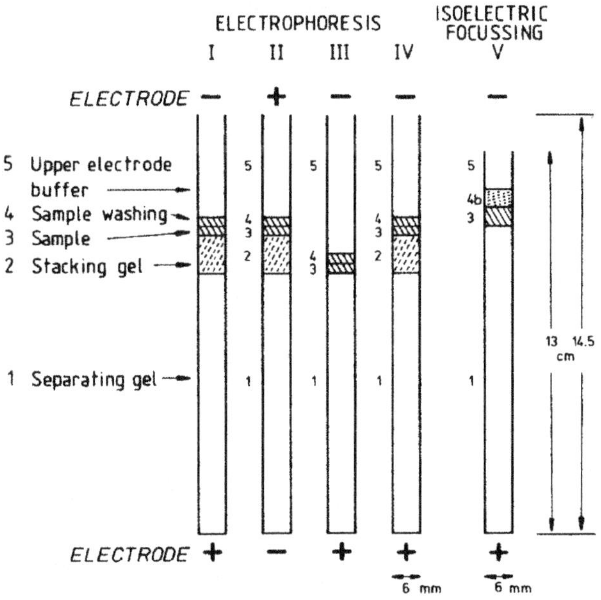

Fig. 6.1. The construction of the composite gels used for the different techniques of electrophoresis in polyacrylamide gel. (I) alkaline-urea gel (Section 6.4.2.3). (II) acidic-urea gel (Section 6.4.2.3). (III) sodium dodecyl sulphate gel: Weber and Osborne method (Section 6.4.2.4). (IV) sodium dodecyl sulphate gel: Neville's method (Section 6.4.2.4). (V) analytical isoelectric focussing (Section 6.4.2.6). The order in which the component gels are added to the tubes is shown by the adjoining numerical key: 4b is the layer of ampholyte solution placed over the sample in isoelectric focussing.

TABLE 6.6
Electrophoresis in alkaline gels (Kane, 1973)
Composition of reagents for alkaline polyacrylamide gel

	Solution						
	A	B	C	D	E	F	G
Tris base		24.23 g			2.907 g		
Acrylamide	40 g			13.3 g			
Bisacrylamide	1.1 g			1.1 g			
TEMED		0.30 ml			0.13 ml		
Riboflavin						2.7 mg	
Ammonium persulphate			143 mg				103 mg
Adjust to pH with ( )		9.1 (HCl)			6.7 (H_3PO_4)		
	Make all these solutions up to 100 ml with distilled water						

Solutions C, F and G must be freshly made. All the other reagents may be kept for up to 7 days in the refrigerator.
Electrode buffers:
Cathode solution (upper compartment): 2.574 g of Tris base
1.745 g of glycine
Water to 500 ml. Adjust to pH 8.91
Anode solution (lower compartment): 7.268 g of Tris base
Water to 500 ml
Adjust to pH 8.07 with HCl
Note that when proteins are te be eluted from the gel for aminoacid analysis, it is important to avoid contamination with glycine. It is then better to use a cathode buffer made as follows: dissolve 2.574 g of Tris base in 500 ml of water and adjust to pH 8.91 with boric acid.

Procedure. The sequence of operations is as described in Section 5.5.1.

Separating gel. Use analytical grade urea to prepare the following two solutions immediately before the gels are to be made (Table 6.6).

(1) 10 ml of soln. A + 10 ml of soln. B + 12.8 g of urea
(2) 20 ml of soln. C + 12.8 g of urea

When the urea is completely dissolved, filter each solution through a fine paper, or a coarse membrane filter (approx. 1 μm pore size; a Millipore Millex filter may be convenient for small volumes). Mix the two solutions in a flask, degas by suction, and pipette 2.5 ml into each prepared 6 mm by 145 mm gel tube.

Stacking gel. While the separating gel is polymerising, prepare the following solutions.

(3) 1 ml of soln. D + 1 ml of soln. E + 1.28 g of urea
(4) 1 ml of soln. F + 1 ml of soln. G + 1.28 g of urea

When the separating gels are ready (Section 5.5.1), filter, degas and mix solutions 3 and 4. Pipette 0.3 ml of the mixture onto each gel, overlayer with water and polymerise under UV light (Section 5.5.1). Be sure to allow ample time for the reaction to go to completion (e.g. 30 min): a detectable scattering of light in the solution does not mark the end of the polymerisation.

Preparation of the sample. The sample should be in a solution of about pH 8 and of low ionic strength. The amount of protein required will depend on the nature and complexity of the mixture to be analysed. About 5 μg of an isolated apo-lipoprotein should be sufficient, but 40–50 μg of apo-HDL will be needed, or up to 100 μg of apo-LDL. The sample may be dissolved in from 10 to 100 μl of solvent but the optimum volume is about 50 μl.

For the electrophoretic characterisation of the soluble apo-lipoproteins on the micro-scale, it is often convenient to use tetramethylurea (TMU) to remove the lipid and the apo-lipoprotein B (Kane, 1973). As in the method of analysis described in Section 6.4.1.2, the TMU must be re-distilled at regular intervals (e.g. once month) and stored in the refrigerator, in a dark bottle. It is also important that the lipoprotein solution should be adjusted to have a pH between 6 and 9, and an ionic strength between 0.05 and 0.15. Mix from 50 to 200 μl of the lipoprotein solution with an equal volume of TMU in a small tube (4 mm by 50 mm) and stand at room temperature for 30 min. If a large proportion of lipid is present, the precipitate may cause the protein bands to be distorted and it is expedient to centrifuge the mixture. Recover the clear aqueous TMU solution and transfer an appropriate amount to the top of the acrylamide gel. With care, the electrolyte buffer can be layered directly onto the sample solution but, if need be, sucrose or urea can be added to increase its density. Kane (1973) recommends the use of one tenth volume of 80% sucrose.

As an alternative to this procedure, the precipitation with TMU can be carried out in the gel tube itself. This is convenient if the sample of lipoprotein is small but carries some risk that the proteins may not migrate evenly. Care must be taken to ensure that the sample and the reagent are properly mixed.

Adequate results can also be obtained by adding urea at the rate of 30–35 mg/50 μl to a solution of the lipoprotein in an alkaline buffer, the mixture then being transferred to the top of the polyacrylamide gel. On passing the current, all the apo-lipoproteins with the exception of B will migrate into the gel. The simplicity of this procedure is attractive when many samples must be analysed, but there seems to be little real evidence that the distribution obtained is quantitatively correct.

Apo-lipoproteins that have been isolated by conventional 'delipidation' (Appendix 3), or have been partially purified, can be dissolved in a buffered urea solution and applied to the gel in the conventional way (Section 5.5.1; Davis, 1964). Cathode buffer to which urea has been added (750 mg/ml) is a suitable solvent which will readily dissolve apo-HDL or isolated apo-lipoproteins of low molecular weight. However, apo-lipoprotein B is virtually insoluble in this solvent and samples of apo-LDL or apo-VLDL must be subjected to a prolonged extraction. Under no circumstances should these apo-lipoprotein mixtures be allowed to dry after they have been wetted with water, for they will then form a horny mass that is almost impossible to deal with. Concentrated solutions of LDL or VLDL may be either freeze-dried and then extracted with solvents, or they may be extracted directly from the aqueous solution after dialysis. Both procedures are described in Appendix 3. When the lipids have been completely removed, the finely divided protein residue can be suspended in buffered urea and allowed to incubate for several hours at room temperature before the supernatant is transferred to the gel.

Loading the gel (cf. Section 5.5.1). Insert the gel tubes into the grommets of the upper electrolyte reservoir before loading them with the protein solutions. If this reservoir is of the usual cylindrical form, the assembly can be conveniently supported on the base of an upturned beaker during the loading procedure. Add the appropriate volume of sample to each gel and carefully overlay it with the cathode buffer to fill the tube. When all the tubes are filled, mount the assembly over the reservoir of anode buffer and check that there are no air bubbles in the ends of the tubes. Finally, fill the upper reservoir with cathode buffer into which 0.2 ml of 0.1% bromophenol blue is well mixed.

Electrophoresis. Connect the upper electrode to the negative pole of the power supply and pass a constant current of 2–4 mA/gel until the band of bromophenol blue tracking dye has entered the separating gel. Then increase the current to 4–8 mA/gel until the dye has reached the pre-determined mark at the end of the tube (Section 5.5.1). Note however, that the maximum acceptable current will be limited by the efficiency with which the tubes can be cooled during the run. Unless the cooling bath can be kept below 20 °C, the gels are likely to be damaged through over-heating, In any event, it is a prudent rule not to exceed a current density of 30 mA/sq. cm., i.e., 4 mA for a gel 4 mm in diameter, or 8 mA for a gel 6 mm in diameter. Since the gels do not all run at the same rate, it will be necessary to stop each one as the tracking dye reaches the end marker. To do this, switch off the power supply, lift the upper reservoir and push the relevant tube up through its grommet until the top is above the level of the electrolyte. Then replace the reservoir and continue the run, remembering to reduce the total current flow to maintain the current per tube at its original value. Repeat this operation for each tube as it becomes necessary.

When the run is completed, switch off the power, extract the gels from their tubes (Section 5.5.1) and stain them by the method (a) of Section 6.4.2.1.

Electrophoresis in acidic gels (adapted from Reisfeld et al., 1962).

The procedures for assembling the gels is essentially the same as that for the alkaline gels described above.

Separating gel. Prepare the following mixtures with analytical grade urea, immediately before they are required (Table 6.7).

(1) 5 ml of soln. H + 10 ml of soln. I + 9.25 g of urea
(2) 17 ml of soln. J + 10.5 g of urea

When all the urea has dissolved, mix the two solutions and degas by suction. Without delay, pipette 2.5 ml of this mixture into each prepared 6 mm × 145 mm glass gel tube and overlayer with water.

Stacking gel. Prepare the following two solutions.

(3) 1 ml of soln. K + 2 ml of soln. L + 1.85 g of urea
(4) 4 ml of soln. M + 1 ml of water + 3.1 g of urea

TABLE 6.7
Composition of reagents for acidic polyacrylamide gel

	Solution					
	H	I	J	K	L	M
Potassium hydroxide		337 mg		337 mg		
Acetic acid		2.14 ml		0.36 ml		
Acrylamide	30 g				18 g	
Bisacrylamide	0.8 g				0.47 g	
TEMED		0.46 ml		0.46 ml		
Ammonium persulphate			164 mg			140 mg
Adjust to pH		4.3		6.8		
	Make all these solutions up to 100 ml with distilled water					

Solutions J and M must be freshly made.
Electrode buffer: the following solution is used for both the anode and the cathode compartments: 31.2 g of β-alanine, 8.0 ml of glacial acetic acid, Water to 1000 ml. Adjust to pH 4.5.

When the urea has dissolved and the separating gels have polymerised, mix these two solutions and de-gas by sunction. While this operation is proceeding, remove the fluid from the top of the separating gels (Section 5.5.1). As soon as the de-aeration is complete, pipette 0.3 ml of the reagent mixture onto each gel, overlayer with water and allow to polymerise for 35–45 min.

Preparation and loading of sample. Proceed as described for the alkaline gels, but add 5 µl of 0.5% Basic Fuchsin as tracking dye.

Electrophoresis. At this low pH, the proteins will migrate towards the cathode and care must be taken that it is the lower electrode that is connected to the negative pole of the power supply. The conditions that determine the permissible current density in acidic gels are the same as for alkaline gels and the same values may be used. When the run is completed, extract the gels and stain them by the method (a) of Section 6.4.2.1.

6.4.2.4. The polyacrylamide gel-sodium dodecyl sulphate (SDS) system. Proteins that are fully complexed with SDS have an electrophoretic mobility in polyacrylamide gel that is inversely related to the logarithm of their molecular weight and is largely independent of their isoelectric point. For this relationship to hold, it is important that the formation of the complex is allowed to go to completion and the sample should be incubated with SDS, either overnight at room temperature, or for 2 h at 37 °C, or for 2–3 min at 100 °C before it is put into the electrophoresis tube. Mercaptoethanol may also be included in the incubation mixture if it is thought desirable to reduce such disulphide bonds as may be present. At least one standard protein mixture must be run with each batch of apo-lipoprotein analyses, to calibrate the gel. Suitable standards are now so readily available from laboratory suppliers that it is rarely worth-while to prepare them in the laboratory.

The conventional apparatus is generally used for this method of analysis, the most widely adopted technique being that of Weber and Osborne (1969). No stacking gel is used in this procedure, which will satisfactorily resolve proteins of 15 000–100 000 dalton in a separating gel of 10% acrylamide. This operational range can be increased to 50 000–250 000 dalton by reducing the gel concentration to 3.3% The more recently developed method of Neville (1971) uses both stacking and separating gels and offers a rather better resolution than the single gel system. The concentration of the separating gel can be varied from 5 to 25% acrylamide, as appropriate, but a value of 15–20% is generally satisfactory for most apo-lipoprotein mixtures. Despite this relatively high gel concentration, the mobility of the proteins is advantageously rapid.

TABLE 6.8
Composition of reagents for polyacrylamide gels that contain sodium dodecyl sulphate

	Solution						
	P	Q	R	S	T	U	V
Tris base						20.68 g	2.62 g
1.0 N HCl						12.3 ml	
1.0 M H_2SO_4							10.7 ml
$NaH_2PO_4 \cdot H_2O$		7.8 g					
$Na_2HPO_4 \cdot 7H_2O$		38.6 g					
Sodium dodecyl sulphate		2.0 g					
Acrylamide	22.2 g			60 g	12.0 g		
Bisacrylamide	0.6 g			0.4 g	0.8 g		
Ammonium persulphate			150 mg				
Water to (ml)	100	1000	100	100	100	100	100

Solutions P, S and T should be filtered before use, and may be stored in the dark, at 4 °C for up to 7 days. Solution R must be freshly made. The final pH of solutions U and V should be 9.18 and 6.14 respectively.

Method of Weber and Osborne (1969)

Electrode buffer. Dilute solution Q with an equal volume of water and use this in both of the anode and the cathode compartments.

Separating gel. To make 10 gels of 10% acrylamide concentration, prepare the following mixtures,

(1) 13.5 ml of soln. P + 15 ml of soln. Q
(2) 1.5 ml of soln. R + 45 μl of TEMED

De-aerate both reagents by suction, then mix thoroughly together and pipette 3 ml into each prepared 6 mm by 145 mm gel tube. Overlay each tube with a layer of water and allow to stand undisturbed until polymerisation is complete. This will take at least 30 min.

To make 10 gels of 3.3% concentration for the analysis of proteins of molecular weight in the range 50 000–250 000, prepare the following solutions

(1) 4.5 ml of soln. P + 15 ml of soln. Q
(2) 1.5 ml of soln. R + 45 μl of TEMED + 9 ml of water

Then proceed as for the 10% gels.

Preparation of sample. Make up the solvent buffer by dissolving 0.8 g of SDS in 100 ml of solution Q.

The amount of protein that is required for each gel is approximately the same as for the urea-gel system. For SDS-gels however, the TMU procedure is inappropriate and the proteins should be extracted with conventional solvent mixtures (Appendix 3). Dissolve the lipid-free protein in the solvent buffer to give a sample volume of 50–100 μl, observing the precautions mentioned in Section 6.4.2.3. When the protein is dispersed, heat the solution for 3 min in a water-bath at 90 °C. If it is intended to reduce the disulphide bonds, add 5 μl of mercaptoethanol before this incubation. It is then wise to perform the operation in a screw-capped vial that can be flushed with nitrogen. When the denaturation is complete, add 50 μl of glycerol, 3 μl of 0.05% Bromophenol Blue, and load the mixture onto the gel as in Section 6.4.2.3.

If the proteins are already dissolved in a buffer that does not contain SDS, modify the procedure as follows. Make up a 4% solution of SDS by dissolving 3.8 g of SDS in 100 ml of solution Q.

Add one volume of this to three volumes of the protein solution and then proceed as above.

If a quantitative estimate of molecular weight is to be made from the experiment (Section 6.4.2.5), samples of the standard protein mixture should be treated in the same way as the apo-lipoproteins.

Electrophoresis. Connect the lower electrode to the positive pole of the power supply and pass a current of 8 mA/gel of 6 mm diameter until the marker dye has reached the predetermined point (Sections 5.5.1 and 6.4.2.3). Note that it may be necessary to run at a lower current density if the equipment is not efficiently cooled.

When the run is completed, switch off the power and remove the gels from their tubes. The 3.3% gels must be handled with care, since they are fragile and can easily be distorted. Stain the proteins by method (a) of Section 6.4.2.1. Because the tracking dye disappears during the staining process and yet the position of the front must be well defined if measurements of mobility are to be accurate, it is recommended that its position is permanently marked before staining is begun. This can easily be done by sticking a needle dipped in India ink through the gel at the dye front (Edelstein and Scanu, 1980; Shore et al., 1980).

Method of Neville (1971)

Electrode buffers. Cathode solution (upper compartment): 2.47 g of boric acid, 4.92 g of Tris base, 1.0 g of sodium dodecyl sulphate, Water to 1 litre. Adjust to pH 8.64. Anode solution (lower compartment): 51.7 g of Tris base, 15.4 ml of 2 M HCl, Water to 1 litre. Adjust to pH 9.5.

Separating gel. The following mixture is sufficient for twelve 15% gels. Stronger gels can be made by appropriately adjusting the proportions of solutions S and U.

(1) 10 ml of soln. S + 10 ml of soln. U

(2) 10 ml of soln. V + 60 μl of TEMED + 10 ml of water

De-aerate both these reagents, then mix them together and immediately pipette 3 ml into each of the prepared glass tubes (Section 5.5.1) and overlayer with 0.1–0.15 ml of isobutanol (cf. Neville, 1971).

These strong gels polymerise very rapidly, and there must be no delay in filling the tubes once the reagents are mixed. After the polymerisation is complete, wash the top of the gel with water before casting the stacking gel.

Stacking gel. This is made from the following two reagents:

(1) 2.5 ml of soln. T + 2.5 ml of soln. V

(2) 1.3 ml of soln. R + 10 μl of TEMED + 3.7 ml of water

De-aerate both reagents by suction, then mix and pipette 0.3 ml onto each separating gel. Overlay with isobutanol, which must be washed away with water after the gel has formed.

Preparation of sample. Refer to the method of Weber and Osborne above, but use a solvent made by dissolving 0.9 g of SDS in 100 ml of the Tris–boric acid cathode buffer just described.

Electrophoresis. As for the method of Weber and Osborne, but pass a current of 1.5 mA/gel until the tracking dye has entered the separating gel and then increase it to 2.5 mA/gel. Some difficulty may be experienced when it comes to removing these rigid gels from the glass tubes. If they have been made in laboratory tubing, it may be as well to smash the glass with a hammer, or with one of the commercial devices made for the purpose, and to wash away the fragments before transferring the gel to its test-tube of stain (method a of Section 6.4.2.1).

6.4.2.5. Estimation of mobility and molecular weight by SDS-gel electrophoresis. Electrophoresis in the presence of sodium dodecyl sulphate is useful, not only as a means of evaluating the heterogeneity of an apo-lipoprotein preparation, but as a way of characterising them by an estimation of molecular weight. The technique is most likely to be used for the study of the 'soluble' apo-lipoproteins that range in molecular weight from about 6000 to 50 000 dalton. The mixture of standard proteins that is used to calibrate the gel must therefore cover approximately the same range. Since apo-lipoprotein B is much larger, with components up to 250 000 dalton, standards of an appropriately higher molecular weight must be used if this apo-lipoprotein is present. It should be noted however, that the apparent molecular weight of apo-lipoprotein B is dependent on the concentration of the acrylamide gel and is therefore not an estimate of the true molecular weight. There is also some evidence that the mobility of apo-lipoprotein AI in this system increases as the load that is applied to the gel is increased.

Two types of standard protein mixture can be used, both of which are available commercially, for example from Bio-Rad Laboratories and BDH Ltd. The first consists of a mixture of purified, well characterised proteins such as serum albumin, ovalbumin, myoglobin and cytochrome C. Other suitable proteins are listed by Weber and Osborne (1969). This type of standard may be the simplest to prepare in the laboratory, although if it is necessary to buy the individual proteins, why not buy them ready mixed? The second type of standard is a mixture of oligomers of a single protein, for example BDH Ltd. prepare a standard containing a protein of M_r 14300 together with dimers, trimers and tetramers of the same protein. This is supplied as a lyophilised powder which can be dissolved in the sample, or in the sample solvent. The re-dissolved protein can be stored in the refrigerator but must be warmed before use, to dissolve the SDS, which precipitates at 4 °C. The same supplier also offers a mixture of oligomers of a protein of monomeric molecular weight 56000 that is useful with 3.3% gels, and a mixture of CNBr fragments of horse heart myoglobin (M_r 2512–16949) that can be used to calibrate 22.5% gels.

If an acceptable estimate of molecular weight is to be made from an electrophoresis in SDS-gel, it is important, because of the variation between batches, that (1) each gel should be run at least in duplicate; (2) the apo-lipoprotein and the standard mixture should be run separately and also when mixed together; (3) these runs should be concurrent and in gels that are prepared from the same reagents, at the same time; (4) the stained protein bands are sharp and undistorted.

After the electrophoresis, the mobility of the proteins must be estimated by measuring the distance both they and the tracking dye have moved from the top of the separating gel. This can be done with a centimetre scale, working either with the gel itself or, perhaps with greater objectivity, from a densitometric scan. Unfortunately, during the staining of the gel, the tracking dye diffuses away and the gels also change in length. Thus, there may be an uncertainty in the measurements which can be reduced in two ways:

(a) The simplest procedure is to cut each gel transversely through the middle of the band of tracking dye, immediately after extruding the gel from its electrophoresis tube. The cut lengths of gel can then be stained. After de-staining, the distance from the top of the separating gel to the cut surface corresponds to the distance the electrophoretic front has migrated. At the same time as this length is measured, also determine the distance from the top of the separating gel to the middle of each band of protein, to the nearest 0.5 mm. The relative mobility (R) of each protein is then given by

$$R = m/f$$

where m is the distance migrated by the protein and f is the migration of the front.

(b) The following procedure is derived from that originally used by Weber and Osborne (1969). At the end of the electrophoresis, measure the total length of the separating gel (u) and the distance (b) from the top of the separating gel to the middle of the band of tracking dye. After the gel has been stained, measure the total length of the separating gel (s) and the distance (m) that each protein has migrated. The migration of the electrophoretic front can then be estimated as

$$f = b \cdot s/u$$

and the relative mobility of the protein is given by

$$R = m/f$$

Plot the mean mobility of each protein of the standard mixture against the logarithm of its molecular weight, on semi-logarithmic paper. Linear interpolation on this curve allows the molecular weights of the apo-lipoproteins to be estimated, with care, with an accuracy of within about 10% for proteins of molecular weight about $1\text{–}70 \times 10^3$.

6.4.2.6. Isoelectric focussing (IEF) in polyacrylamide gel. The ability of the IEF technique to resolve proteins that differ in isoelectric point by as little as 0.01 pH units makes it a powerful tool for the analysis of apo-lipoproteins, particularly those like A–I and E which are electrically diversiform. Moreover, the isoelectric point that can be determined by this method is a valuable physical characteristic of the apo-lipoprotein, especially when taken in conjunction with the molecular weight that can be estimated by SDS-gel electrophoresis.

The detailed principles and practice of the technique are described in many authoritative laboratory manuals (e.g. Smith, 1976; Righetti, 1983) and we shall not pursue them here. Like conventional electrophoresis, IEF can be performed either in cylinders or in slabs of gel but, for the analysis of apo-lipoproteins, the former appears to be the more popular of the two. However, as Righetti (1983) has pointed out, apparatus that is designed for conventional electrophoresis may not always be ideally suited to IEF, notably in the configuration of the electrode compartments and in the efficiency of the cooling system.

For the IEF of apo-lipoproteins, the production of the pH gradient is achieved by the use of carrier ampholytes in the usual way, but it is necessary for the gel to contain 6–8 M urea as well. The sample, which must also contain urea, is usually put on top of the separating gel and is then overlayered with a solution of ampholyte to separate it from the strongly acid cathode solution. As with conventional gel electrophoresis, there are many variants of the same general technique. The one that follows is a modification of the method of Pagnan et al. (1977) in which the analysis is performed in a 7.5% acrylamide gel that contains 6 M urea.

Electrode buffers. Cathode solution (upper compartment): dissolve 7.5 g of glycine in 1 litre of water and adjust to pH 7.0. Anode solution (lower compartment): dissolve 2.38 g of *N*-2-hydroxyethylpiperazine-*N'*-2-ethane sulphonic acid (HEPES) in 1 litre of water and adjust to pH 3.5.

The pH of these solutions should be adjusted with HCl or NaOH as appropriate, to be just beyond the extreme p*I* values of the

TABLE 6.9
Composition of reagents for IEF in polyacrylamide gels

	Solution			
	W	X	Y	Z
Urea	22.9 g			36 g
Ethylmorpholine				0.25 ml
Sucrose				20 g
Ampholytes (40% soln.)		3.0 ml		
Acrylamide	4.5 g			
Bisacrylamide	0.12 g			
TEMED (5% in water)		0.06 ml		
Ammonium persulphate			375 mg	
Water to (ml)	60		10	100

Ampholytes are obtainable from a number of sources e.g. LKB Produkter, Bio-Rad Laboratories, Pharmacia. Shore et al. (1980) recommend that, if the nature of the apo-lipoprotein sample is unknown, a preliminary analysis should be made in ampholytes of wide pH range. If necessary, a suitable ampholyte can be made by mixing equal quantities of two of the narrow range examples.
Solution Y must be freshly prepared. Solutions W, X and Z should be filtered and preferably used at once but, if need be, they may be stored for up to 7 days at 4 °C.

ampholyte. This will ensure that the maximum length of gel is available for isoelectric focussing to take place.

Separating gel. Mix together the 60 ml of soln. W and the 3.06 ml of soln. X, and de-aerate by suction. Then add 0.5 ml of soln. Y, briefly de-gas once more and immediately pipette 3 ml of the mixture into each ready-prepared precision-bore glass tube (6 mm × 130 mm). Overlay with water to a depth of about 3 mm and stand undisturbed until the gel has formed (ca. 45 min).

Preparation of sample (cf. Section 6.4.2.3). From 100–150 μg of a crude 'delipidised' apo-lipoprotein mixture (Appendix 3) will be needed, but substantially less of an isolated apo-lipoprotein can be used. Dissolve the appropriate amount of protein in solution Z, to give a sample volume of 50–75 µl. Transfer this amount to the top of the separating gel and carefully overlay with a solution made by diluting 0.5 ml of 40% (w/v) ampholyte with 10 ml of 5% sucrose. This in turn must be overlayered with the cathode buffer. Also prepare control gels, for which the sample contains no protein, on which to determine the shape of the pH gradient. Finally, assemble the apparatus and pass cooling water to maintain the temperature at 10 °C.

Electrophoresis. When the power is switched on, a comparatively high current will pass until the carrier ampholytes have focussed to form the pH gradient. As this equilibrium is established, the electrical conductivity of the gel decreases and overheating may occur if too high a potential is applied. To avoid this misfortune, limit the power dissipation per gel to about 1.5 W. The most convenient way of doing this is to use a power supply that will operate in a constant power mode. However, if equipment of this kind is not available, a constant voltage power supply can also be used. Connect the upper electrode of the electrophoresis cell to the negative pole of the supply and pass a current of 1 mA/gel until the potential reaches 400 V, at which value it should be maintained for 5.5 h. When the run is completed, switch off the power, extract the gels from their tubes and stain those in which proteins have been focussed by method (b) of Section 6.4.2.1.

Because the focussing of apo-lipoproteins is carried out in concentrated urea solution, in which the p*I* of both the proteins and the ampholytes differs from their values in dilute salt solution, it is important to determine the configuration of the pH gradient in the gel whenever the p*I* of the apo-lipoproteins is to be measured. This can be done in several ways: (1) Proteins of known p*I* can be run in parallel with the apo-lipoproteins. (2) Cut serial sections 2.5 mm thick along the length of the control gels. Macerate each one with 0.2 ml of freshly boiled water and measure the pH of the extract. For the highest accuracy, the measurement should be made at the temperature at which the focussing was performed. Note that this procedure assumes the use of a micro-electrode for the pH determination. If this type is not available, the thickness of the gel section and the volume of the eluate may be slightly increased, but this is undesirable since it leads to a loss of resolution. (3) A combined micro-electrode which is specially designed to be applied to the surface of the gel (e.g. Bio-Rad Laboratories) can be used to measure the pH directly.

From the data obtained by one of these methods, plot a curve to relate pH with distance along the gel, expressed as a fraction of the total length of the gel. This will eliminate the differences that will arise from the varying shrinkage of the gels during the staining process. The position of an apo-lipoprotein in the gel can then be measured with a scale and the p*I* interpolated from the calibration curve.

6.4.3. The identification of apo-lipoproteins by electrophoretic analysis

The strategies for the qualitative analysis of the apo-lipoproteins were outlined in Section 6.4, where it was pointed out that, in many laboratories, this depends heavily on the comparison of their relative positions after electrophoresis in several different systems. The bands of apo-lipoprotein appear on the gels in a more or less reproducible order which is shown in the generalised sketches of Fig. 6.2, 6.5 and 6.6. However, whereas the relative retention times of peaks on a gas chromatogram are often sufficiently reproducible to be used as an identifying characteristic of a substance, the mobilities of the apo-lipoproteins during gel electrophoresis are too variable to be used in this way. An example of this variation in relative mobility is shown in the middle photographs of Fig. 6.3, which are the result of running un-reduced apo-HDL on gels prepared at different times. Differences of this magnitude can easily be confusing if there are only one or two proteins in the sample, or if extra bands are present, either in the form of unusual 'apo-lipoproteins' (Table 6.5), or because of contamination with plasma proteins. There can also be marked differences in the relative proportions of the apo-lipoproteins from different individuals. An example of this can be seen by comparing the C-I, C-II

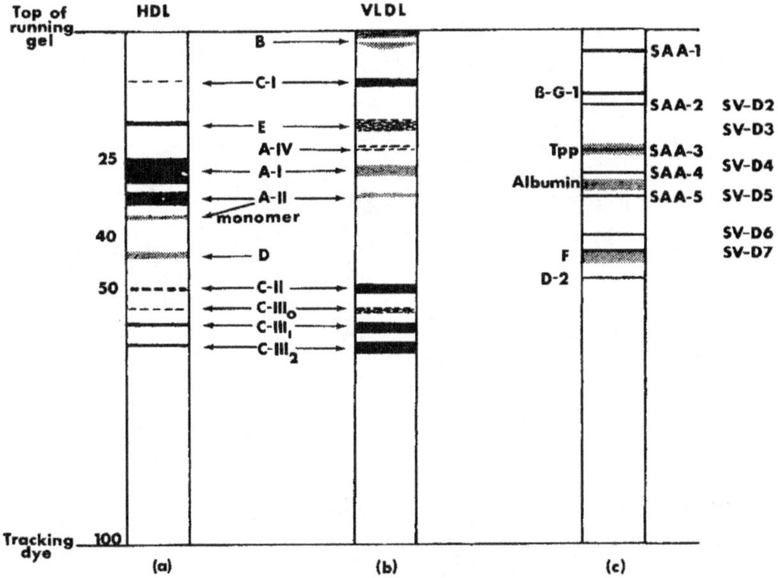

Fig. 6.2. A schematic representation of the relative mobilities of apo-lipoproteins during electrophoresis in alkaline-urea gel. The major apo-lipoproteins are shown in sketches (a) and (b) which illustrate the approximate quantitative distribution to be expected in apo-HDL and apo-VLDL respectively. The relative positions of minor proteins and those which have been reported in unusual pathological conditions are shown in (c). Abbreviations: β-G-1, β_2-glycoprotein-1. Tpp, threonine-poor protein. SAA, serum amyloid A protein. SV-D, Sephadex-fraction V-D.

and C-III bands in the middle two gels of Fig. 6.3. An unusual protein distribution is also shown in the first gel of the series, which is of apo-VLDL. Although from a hypertriglyceridaemic subject, there is little of the C-I and E apo-lipoproteins, but a protein with the mobility of A-I is present.

The determination of apparent p*I* by electrofocussing is also open to uncertainties, as can be seen from a study of the published values for the 'polymorphs' of apo-lipoprotein E. There are at least four of these, with isoelectric points that fall within a range that has been variously stated to be 5.3–5.55 (Utermann, 1975), 5.26–5.75 (Beisiegel, 1979), 5.97–6.03 (Pagnan et al., 1977) or 5.7–6.2 (Marcel et al.,

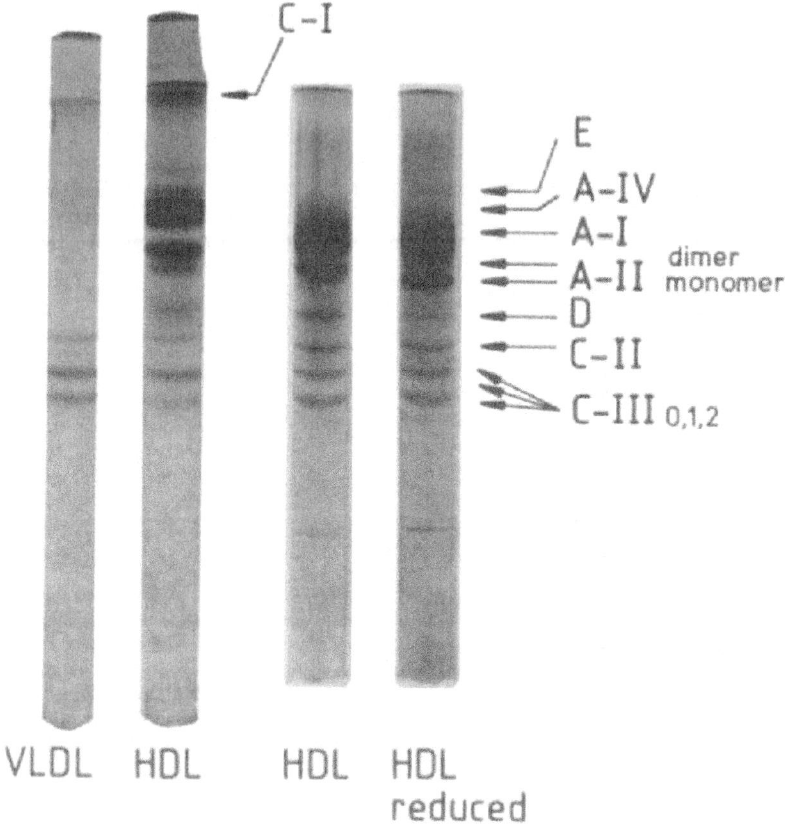

Fig. 6.3. apo-VLDL and apo-HDL analysed by electrophoresis in alkaline-urea gel (Section 6.4.2.3) and stained with Coomassie Blue R250. The lipoprotein fractions used were: VLDL 1.019 g/ml and HDL 1.070–1.210 g/ml. A load of 100 μg of protein was applied to each gel. The two gels to the right show the same preparation with and without the addition of β-mercaptoethanol. The increased mobility of A-II in the presence of the reducing agent can be seen in the rightmost gel.

1974). Clearly, by this criterion alone, at least some of the proteins studied by Utermann and Beisiegel would not be regarded as apo-lipoprotein E by the other investigators.

An unequivocal identification of the apo-lipoproteins by electrophoresis therefore requires the following precautions to be taken: (1) The sample must be analysed at least in duplicate, in at least two different systems. (2) At least two control samples of fresh apo-HDL or apo-VLDL must be run in parallel with the unknown. (3) All the gels must be prepared from the same reagents, at the same time.

For a more positive identification still, it may be advantageous to co-electrophorese the sample with an authentic apo-lipoprotein (if this can be obtained), or to fix the positions of those apo-lipoproteins that contain carbohydrate by staining with Schiff reagent (Appendix 4). The latter can provide a useful confirmation of the identification made by electrophoretic mobility, but has the disadvantage that the method is comparatively insensitive. For the most certain confirmation however, it may still be necessary to resort to immunochemical analysis (Section 6.5).

If the apo-lipoprotein of interest is one that has been substantially purified, it may be possible to observe its behaviour in the different electrophoretic systems without difficulty. To follow one protein in the mixture that constitutes apo-HDL, for example, may be more of a problem and may require the protein to be eluted from one gel and then run in a second. This can be done by cutting out the relevant slice of the first gel, chopping it up and steeping it in about 0.2 ml of the solvent to be used for the next electrophoresis. In view of the undesirable reaction that may occur between the protein and urea during the process, it should be performed at 4 °C for the shortest practicable time, and for no longer than overnight. For the same reason, it is prudent to run urea-gels before SDS-gels, if possible. After the maceration, transfer both the original slice of gel and its extract onto the gel to be used for the next stage of the analysis.

Proteins can also be recovered from slices of gel by electrophoresis in the apparatus described by Allington et al. (1978), which is sold by the Instrumentation Specialities Co. Alternatively, the pieces of gel can be supported on a porous plug in a glass or plastic tube that has a dialysis membrane fixed over the bottom end (Fig. 6.4). The whole is filled with electrolyte and inserted into the disc electrophoresis

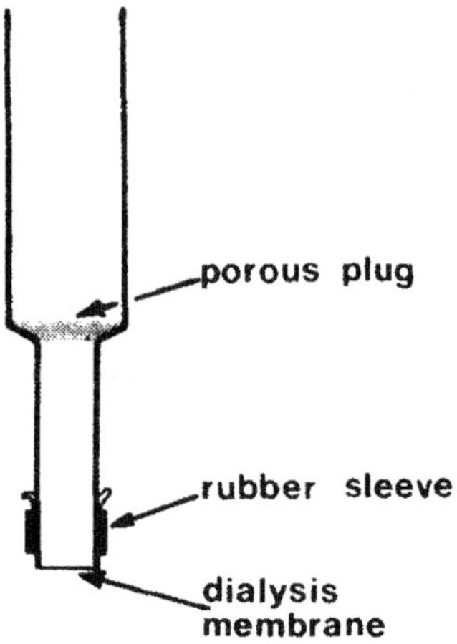

Fig. 6.4. Sketch of an apparatus for recovering apo-lipoproteins from plugs of polyacrylamide gel by electrophoresis.

apparatus. In this way, the apo-lipoproteins can be recovered quite expeditiously, but low yields have been reported by some laboratories. There may be two reasons for this varied experience. Firstly, it is desirable that the electrolyte that surrounds the pieces of gel should be of slightly lower concentration (conductivity) than that within the gel. Secondly, the apo-lipoproteins may be adsorbed by the gel to an extent that varies with the source of the monomer. In this context, the product marketed by Bio-Rad Laboratories has been found to be one of the most satisfactory.

Procedure

A-1. Add an equal volume of tetramethylurea (TMU) to a solution of the lipoprotein, as for electrophoresis in alkaline-urea gel (Section 6.4.2.3). After incubating for 30 min, remove the precipitate by centrifugation. This may contain apo-lipoprotein B and should be examined as in B-1 below.

Complete the electrophoresis of the TMU-soluble fraction in replicate, with at least two parallel samples of authentic apo-HDL or apo-VLDL. These reference samples should give rise to a pattern of bands like those illustrated schematically in Fig. 6.2 (a) or (b). Note however that the relative positions of the bands may differ slightly from those because of local variations in technique. This will be the more likely if the procedure used is not that described in Section 6.4.2.3. In addition, the pattern actually observed may include traces of plasma proteins, or the minor 'apo-lipoproteins' that are shown in Fig. 6.2 (c). These are unlikely to be present in significant amounts in reference samples of apo-HDL or apo-VLDL that have been carefully prepared from normal plasma, but may become concentrated in sub-fractions, or be present in preparations from pathological sera.

From the electrophoretic patterns obtained, make a preliminary identification of the proteins present in the unknown sample, bearing in mind the following:

(1) The impurity that is most likely to be present is serum albumin, which runs slightly faster than apo-lipoprotein A-I.

(2) There is sometimes a partial resolution of the 'polymorphs' of apo-lipoprotein A-I in this system, which may add to the difficulty of interpretation.

(3) Additional evidence for the identity of apo-lipoprotein A-II may be obtained by reducing the sample to be analysed with β-mercaptoethanol. This yields the monomeric form of A-II, which has a slightly greater mobility than the native dimer.

(4) Since apo-lipoproteins B, D, E and C-III contain carbohydrate components, they can be differentiated by their ability to react with the Schiff reagent (Section 6.4.2.1). However, we have already pointed out that this test is relatively insensitive and it is therefore likely to be of practical value only when the protein in question is partially purified and can be applied to the gel at a comparatively high concentration.

Having made a provisional identification of the proteins that are found by this electrophoresis, it remains to confirm (a) that the identification is correct, and (b) that each band in the gel corresponds to a single protein only.

The latter is important, for example, in the case of apo-lipoproteins F and D which are difficult to resolve in alkaline-urea gel.

The most satisfactory confirmation of identity is immunochemical, using the technique of immunodiffusion that is described in Section 6.5. This can be especially useful for the detection of albumin and β_2-glycoprotein-1, for both of which commercial antisera are readily available. However, it can still be difficult to make a confident identification if several proteins run close together, since the apparent resolution that is obtained by immunochemical detection is often less than that given by the less sensitive staining technique. Under these circumstances, it may be necessary to cut out the band of interest and to electrophorese it a second time before the immunodiffusion.

When authentic antisera are not available, confirmatory evidence must take the form of estimates of molecular weight and isoelectric point (p*I*), as follows:

A-2. Electrophoresis in SDS-gel (Section 6.4.2.4).

Since the analyses must be run in duplicate, both with and without reduction by β-mercaptoethanol, four alkaline-urea gels will usually be needed to provide the material for this stage. Cut out the bands of interest and macerate each slice of gel overnight in 0.2 ml of 1% SDS buffer (with 5 µl of mercaptoethanol when the protein is to be reduced) at 4 °C. Then heat the mixture to 90 °C for 3 min and transfer it to the top of the SDS-gel for electrophoresis in the usual way. At the same time, analyse an appropriate mixture of reference proteins, also in duplicate.

Determine the M_r of the apo-lipoproteins as described in Section 6.4.2.5, and identify them by reference to Table 6.4 and Fig. 6.5. As the figure indicates, the resolution of the major apo-lipoproteins in this system is such that most of them can be distinguished, although it may be difficult to separate the low-molecular weight C proteins. It should also be noted that F cannot be resolved from A-I and that superposition of bands may occur if 'apo-lipoproteins' from pathological sera are present.

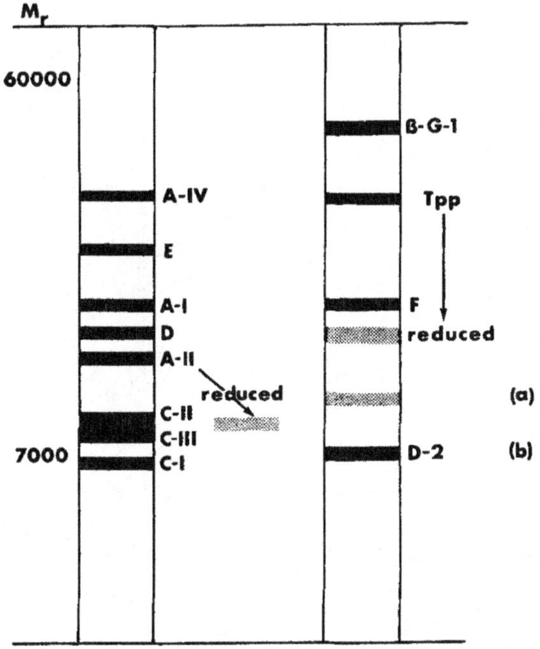

Fig. 6.5. Schematic representation of the relative mobilities of apo-lipoproteins during electrophoresis in SDS-gel. Abbreviations as in Fig. 6.2. In the presence of β-mercaptoethanol, A-II and Tpp are reduced to monomers that run with the mobilities indicated. Band (a) represents the SAA proteins and some of the SV fractions. Band (b) represents smaller proteins of the SV group.

A-3. Isoelectric focussing (Section 6.4.2.6).

In contrast to the SDS-gel system, IEF gives rise to a pattern of bands (Fig. 6.6) in which there is much superposition of proteins, and which is complicated by the overlapping of the ‘polymorphic’ forms of different apo-lipoproteins. In general, the resolution given by a pH gradient from 4 to 8 is adequate and there is probably little to be gained from the use of a shallower gradient except, perhaps, for the more acidic proteins. For these, a gradient of pH 4–6 may offer some advantages but, in most cases, a confident identification is only

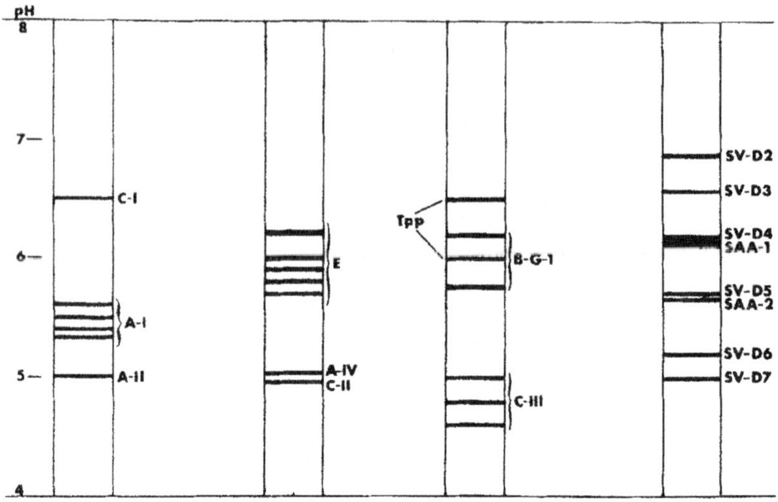

Fig. 6.6. A schematic summary of the isoelectric points of the apo-lipoproteins as found by analytical isoelectric focussing. Abbreviations as in Fig. 6.2. Because many of the apo-lipoproteins are superimposed, or lie close together on the gradient, they are illustrated here in four groups. For the same reason, it may be necessary to isolate a protein of interest by a preliminary separation, using electrophoresis in alkaline-urea gel. N.B. the p*I* of apo-lipoprotein F is 3.7 and this protein can not be observed on the pH gradient illustrated.

possible when a single protein has been applied to the IEF gel. Note however that apo-lipoprotein F is said to be highly acidic and its p*I* must be estimated on a gradient of pH 3–5.

Macerate the appropriate slices of gel from an alkaline-urea electrophoresis in 0.2 ml of buffer Z (Table 6.9) at 4 °C, for a period not exceeding 18 hours. Transfer the whole mixture to the prepared IEF gels, carry out the electrophoresis and estimate the p*I* of each protein band in the normal manner.

The combined evidence of relative mobility in alkaline-urea gel, relative molecular weight and p*I* will usually allow an unequivocal identification of the important apo-lipoproteins but, in view of the uncertainty attached to the published estimates of p*I*, added confidence will result if the unknown protein can be focussed concurrently with an authentic sample. In general, the technique of isoelectric focussing will provide better evidence of identity than can be obtained by electrophoresis in acid-urea gels, and the latter are not now much used for the characterisation of apo-lipoproteins. Nonetheless, they may be of some value for the identification of apo-lipoprotein C-I, which has a low mobility in alkaline-urea gel but is the fastest moving protein in acidic gels (Section 6.4.2.3).

Under certain circumstances, it may also be possible to obtain chemical evidence to support an identification by electrophoretic analysis. We have already commented on the use of the Schiff reaction as a means of confirming the identity of those apo-lipoproteins that contain carbohydrate, e.g. C-III, E and D. Additional evidence can be obtained by treating the protein with neuraminidase. By removing sialic acid residues, this will lower the p*I* of the protein and thus reduce its mobility in alkaline-urea electrophoresis. Perhaps the most useful application of the technique is as a means of confirming the identity of the polymorphs of C-III. Since C-III_1 and C-III_2 contain one and two residues of sialic acid respectively, treatment with neuraminidase will convert these glycoproteins into C-III_0, which is not normally prominent among the plasma apo-lipoproteins.

A-4. Digestion with neuraminidase.

In the process we shall describe, the substrate is the native lipoprotein but the technique can be applied to an isolated apo-lipoprotein if it is available in sufficient quantity (250–500 μg).

Materials. Protease-free neuraminidase (ex. *Clostridium perfringens*) that is co-valently bound to agarose (Sigma Chemical Co.).

Incubation buffer: 6.8 g of sodium acetate trihydrate; 9.0 g of sodium chloride; 0.1 g of EDTA; Water to 100 ml; adjust to pH 5.0.

Procedure. Transfer an amount of the agarose-neuraminidase equivalent to 0.04 units/1 mg of apo-lipoprotein to a small centrifuge tube and wash twice by centrifugation with a 1:10 dilution of the incubation buffer. To the washed agarose in the centrifuge tube, add 1 ml of the lipoprotein solution containing at least 1 mg of protein and 0.1 ml of the concentrated incubation buffer. Shake gently to re-suspend the agarose and incubate at 37 °C. Every 15 min, centrifuge the tube and remove a sample of the supernatant equivalent to 100–200 μg of protein, then re-suspend the agarose and return the tube to the incubator. Continue until four samples have been collected, agitating the tube periodically between-times. Store the first three samples in an ice-bath until the last one has been collected. Finally, analyse each sample, in parallel with a specimen of the undigested lipoprotein, by electrophoresis in alkaline-urea gel (Section 6.4.2.3). Stain the gels with Coomassie Blue R250 and compare them visually, or densitometrically. The bands due to susceptible glycoproteins will progressively dwindle in intensity as the digestion progresses, to be replaced by others of lower mobility.

If the apo-lipoprotein under investigation has been purified, further chemical evidence of its identity can be obtained from aminoacid and/or end-group determinations. Modern aminoacid analysers do not require unrealistic amounts of sample, and it is even possible to recover sufficient material for an analysis from slices of polyacrylamide gel, if the eluates from several gels are combined. However, it may be difficult to ensure that these eluates are not contaminated with, for example, glycine derived from the electrophoresis buffer. (In this case, the best solution may be to substitute β-alanine.) The aminoacid composition of many of the major apo-lipoproteins is now well established (and for several of them is made more certain by a knowledge of the sequence), but an analysis is probably most useful when the protein of interest lacks a particular aminoacid or combination of acids (Table 6.4). However, because the method is not of wide application and the details of the analytical procedures are easily available (e.g. Allen, 1981), we shall not describe them here.

B-1. The identification of apo-lipoprotein B.

Because apo-lipoprotein B is insoluble in all but the most powerful solvents, it is inaccessible to most physico-chemical methods of identification. The easiest and most satisfactory way of detecting it is therefore by immuno-analysis of the native lipoprotein (Section 6.5).

It is fortunate, in this context, that good quality antisera to LDL react only with apo-lipoprotein B, and are readily available commercially (Hoechst AG; Nordic Immunological Laboratories). However, the B protein is soluble in sodium dodecyl sulphate solution (SDS), in which it is dissociated into components of characteristically high molecular weight. Evidence for the presence of apo-lipoprotein B may therefore be obtained by electrophoresis in the SDS-gel system. At the same time, it may be noted that the smaller components of the B complex (which may actually be artifacts that result from mishandling of the preparation) are slightly soluble in 8 M urea solution and may appear at the junction of the stacking and running gels in urea-gel electrophoresis (Fig. 6.2 b).

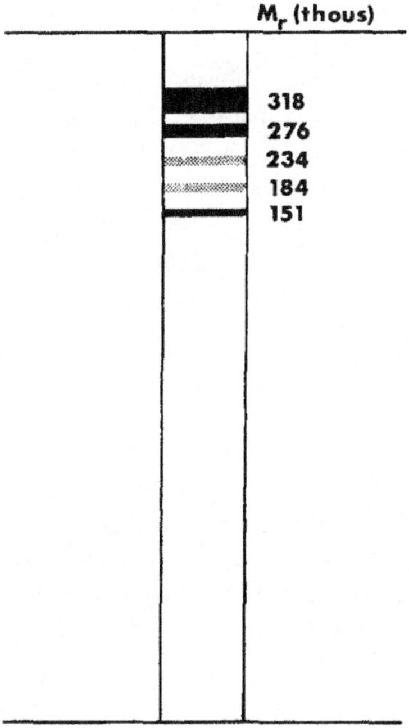

Fig. 6.7. The analysis of the protein components of apo-lipoprotein B by electrophoresis in SDS-gel.

Although it is, in principle, possible to take the TMU precipitate from stage A-1, extract the lipids with ethanol–ether and dissolve the protein residue in SDS solution, this can be a delicate manipulation if the amount of protein is small, as in a VLDL for example. It may therefore be simpler, if there is sufficient material available, to take a fresh portion of the lipoprotein and to extract it with ethanol–ether in the conventional way (Appendix 3). The protein having been isolated by one of these methods, it can be dissolved in the SDS solution and subjected to electrophoresis in a 3.3% polyacrylamide gel by the method of Weber and Osborne (Section 6.4.2.4). Under these conditions, undegraded apo-lipoprotein B yields five components which range in molecular weight from 150 000 to 320 000 (Fig. 6.7). Of these, the two of lowest mobility (M_r 275 000 and 320 000) form the greater part of the mixture.

If the attempt is made to obtain confirmatory evidence in the form of end-group or aminoacid analyses, note that the apo-lipoprotein B must first be purified by electrophoresis or filtration chromatography in SDS solution (Herbert et al., 1974). Equally, the reference analysis must be that of apo-lipoprotein B and not of apo-LDL. Probably the most reliable analyses of the B protein that are at present available are those of Lee and Alaupovic (1974).

6.5. Qualitative analysis of apo-lipoproteins by immunodiffusion

This technique can be used to detect the presence of a specific apo-lipoprotein either in the native lipoprotein, or in the protein residue after the lipids have been extracted. The experimental procedures are essentially standard (Ouchterlony, 1953) and details will be found, for example, in Clausen (1974). Detailed working instructions are also given with the Gelman equipment (Gelman Sciences Inc.). We shall confine ourselves here mainly to those technical considerations that are pertinent to the analysis of apo-lipoproteins.

Despite the apparent simplicity of the method, immunodiffusion is practised in a number of variations. For example, some workers prefer to make their gels from pure agarose, whereas others add a little agar. In agarose, the precipitin lines tend to be sharper but fainter than they are in agar, and some laboratories examine their samples by both systems in parallel. The agarose–agar mixture that is described in Section 5.4 is satisfactory for routine use, since it gives an enhanced precipitin reaction without the artefacts that are liable to occur in gels of pure agar. For refined experiments however, pure agarose is probably more suitable. But no matter what gel is used, the agar or agarose should be of the highest available quality.

There have been several reports (e.g. Kostner and Holasek, 1972) that precipitin reactions with lipoproteins can be enhanced by the addition of such substances as dextran or polyethylene glycol to the gel. In our experience, these substances are most effective when the

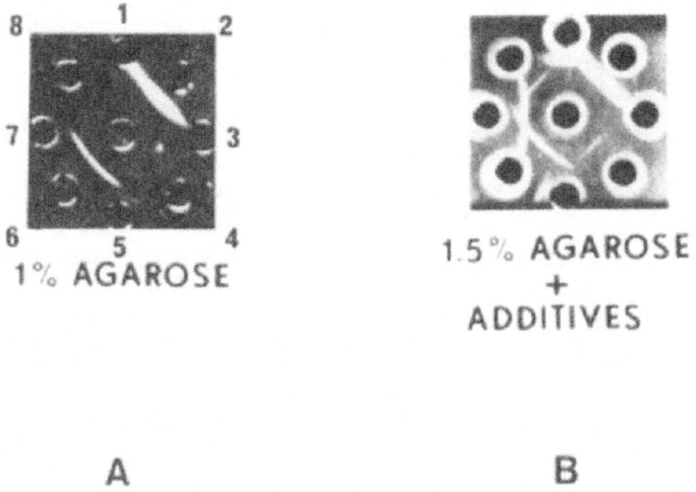

Fig. 6.8. The increased immunoprecipitation of apo-lipoproteins that can result from the addition of 10 mM $CaCl_2$, $MgCl_2$ and $MnCl_2$ to the agarose gel. The two experiments were identical, with the antisera arranged in the following order: (1) anti-A-II; (2) anti-A-I; (3) anti-C-III; (4) anti-C-II; (5) anti-C-I; (6) anti-B; (7) anti-E; (8) anti-D.

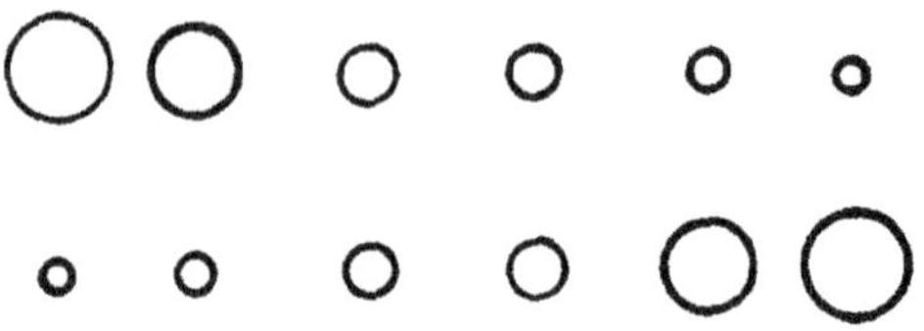

Fig. 6.9. An array of wells suitable for the immunoassay of apo-lipoproteins when the antigen and antibody may not be in balance.

antibody-antigen balance of the system is not at the optimum. It is therefore better to ensure that the balance is correct than to rely on the addition of these 'stimulants' (cf. below). However, a useful increase in the sensitivity of the precipitin reactions of at least some apo-lipoproteins can be obtained by including 10 mM $CaCl_2$, $MgCl_2$ and $MnCl_2$ in the agarose. An example of this effect is shown in Fig. 6.8. However, when this system is used, the agarose should be dissolved in 0.15 M NaCl and not in the veronal buffer that is proposed below.

Equipment. The apparatus marketed by Gelman Sciences Inc. is convenient and allows six microscope slides to be handled at one time. However, no-one need be deterred by the lack of this apparatus or some similar commercial product, since it is easy to coat slides singly and to punch wells in the gel with a suitably abbreviated Pasteur pipette. Moreover, although the commercial equipment is widely used, its fixed geometry gives it an inflexibility that is sometimes a disadvantage. For example, the precipitin lines may form too near to one of the wells for easy examination, or may be very weak if the antibody–antigen balance is too far from the optimum, a situation that can easily arise when the concentration of the antigen solution is not well defined. The simplest way to overcome this difficulty is to use a system of variable geometry in which either the distance between the wells, or the diameter of the wells can be changed. An example of the latter arrangement is shown in Fig. 6.9. A suitable array of wells would be as follows.

Diameter (mm)	Approx. volume (μl)
1.5	3
2.0	5
3.0	11
4.0	19
6.0	42
9.0	95

These can be cut with any suitably sized pieces of thin-walled tubing, for example, cork borers can be used for the large ones, while the smaller can be cut with cannulas. A full-sized master drawing of the desired pattern of wells must first be made as a guide, over which the agarose slide can be placed to ensure that the holes are punched in exactly the required places. With this system, the precipitin reactions can be observed under conditions close to the optimum, without the need to titrate the antigen with the antibody beforehand. Even with this simple equipment however, an instrument will be needed for cutting the troughs in the gels for immunoelectrophoresis. This can be made from two thin blades, about 65 mm long, clamped against a spacer to form parallel cutting edges either 1.0 mm or 2.0 mm apart.

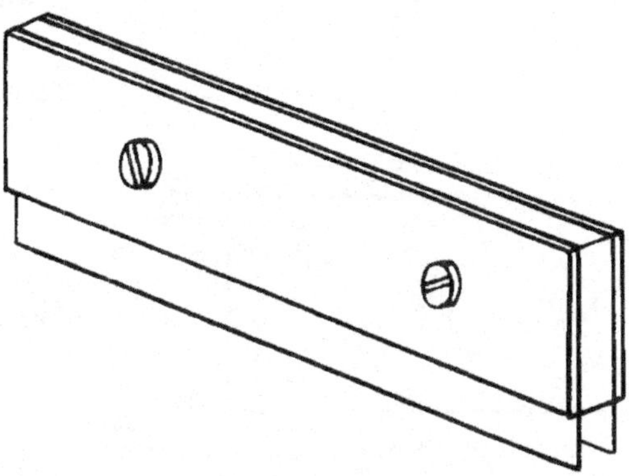

Fig. 6.10. Cutter for making the antiserum trough in immunoelectrophoresis gels. The two sharp blades are clamped onto a metal spacer 1–2 mm thick.

This assembly needs to be mounted in a spring-loaded frame so that the cutters can be aligned with the axis of the slide (Fig. 6.10).

A valuable piece of equipment is a light source that will allow the precipitin lines to be viewed with oblique illumination against a dark ground (Fig. 6.11). This enables faint lines and details such as spurs to be seen more easily. Although a permanent record of the final stage of the analysis can be obtained by staining the precipitin lines, it is often helpful to record the intermediate stages in their development. This can be done by making pencil drawings, but it is better to take photographs as described in Appendix 5.

Materials (refer to Section 5.4). Antisera to human serum and to the sera of certain other animals are commercially available and can be used to detect the presence of plasma protein contaminants in isolated lipoproteins. Antisera to human α- and β-lipoproteins can also be obtained from commercial sources (e.g. Hoechst AG; Nordic Immunological Laboratories) but the anti-α-lipoprotein is often of low titre. For refined studies with other lipoprotein fractions, or with mono-specific antisera to the individual apo-lipoproteins, it will be necessary to raise antisera to the specific proteins that have been isolated as described in Section 8.1.

The detailed technique for the production of antisera is readily available in Clausen (1974). The rabbit is a suitable animal for the

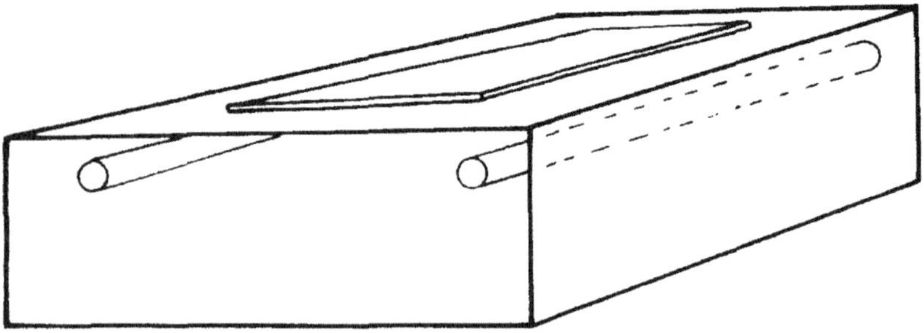

Fig. 6.11. A source of oblique illumination for viewing precipitin reactions. Two small fluorescent tubes provide the light but are arranged so that they are not directly visible through the apperture in the top of the box. The slides are supported on a piece of glass over this hole.

production of modest quantities of antiserum. Moreover, as Clausen (1974) points out, the precipitin lines are not soluble in excess of rabbit antibody and faint lines therefore do not tend to disappear on continued diffusion, as they do with horse antibodies, for example. Intradermal, subcutaneous and intra-peritoneal administration have all been used satisfactorily, but the first of these has the merit that adequate titres can often be obtained with relatively small amounts of protein.

The procedure is as follows: Inject 0.25–0.5 ml of 5% antigen solution into a rabbit, either intradermally or subcutaneously. The first injection should be emulsified with 0.5 ml of Freund's complete adjuvant. Follow this with four more injections of the antigen, at intervals of three weeks. Take a trial sample of antiserum one week after the last injection.

Method. Prepare the agar–agarose solution described in Section 5.4, but do *NOT* add albumin. However, it is advisable to add 1:1000 v/v of sodium azide to maintain the sterility of the gel during what may be an incubation of several days. It is essential to continue the heating of the agarose for long enough to ensure that it is all dissolved; some workers filter through gauze to make certain that any undispersed material is removed. If the gels are to be made of pure agarose, the procedure is essentially the same.

The slides on which the gels are to be cast must first be thoroughly cleaned with chromic acid solution, followed by an exhaustive washing with clean water. The slides may then be dried either in an oven, or by immersion in acetone. Smear each of these clean, dry slides with a drop or two of freshly prepared 1% agarose solution and allow this to dry under cover (e.g. an upturned Petri dish), at room temperature, for 24 hours. This gives good adhesion to the layer of gel in which diffusion will take place. Set the slides on a level surface (check with a spirit level) and pipette the warm agarose solution onto them at the rate of 3 ml for a standard 75 mm × 25 mm slide (i.e. a gel thickness of about 1.5 mm). Allow to set for about 15 min and then transfer to a humid chamber for at least 1 hour, but not more than 24 hours, before use. The humid chamber can easily be made from a large Petri dish, in the bottom of which is a disc of filter paper soaked in water.

If the wells are cut individually, as described above, place the slide over an accurately drawn template as a guide. Remove the cores from the wells by suction and fill them, as appropriate, with sample or antiserum. This can be done with the aid of a piston pipetter fitted with disposable plastic tips, taking care not to splash the solutions outside the wells. Finally, return the slides to the humid chamber, where the diffusion is allowed to proceed. This can take place either at room temperature, or at 4 °C, but it is important that the temperature should be constant, as this results in sharper precipitin lines with less branching and other artefacts. Since the temperature should also not be allowed to exceed 20 °C, it is in practice easier to put the gels into the refrigerator. Keep them under observation for at least two days and record the appearance of the precipitin lines at intervals.

When the reaction is complete, immerse the slides in 0.9% NaCl solution (use about 3 litres for 24 slides) for at least 48 hours. Next wash the slides in distilled water for 60 min to remove the NaCl and then fill the wells with molten agarose solution. When this has set, flood the slide with water and plant on the surface a strip of Whatman No. 1 filter paper. Allow the covered gel to dry on the bench, then moisten the filter paper and peel it off. Finally, rub the slide gently with a finger, under a stream if running water, to remove any fibres etc. that may adhere to it.

Stain the precipitin lines for lipid by putting the slides in a bath of Sudan Black (Appendix 4) for 1–2 hours, and then wash in several changes of 50% ethanol until the background is clear. Proteins can be stained with Amidoschwarz (Appendix 4); immerse the slide in the stain for 10–30 min according to the strength of the precipitin line. Transfer the stained slide into the de-staining mixture of glycerol in 2% acetic acid as described in the Appendix. Finally, wash under the tap and dry at room temperature.

The diffusion technique can also be used to characterise the protein bands that are separated by electrophoresis in polyacrylamide gel (Section 6.4.2). To do this, cut the rod of gel longitudinally into halves, one of which is then stained in the usual way (Section 6.4.2.1).

Alternatively, the rod may be cut transversely into short sections, using a second, stained gel as a guide to the positions of the bands. If the bands are both few and strong, it is sometimes possible to reveal their positions by a *brief* treatment with stain which does not denature all of the protein that is present. This residual protein can then be identified immunochemically.

Arrange the slices of unstained gel along the axis of microscope slides, and pour warm agarose–agar mixture round them. The small disc-like slices of gel tend to float about while this is done, so stick them to the slide beforehand with a dab of molten agarose. Alternatively, insert them into holes of the appropriate size that are punched with a cork borer in a previously prepared agarose-agar gel. These prepared slides must then be kept in the humid chamber for 48 hours at 4 °C, to allow the protein in the polyacrylamide gel to diffuse into the agarose. After this time, cut troughs or wells in the agarose as appropriate, load them with antisera and return the slides to the humid chamber for at least 48 hours. When immunoprecipitation is complete, carefully detach the polyacrylamide from the agarose and process the slides in the usual way.

6.6. Identification and estimation of carbohydrates

Some of the apo-lipoproteins (Table 6.4) are conjugated with carbohydrate residues. Little work has been done on this carbohydrate moiety and, because carbohydrate assays tend to be less reliable than those for aminoacids, the aminoacid composition is the more firmly established. Despite this, a carbohydrate analysis is not without some value as a means of characterising the apo-lipoproteins. Moreover, it may become of interest in the light of evidence that carbohydrates have a role in the antigenic behaviour of the apo-lipoproteins, or in view of their possible significance for the metabolism of lipoproteins. However, because the apo-lipoproteins do not all contain carbohydrate and the content of the lipoproteins is therefore not constant, carbohydrate analysis is better for the characterisation of apo-lipoproteins than of the lipoproteins.

As we have already remarked, it may be helpful to use the periodic acid–Schiff reaction to test qualitatively for the presence of carbohydrate after electrophoretic or immuno-diffusion analysis (Section 6.4.2.1; Appendix 4). However, a better characterisation will result if the individual carbohydrates present are identified and estimated. The conventional colorimetric methods are generally quite specific but often require an inconveniently large amount of protein if all the carbohydrates are to be estimated. Since partial analyses are of little or no value, it is better to take advantage of the greater sensitivity offered by gas chromatographic analysis of silylated carbohydrates. The following procedure, which allows the analysis of microgram quantities of carbohydrate, is that of Clamp et al. (1971). Because of its high sensitivity however, interference from substances leached from the support phase may be experienced if the apo-lipoprotein is prepared by chromatography or electrophoresis. Moreover, spurious peaks are produced on the chromatogram by a number of widely used buffer salts, which must be removed before the analysis.

Reagents: silver carbonate, acetic anhydride, hexamethyl-disilazane and trimethyl-chlorosilane. Dry pyridine (re-distilled and stored over solid KOH).

Dry methanol, prepared as follows: stand methanol over calcium oxide for at least 24 h. Reflux 500 ml of this partially dried methanol with 2.5 g of magnesium turnings and 0.1 g of iodine for 1 h and then distil into a clean dry flask. Keep tightly stoppered.

1.0 N HCl in dry methanol: produce dry HCl gas by adding concentrated H_2SO_4 dropwise to solid NH_4Cl. Pass the gas through concentrated H_2SO_4 into 100 ml of dry methanol until the weight has increased by about 3.6 g. Check the concentration of the solution by titration with standard NaOH. Note that HCl gas is also available commercially.

Pure mannitol is required for use as an internal standard for the gas chromatograph, and appropriate solutions of fructose, mannose, galactose, *N*-acetylglucosamine and *N*-acetylneuraminic acid as standards for the calibration of the chromatograph.

Equipment. The analysis is carried out on a gas chromatograph that is equipped with a flame ionisation detector and a 2 metre column packed with 4% SE-30 on Chromosorb W-HP (100–120 mesh). The temperature of the injector port is set at 200 °C and the detector at 250 °C. The temperature of the column is programmed from 120–200 °C at 1 °C/min, and is held at 200 °C until the last peak has emerged.

Procedure. The apo-lipoprotein must be freed from lipid by extraction with organic solvents (Appendix 3) and dried from ether or acetone in the form of a powder. Note that this preparation must also have been freed from salt by a preliminary exhaustive dialysis.

(1) Weigh a sample of 0.1–5.0 mg of apo-lipoprotein (equivalent to about 5–250 µg of carbohydrate) into an ampoule, or a Reacti-vial (Pierce Chemical Co.) with a Teflon-lined screw-cap. Add a volume of standard mannitol solution equivalent to about 10% of the expected total carbohydrate (0.5–25 µg of mannitol) and dry in a vacuum desiccator over P_2O_5. Add 2 ml of 1 N HCl–methanol, purge with nitrogen, seal the tube and heat for 24 h at 85 °C. Cool the tube and neutralise the contents with silver carbonate.

(2) Re-acetylate the amino sugars and sialic acid by adding 0.3 ml of acetic anhydride to the vial, sealing it and keeping the mixture at room temperature overnight. Centrifuge down the solid residue and transfer the supernatant to a small stoppered tube. Triturate and centrifuge the residue twice with dry methanol, and add both washings to the original supernatant. Finally, evaporate the solution of methylglycosides to dryness under nitrogen at 30–35 °C and keep in a vacuum desiccator for 12 h over P_2O_5.

(3) To prepare the silylating reagent, mix 0.5 ml of dry pyridine, 0.2 ml of hexamethyl disilazane and 0.1 ml of trimethylchlorosilane in a small stoppered tube. Note that these silicon derivatives are noxious substances that react with moisture and should be handled accordingly. Allow the reagent mixture to stand at room temperature for 15 min, then centrifuge and add 0.1 ml of the supernatant to the dry methylglycosides from (2). Stand the mixture at room temperature for 30 min and then centrifuge.

(4) The pyridine solution of silyl ethers can be injected directly onto

the chromatography column. A volume of up to 5 μl can be tolerated at the usual sensitivity settings and, if the amount of protein taken was greater than about 1.0 mg, a sample can be taken directly from the reaction mixture. If the amount of protein was less, a larger aliquot must be transferred to a small conical tube (vol. approx. 0.5 ml) and concentrated to 2–3 μl by evaporation in a vacuum desiccator over P_2O_5.

Calculation. As is shown in Fig. 6.12 this chromatographic system will resolve each sugar into several peaks which have the approximate retention times, relative to mannitol, that are given in Table 6.10. For each sugar, each of these peaks is a roughly constant proportion of the total area for that sugar, and these proportions are also listed in the table. However, it must be emphasised that both the retention time and the relative proportions of the peaks are not constant and must be regularly checked by the analysis of standard mixtures. Moreover, the different sugar derivatives do not give the same molar response in the ionisation detector. The relevant correction factors that are

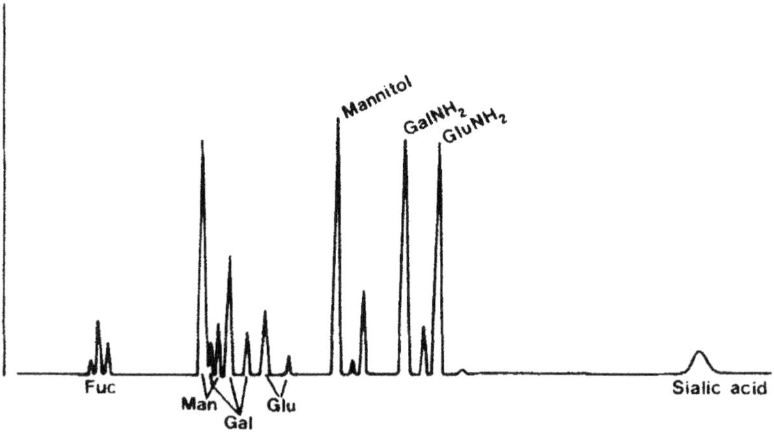

Fig. 6.12. Schematic chromatographic elution pattern of the carbohydrates commonly found in apo-lipoproteins. Mannitol is added as a standard. Fuc, fucose. Man, mannose. Gal, galactose. Glu, glucose. $GalNH_2$, acetylgalactosamine. $GluNH_2$, acetylglucosamine.

quoted in Table 6.10 are taken from Clamp et al. (1971) but these also should be regularly checked for each individual system.

To calculate the amount of each carbohydrate in the sample injected: (1) determine the area of each peak on the chromatogram (e.g. by triangulation); (2) find the total area for each carbohydrate by summing the areas for the relevant isomers; (3) find the ratio (R) of each of these areas to that for mannitol; (4) the molar concentration of the carbohydrate in the sample injected is then given by the relation: $R \times F \times C$, where F is the molar correction factor (Table 6.10) and C is the concentration of the internal standard in µmoles.

TABLE 6.10
Chromatographic properties of silyl methylglycosides on SE-30

Carbohydrate	Retention relative to mannitol	Approx. relative proportions of peaks (%)	Molar correction factor (F) relative to mannitol
Fucose	0.25	11	
	0.28	54	1.4
	0.31	35	
Mannose	0.59	91	1.0
	0.65	9	
Galactose	0.61	12	
	0.68	60	1.0
	0.74	28	
Glucose	0.79	73	1.1
	0.86	27	
N-Acetylgalactosamine	1.08	27	1.6
	1.21	73	
N-Acetylglucosamine	1.05	5	
	1.17	14	1.3
	1.31	78	
	1.38	3	
Sialic acid	2.22		2.1

6.7. Quantitative analysis of lipoproteins by micro-scale thin-layer chromatography

This is a comparatively new technique in which the components of a mixture are resolved by thin-layer chromatography and are then estimated by quantitative combustion in a flame ionisation detector (Ackman, 1981). The high sensitivity of the detector allows very small amounts of material to be analysed. Moreover, the method is very rapid. In the context of lipoprotein analysis, the technique makes it possible to analyse a few tens of micrograms without the need to separate the lipids from the protein first (Mills et al., 1979). It is therefore easily possible to analyse fractions from a density gradient centrifugation, or samples from small laboratory animals.

However, the method is not without its disadvantages. In the first place, the only commercial equipment at present available (the TH10 analyser: Iatron Laboratories Inc.) is comparatively expensive. Secondly, the detector is non-specific. Consequently, only pure isolated lipoproteins can be analysed. Even the salt or sucrose etc. used in the isolation of the lipoprotein must be removed before the combustion is carried out. The third draw-back to the method, at least in its present stage of development, is its relatively low reproducibility. This makes it necessary to replicate the analyses several times if the results are to be of acceptable precision. A substantial improvement in precision could be obtained by the use of an internal standard, if the conflicting requirements for such a substance could be satisfied. At the time of writing however, a compound that is soluble in water but chromatographs like a lipid, and runs to an unoccupied part of the chromatogram is not available. Under the existing circumstances therefore, conventional chemical analysis will probably give the more precise results if there is sufficient material available to permit of adequate replication. However, this advantage is lost if, as if often the case, the precision is not reported.

The essential feature of the micro-method is the so-called 'chromarod' (Iatron Laboratories Inc.), which consists of a silica rod (15 cm × 0.9 mm diam.) that is coated with a porous layer of silicic acid about 100 μm thick. This layer is fused to the rod and can be used many times before its performance deteriorates. A small sample of lipoprotein is applied near one end of the rod and is then developed chromatographically in a sequence of solvents that leave the separated components distributed up the rod. This is then scanned in the flame ionisation detector, the output of which can be recorded in both the differential and the integral modes. Because the response of the detector is different for each component, an appropriate range of calibration curves must also be prepared. However, these are linear up to at least 12 μg of each component. Much the most laborious part of the procedure is the measurement of the recorder traces, the calculation of the results and their subsequent statistical analysis. This computational load can be greatly eased if the calibration curves can be incorporated into a programme to run on a desk-top computer.

Equipment and reagents. In addition to the analyser itself and a supply of chromarods, a two-channel recorder is needed, and a plunger-type micropipette to apply the samples to the rods. The latter must be capable of depositing 1–2 μl of solution in controlled droplets of about 0.5 μl. In our experience, the Drummond Microdispenser has proved very satisfactory. A supply of hydrogen is also required for the flame detector.

Since the chromarods are very fragile, they are handled in special frames that hold ten rods at a time. These frames can be put into the chromatography tank, so that all the rods are developed simultaneously. However, the resolution is somewhat better if each rod is developed separately in a small vessel like that shown in Fig. 6.13.

The following solvents are required for lipoprotein analysis: Diethyl ether, acetic acid, methanol, ethanol, chloroform and petroleum ether of boiling range 40–60 °C. All these should be of the best quality and should be purified as described by Kates (1975), who also defines the term ‘petroleum ether’.

For the calibration of the instrument, pure specimens of cholesterol, cholesteryl stearate, tripalmitin (or tristearin), lecithin and bovine serum albumin are needed. A comprehensive list of suppliers is given by Kates (1975).

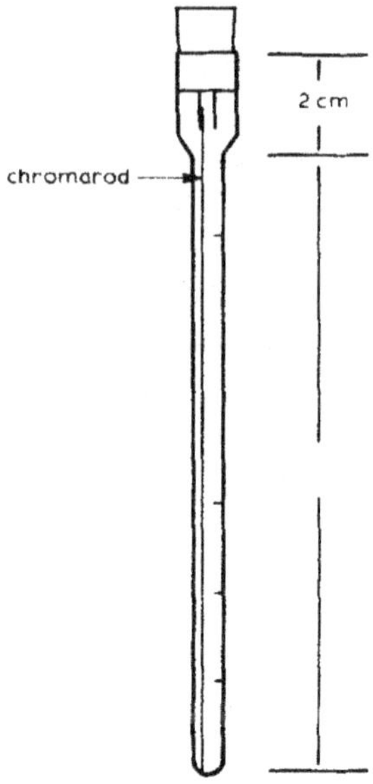

Fig. 6.13. Tube for the development of individual 'chromarods'. The bore of the tube is 5–6 mm diam. and the chromarod is kept in an axial position by the short length of tubing in the stopper. The lower index mark – 2 cm from the bottom of the tube – marks the point of application on the rod. The upper three marks – respectively 4, 6 and 12 cm from the bottom of the tube – are the points to which the different developing solvents are allowed to run.

Preparation of samples. The lipoproteins to be analysed should be fresh. Old samples will give spurious peaks on the chromatogram, the identity of which is uncertain. The lipoproteins may be isolated by any of the procedures given in Chapter 2 but, because high concentrations of salt, sucrose, etc. will interfere with the running of the chromatogram, the preparations must first be dialysed against 0.15 M NaCl overnight at 4 °C. The resulting solution may itself be sufficiently concentrated to be applied directly to the chromarod. If this it not the case, the sample can be concentrated by ultracentrifugation in physiological saline overnight, in a 2 or 3 ml tube. An LDL or HDL will sediment to the bottom where it will form a syrupy deposit. Carefully remove the supernatant and dissolve this syrup in a drop of fresh saline. Very-low-density lipoproteins must be drawn directly from the top of the tube into the micropipette.

The amount of lipoprotein applied to each rod must be sufficient to allow the weakest component to be determined, without overloading the rod with the strongest component. This consideration makes it impractical to analyse chylomicrons in this way. It must also be remembered that, if too much lipoprotein is used, some may not be burnt in the flame. In most cases, 15–25 μg of material should be sufficient. The easiest way to determine the appropriate volume of sample solution is to run a trial rod.

Procedure. It is vital that the chromarods are kept scrupulously clean since any contamination will produce a spurious response in the detector. Moreover, the life of the rods is greatly prolonged if all inorganic salts are washed away before the rods are scanned. The rods must be activated before use by passing them through the flame of the detector and this allows each one to be checked for the presence of persistent contaminants. If they are dirty, it may be possible to clean them by soaking in chromic acid cleaning mixture, followed by exhaustive washing in distilled water.

It is equally important that the chromatography tanks, the solvents and the air supply to the flame should be clean and free from particulate matter.

To apply the samples, set a rack of rods over a small hot-plate that is adjusted to about 40–45 °C. Do not attempt to dry the rods by blowing air over them, for they will be contaminated with dust. Fit a clean capillary to the micropipette and adjust it to deliver the required volume of about 2 μl. Fill the pipette with the sample solution and apply 0.25–0.5 μl to a point about 2 cm from the end of the first rod. Allow this to dry before applying a similar volume to the same point. Continue this process until all the sample has been transferred to the rod. This loading of the rods is a delicate process in which the adsorbent coating of the rod is easily damaged: a good light and a support for the hand are desirable. Repeat the procedure for the required number of replicate analyses.

The chromatographic resolution is brought about by five consecutive developments, at room temperature, as follows:

(1) Ethanol/diethyl ether (3:1 v/v). Put about 0.3 ml of this solvent in the bottom of the chromatography chamber (Fig. 6.13). Transfer a prepared rod (with forceps) from the rack to the chamber, which is then closed with the special stopper. Continue the development until the front is 2 cm beyond the point of application. Then remove the rod from the chamber with forceps and allow it to dry at room temperature for 5 min. This leave the neutral lipids in a sharp band at the position of the front.

(2) Petroleum ether/diethyl ether/acetic acid (90:10:1 v/v/v). It is preferable to use a separate chamber for each solvent, but the same one can be used if great care is taken to rinse out the preceding solvent with the fresh one at each change. Allow the second solvent to rise up the rod until the front is 10 cm beyond the point of application. Remove the rod and dry it as before. At this stage, the cholesteryl esters, triacylglycerols and unesterified cholesterol are resolved in the upper part of the rod, leaving the protein and phospholipid at the start.

(3) Methanol. Allow the front to rise 4 cm beyond the point of application and then dry the rod as before. Repeat the process a second time, which completes the resolution of the phospholipid from the protein.

(4) Finally, drop the dry rod into a tube containing double-distilled water, to remove small amounts of salt which would otherwise give an unwanted response in the detector. This stage is necessary because lipoproteins will not dissolve satisfactorily unless there is salt in the solvent. Double-distilled water gives a noticeably better baseline than that which has only been distilled once. After soaking for at least 5 min, transfer the rods back to the rack and dry them for 20 min at 100 °C.

This chromatographic protocol has been designed to give an adequate separation of the different components of the lipoprotein. But it must be closely adhered to. If the ethanol/ether is allowed to advance too far up the rod, the length that is available for the resolution of the neutral lipid is reduced. Moreover, if the methanol is run too far, it will displace the free cholesterol. Apart from these mechanical considerations, the isotherms for the adsorption of lipids by the chromarods are apparently non-linear. This leads to a loss that is roughly proportional to the distance that each substance has moved up the rod and is another reason for standardising the conditions of the analysis very precisely. In this context, it is a great advantage to use the individual chromatography chambers shown in Fig. 6.13, as it is then possible to ensure that each solvent front moves the same distance on all the rods. As the rods do not 'run' at the same rate, this is difficult to arrange in a single large tank. It is also worth noting that different batches of rods can differ widely in their average running time, for reasons that are not at present clear. Sluggish rods may run more rapidly after a period of use, and may be improved by exhaustive pickling in chromic acid, but they will always tend to be 'slower' than average.

Quantification procedure. While the chromarods are drying, light the flame ionisation detector in the analyser and check that it is properly adjusted in accordance with the manufacturer's instructions. The detector should be fed with hydrogen and air at the rate of about 165 ml/min and 2 litres/min respectively.

When the rods are thoroughly dry, instal them in the scanning frame of the analyser and scan them at the rate of 20 cm/min. Note that throughout all these operations, the rods must be handled with forceps as finger-prints can be detected by the instrument. Try to adjust the zero of the instrument so that the integrator trace is parallel to the baseline when no peaks are emerging. This will simplify the subsequent estimation of the areas under the peaks, for which full instructions are provided by the makers of the analyser.

Calibration of the analyser. Prepare a standard solution of each of the following substances at the rate of 2 mg/ml. Cholesterol, cholesteryl ester and tristearin or tripalmitin, in chloroform. Lecithin in chloroform/methanol (2:1 v/v). Bovine serum albumin in 0.025 N ammonia solution.

Apply 1 μl of the first solution to each of ten chromarods and carry out the chromatographic development and quantification just described. Repeat the process with 4 or 5 larger volumes of solution up to 6 or 7 μl. Use the least squares procedure to obtain the equation of the best fitting straight line to the resulting 50 or 60 pairs of values that relate the weight of lipid with the area under the peak. The method of computation can be found in any manual of statistics, and is also a library programme on many computers or programmable calculators. Go through the same procedure for each of the standard solutions. Note that the resulting calibration curves will apply only to the batch of rods that was used. Each new batch will require a new calibration.

Note that enough determinations of each lipoprotein component are made during the construction of these curves to permit their standard deviation to be calculated. With this information, it is then possible to calculate the number of analyses that must be made to determine a component of an unknown sample with a desired precision. For example, if the estimation of cholesterol has a standard deviation of ± 0.3 μg, the number of rods (*N*) that must be run in order to determine the cholesterol content of a lipoprotein with a precision of ± 0.25 μg, with a confidence of 95%, is

$$N = \left(\frac{1.96 \times 0.3}{0.25}\right)^2 = 5.5$$

where the factor 1.96 is the value of the standardised variable (z) at which the area under the normal distribution curve is 0.95.

In practice, the variance of the analyses is such that at least five rods must be used for each sample of lipoprotein, which will require a minimum of about 75 μg of material. However, it is better to use all the ten rods that constitute a batch and thereby gain the advantage of higher precision. The raw data from this number of rods will amount to 50 values of peak area, from which the equivalent weights of protein or lipid can be calculated by means of the known equations of the calibration curves. From these results, calculate the mean weight of each component of the lipoprotein, with its standard error;

the average weight of sample applied to the rods and the mean percentage composition of the lipoprotein, with its standard error. Clearly, this task is laborious and is well suited to a small computer or programmable calculator. Fortunately, it is a simple matter to write a suitable programme for the purpose.

The estimation of lipoproteins

This chapter deals with the quantitative determination of intact lipoproteins, with particular reference to their estimation in plasma, which has come to be important to the clinician in the context of atheromatous disease. The analysis can be considered as having three parts:

(1) The isolation of the lipoproteins. This step may not be necessary if a specific method of estimation is to be used, e.g. the determination of cholesterol, or an immunoassay. Centrifugation and precipitation are the most widely used procedures.

(2) The resolution of the lipoproteins. This stage is often combined with the first. Centrifugation is the standard method and has the advantage that it will handle comparatively large samples at high resolution. Precipitation is a cheap and convenient alternative, but its limited resolution has restricted its application mainly to the estimation of HDL. Neither of these techniques is very suitable for micro-scale analysis, for which electrophoresis has sometimes been used. However, it is difficult to measure the small samples reproducibly, and electrophoretic analysis is therefore often used to estimate the relative proportions by weight of the main classes of lipoprotein. Their absolute concentrations must then be derived from these by means of an independent estimate of the total lipoprotein concentration.

(3) The estimation of the lipoproteins. This step is made difficult by the physical and chemical heterogeneity of the lipoproteins which, as was said in Chapter 6, is such that there is no known property that is a constant proportion by weight of every lipoprotein particle.

Consequently, unless the lipoproteins are determined gravimetrically, it is necessary to assume that each class has a characteristic mean value of the parameter that is being measured. This mean is often ill-defined and depends on the distribution of the lipoproteins in the sample, which is usually unknown. A considerable uncertainty is therefore attached to many lipoprotein estimations and especially to those made by the chemical determination of a constituent like cholesterol or protein. Such methods may be no better than semi-quantitative, although this may not disqualify them for use in clinical or epidemiological studies that make no pretence to be accurate.

Physical methods of estimation have the advantage that they measure properties that are a function of the whole lipoprotein particle and are more directly related to lipoprotein mass than measurements of chemical constituents. However, they require a comparatively large sample and, because they are nonspecific, the lipoproteins must be isolated in a highly pure state.

The gravimetric estimation of lipoproteins has been described in Section 6.1, and deserves to be more widely used in academic studies, although it is unsuited to the routine estimation of lipoproteins in plasma. The two other physical methods of estimation that are available are based on refractometry and light scattering respectively. The most sophisticated of the refractometric methods is analytical ultracentrifugation. This has the outstanding advantage that the quantitative estimation is combined with the highest available resolution, but the extraction of quantitative results from the raw data is a mathematical process that some may find daunting. Despite some reservations about the theory and practice of the method, it remains the most convenient and most powerful standard of lipoprotein analysis.

The estimation of isolated lipoproteins by conventional refractometry has been described in Section 6.1.1. The original proposal that this technique could be used to estimate the conventional VLDL, LDL and HDL classes in plasma (Lindgren et al., 1964) seems to have attracted little attention. This disregard is probably to be regretted since the chemical methods that are more often used, e.g. the determination of cholesterol or protein, are technically inferior. Nonetheless, it has to be said that the specific refractive increment of the different lipoproteins has not been established with a convincing certainty.

The estimation of lipoproteins by measurements of light scattering is an ingenious technique that acquires an artificial specificity from the fact that the lipoproteins are the largest particles normally present in plasma. It is therefore not necessary to isolate the lipoproteins. However, in its simplest form, the method can only detect the larger particles and has the defect that it estimates them in terms of an arbitrary unit. Nonetheless, it has a place as a quick and economical screening method when interest is focussed on VLDL and chylomicrons.

By contrast with these techniques, the chemical estimation of a component of the lipoprotein is more sensitive and often more specific. But, to determine the true mass of lipoprotein in the sample, all the constituents must be measured, which is not usually feasible if conventional methods of analysis are used. However, it can be done readily enough (on isolated lipoproteins only) by micro-scale thin-layer chromatography as described in Section 6.7, thus obtaining not only the amount of lipoprotein but also its composition. Despite their rough-and-ready nature as indices of lipoprotein concentration, chemical estimates of a single constituent are attractive because they are relatively speedy and can be made by easily accessible methods. This has often led to their uncritical use in studies for which they are inadequate. Although the technique can be used on lipoproteins that have been resolved by any method, perhaps the most useful current example is the estimation of 'HDL'-cholesterol after the 'low-density' lipoproteins have been removed from the plasma by precipitation (Section 7.2.1).

The estimation of lipoproteins that have been resolved by electrophoresis can also be made by determining the cholesterol content of the separated fractions. Alternatively, the lipoproteins can be stained under controlled conditions and then estimated spectophotometrically. This is a sensitive technique but, since it does not depend on a defined chemical reaction, it must be the subject of a laborious

TABLE 7.1
Index to methods of lipoprotein analysis

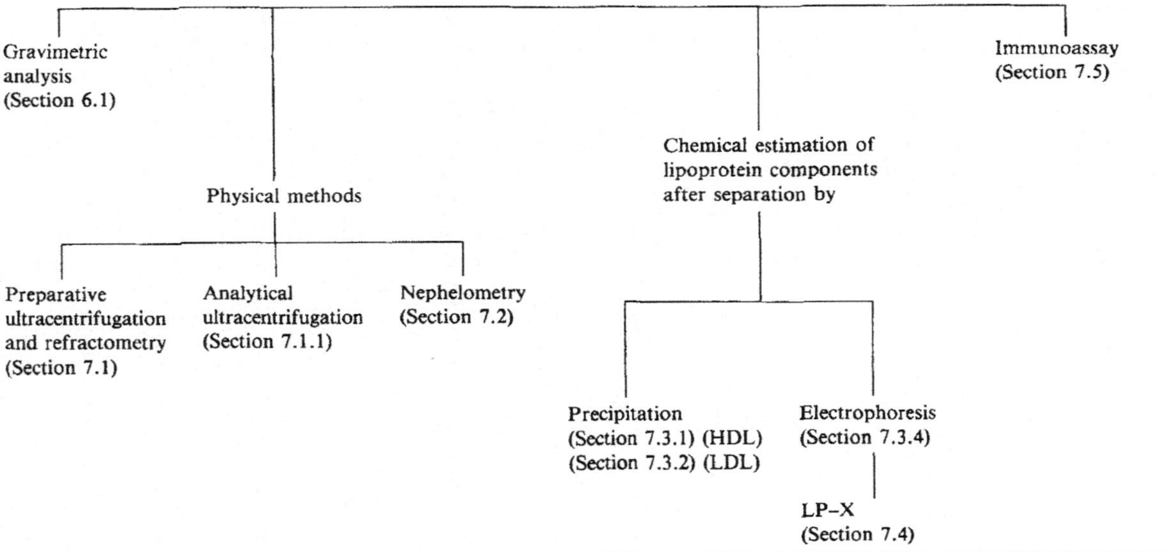

verification and calibration when it is first mounted. It is therefore not an appropriate method for a short series of analyses. Moreover, the verification must be continuously checked, either by comparison with a reference method of analysis, or by the repeated analysis of a standard lipoprotein mixture (e.g. a reference serum). In view of the well-established instability of lipoproteins during storage (Appendix 1; Section 7.3.1), the maintenance of an appropriate standard for an adequate length of time is a difficult, uncertain and poorly understood operation.

Immunochemical analysis offers an attractive combination of sensitivity with specificity, but is requires the use of a reference lipoprotein. It therefore suffers from the same handicap as the quantitative electrophoretic analysis.

To summarise, it must be said that, despite the great interest in lipoproteins since 1950, little work has been done on improving the quality of methods for their quantitative estimation. In this admittedly difficult field, the emphasis has been on the development of fast, facile methods, and this has been achieved only at the cost of real accuracy and precision. Where the volume of observations is not large, these methods should be avoided and preference given to gravimetric or refractometric procedures. However, if a chemical method of estimation must be used, we recommend that unesterified cholesterol should be chosen measure, since the proportion of this constituent is less variable, within a given lipoprotein class, than that of the others. Whatever procedure is adopted, the precision of the results should be determined and, particularly in the case of chemical methods, the accuracy should be verified by reference to a gravimetric procedure.

7.1. Estimation of lipoproteins by refractometry

This is the basis of the classical technique of measung plasma lipoprotein levels by analytical ultracentrifugation, that was developed by Gofman and his collaborators around 1950, and which is described in Section 7.1.1. Where the high resolution of the analytical centrifuge is not necessary, e.g. for the simple estimation of total LDL or HDL, the refractometric method of Section 6.1.1 can be used. The defects of this technique were discussed in Section 6.1.1, where it was pointed out that there is a lack of reliable data for the refractive increment of lipoproteins in solvents of differing composition. In the context of plasma analysis, this tends to discourage the use of the method for fractions other than the conventional lipoprotein classes i.e. VLDL, LDL and HDL. For fractions that are isolated at other densities, it may be necessary to make specific determinations of the relevant values of the refractive increment.

Although it is possible to measure the refractive index of a single drop of solution, it is impracticable, with the present techniques, to isolate the purified lipoproteins quantitatively in a volume of less than about 0.5 ml. For the routine analysis of plasma it is therefore convenient to use the refractometer in conjunction with the preparative ultracentrifuge. However, centrifugation leads to a re-distribution of the salt molecules in the solvent and thus to a gradient of refractive index down the tube, the magnitude of which must be determined separately. This can be done by centrifuging control tubes at the appropriate density in parallel with the adjusted plasma. However, if the procedural details are exactly maintained, the re-distribution of salt is highly reproducible and Lindgren et al. (1964) have used this fact to predict the refractive index of the solvent in the lipoprotein layer from that of the layer immediately beneath. The procedure can, in principle, be carried out on a single sample by serial centrifugation, but it may be easier, as well as quicker, to perform parallel centrifugations at different densities as we shall describe below.

Equipment. The specification of the refractometer and its method of use are described in Section 2.1.2.5. Note that it is important NOT to use a centrifuge tube with a skirted cap. For analytical purposes, the Beckman 40.3 rotor, with a 6 ml load in each tube, is satisfactory.

Procedure. To determine the re-distribution of salt, prepare a salt solution of exactly the same density and composition as that in which the lipoproteins are to be isolated (Section 2.1.1.1) and centrifuge it under conditions identical to those to be used for the isolation. Use the pipetting technique (Section 2.1.1.7) to remove the contents of the tubes in successive volumes of 1 ml. Do not delay in this or the salt gradient will diffuse. Then determine the refractive index of each sample as described in Section 2.1.2.5. These measurements should be made at 26 °C, since this is the only temperature at which the specific refractive increment of lipoproteins is known.

It is a disadvantage of this method that it involves the measurement of small changes in refractive index that are superimposed on the relatively large constant value for water. This leads to unwieldy computations that can be simplified by subtracting the refractive index for water (n_o) from the values determined for each of the fractions from the gradient. Thus, $\Delta n_1 = n_1 - n_o$; $\Delta n_2 = n_2 - n_o$ etc. where n_j is the refractive index of the *j*th fraction. Under a given set of experimental conditions, the ratio n_1/n_j is a constant (R_j) which can be used to estimate n_1 from a measurement of n_j.

Estimation of VLDL. Pipette 6 ml of plasma into a 13 mm × 63.5 mm tube and centrifuge it at 40000 rev/min for 18 hours at 18 °C (Section 2.1.1). If less than 6 ml of plasma is available, the volume can be made up by the addition of 0.196 molal NaCl solution. Use a smaller volume of plasma if there is reason to think that the level of VLDL is high, otherwise there may be difficulty in recovering the lipoproteins quantitatively. After the centrifugation, pipette off the supernatant lipoproteins quantitatively in a volume of 1 ml (Section 2.1.1.7), then recover the next 1 ml separately, using a clean pipette. If there is too much VLDL to take off in 1 ml, combine the top two fractions and recover the third separately for the estimation of the salt re-distribution. Finally, measure the refractive index of each fraction (Section 2.1.2.5) in the following sequence: (1) distilled water (n_o); (2) the VLDL concentrate (n_1); (3) the sample below the VLDL fraction (n_2).

Estimation of LDL. As originally described by Lindgren et al. (1964), the plasma LDL level is determined by deducting the value for VLDL from that for the total low-density lipoproteins. To isolate the latter, adjust the density of an appropriate volume of plasma to $d_{20} = 1.063$ g/ml as described in Section 2.1.1.1, e.g. by adding 3 ml of 3.162 molal NaCl solution to 3 ml of plasma. Centrifuge this mixture for 18 hours at 40000 rev/min at 18 °C. Recover the lipoproteins and measure the refractive index of the fractions as described for the VLDL.

Method of calculation. To convert the observed values of refractive index into concentration of lipoprotein it is necessary to know their specific refractive increment (S) under the conditions of the experiment. For a discussion of these data refer to Section 6.1.1.

In most cases the contribution of the solvent to the refractive index of the lipoprotein solution can be estimated from the refractive index of the subjacent layer (n_2) by means of the constant R_2 that was determined in the control experiment, i.e. the required value = $R_2\ (n_2 - n_o)$. The contribution of the lipoprotein to the refractive index of the solution is then given by: $(n_1 - n_o) - R_2\ (n_2 - n_o)$. The concentration of lipoprotein in the original plasma (in mg/dl) is then:

$$\frac{(n_1 - n_o) - R_2\ (n_2 - n_o)}{S} \times \frac{1000}{C}$$

where C is the volume of plasma centrifuged, in ml.

7.1.1. Analytical ultracentrifugation

It will be evident from what has been said in Chapter 4, that the analytical centrifuge is a clumsy and expensive way of setting out to determine merely the total concentration of a major lipoprotein class such as LDL or HDL. The greatest value of this technique lies in its ability to display and to measure the subtle differences in the distribution of the lipoproteins within these classes, which reflect the finer aspects of their metabolism. Unfortunately, interest has turned towards simpler but less informative methods of analysis, with the result that the development of the centrifugal technique has remained static for some 30 years. Consequently, as we remarked in Chapter 4, it does not incorporate the newest advances in either the theory of centrifugation, or in our understanding of the physical chemistry of lipoproteins. Nonetheless, despite this obsolescence, the analytical centrifuge is the method of choice for serious studies of the quantitative distribution of lipoproteins. However, it requires a considerable investment in both time and money to establish the method in the refined forms with which we are concerned here, and this effectively puts it out of the reach of the casual user. Moreover, there are different versions of the technique from which a choice must be made according to the circumstances and intentions of the investigator. In order to make this choice, it is necessary to consider the options that are open at each stage of the analysis.

The procedure can be divided into two main parts:

(1) The centrifugation itself and the recording of the refractive index gradient within the cell, which is usually done by photographing the schlieren pattern. These procedures are quite standard, but they include an exacting and time-consuming calibration of the centrifuge that must be performed at regular intervals.

(2) The extraction of quantitative data from the record. The labour that is involved here is largely taken up by reading data from the record and with the correction of the analysis for the effects of the concentration dependence of flotation rate. The higher the resolution that is desired, the greater the number of readings that must be taken, with a corresponding increase in the computational load. By performing the calculations on a digital computer, as was first done by Ewing et al. (1965), it becomes feasible to take a large number of data points and thus to achieve a relatively high resolution and accuracy. By using digitising equipment to reproduce the data from the film directly in a computer-readable form, it is possible to handle even more data. The first mechanised film reading equipment to be used in this context was made by P.C.D. Ltd., the data being recorded on punched paper tape. However, this apparatus has been made obsolescent by recent technical developments that have led to equipment of better capabilities. This use of the computer has evident advantages when the number of analyses to be done is likely to be large but, apart from the expense of the reading equipment and of computer time, it has

the objection that it may be necessary to write, or at least to modify a programme for the computer. Variants of the comprehensive programme that was developed by Ewing et al. (1965) have subsequently been used in other laboratories, but they may not be directly interchangeable either because they require the data to be presented in different ways, or because they are in different programming languages. Although these programmes have been developed and run on large main-frame computers, it would be possible to use a relatively cheap modern micro-computer for the purpose, with some sacrifice in speed of operation, provided that it had a sufficiently large memory. However, it would then almost certainly be necessary to write a new programme.

Clearly, it may be unrealistic to consider using this method of quantification if the number of analyses will be small. However, by reducing the volume of the data input and by approximating the correction for the concentration dependence (Johnston–Ogston) effect, the computations can be brought within the scope of a desk calculator. A procedure of this kind has been described by deLalla et al. (1967). This exemplifies the way in which simplification can be achieved but, since the Johnston–Ogston correction is applied only to the S_f 0–12 and S_f 12–400 classes, the result is effectively in terms of LDL and VLDL only. Consequently, the method makes inefficient use of the equipment and of the information that is recorded in the schlieren pattern. In this respect therefore, a less radical simplification of the computational procedure would be more satisfactory and is not difficult to arrange. The draw-back to all such 'simplified' procedures is, of course, that the manual component of the work is much increased and leads to a greater use of approximations. Moreover, a considerable amount of work must be expended on the preliminaries and it is unrealistic to consider even these methods unless it is proposed to run at least 25 analyses, or a series of infrequent analyses over a considerable period.

Since we have not the space to describe all these procedures in detail, we shall confine ourselves to one that requires a minimum of special equipment but which takes advantage of the great flexibility that results from the use of a programmable calculator or, better still, a micro-computer.

The procedure depends on the fact that the concentration of lipoprotein in the ultracentrifuge cell is proportional to the area under the schlieren curve. In these analyses however, the curve is usually grossly asymmetrical and may represent lipoproteins that cover a range of more than 100-fold in particle weight (Figs. 1.1, 1.3). To make it possible to determine quantitatively the relative amounts of these different particles, the schlieren pattern is divided into sections by boundaries that are set at arbitrary points, e.g. at S_f 0, S_f 12, S_f 20 and S_f 100. These boundaries represent the positions to which particles of flotation rates 12, 20, 100, etc. would have moved if they started at the bottom of the cell and floated under ideal conditions, at zero concentration, in the absence of diffusion. The area lying between two boundaries is then considered to represent the concentration of lipoproteins with flotation rates lying between these values, e.g. greater than S_f 12 but less than S_f 20. In practice, of course, an ideal, clear-cut distinction between the faster and slower particles is not achieved at these boundaries, but it is assumed that the errors are approximately self-compensating, for example that the amount of lipoprotein of $S_f > 20$ which is found on the 'slow' side of the S_f 20 boundary will be about the same as the amount of material of $S_f < 20$ that is on the 'fast' side.

In principle therefore, it is only necessary to measure the area of the schlieren curve between the selected boundaries and to convert these values into concentration of lipoprotein by means of the predetermined calibration factor for the centrifuge. In reality, the procedure is much less straight-forward:

(1) because the schlieren curve is highly asymmetric, some areas are measured close to the base of the cell, whereas others are measured much nearer to the meniscus. But the cell is wedge-shaped and the lipoproteins become concentrated as they move towards the centre of the rotor. Therefore, it is necessary to correct each measured concentration to the conditions at the start of the analysis, i.e. to the base of the cell. This is done by multiplying each measured value by the

factor $(r/r_0)^2$, where r is the mean distance from the centre of the rotor at which the area is measured, and r_0 is the radial distance of the base of the cell.

(2) the apparent flotation rate of a given lipoprotein is negatively correlated with its concentration, roughly in accordance with the equation $S_f = S_f^o\,(1-KC)$, where C is the concentration of the lipoprotein and K is a constant. From this, it follows that the exact position of the S_f boundaries along the schlieren pattern cannot be calculated in advance of the analysis but must be determined each time. Two strategies can be employed. The first requires a graph that relates the distance of each arbitrary delimiting boundary from the base of the cell (x in Fig. 7.2) to the concentration, measured as the area under the schlieren curve between the boundary and the base of the cell (Fig.

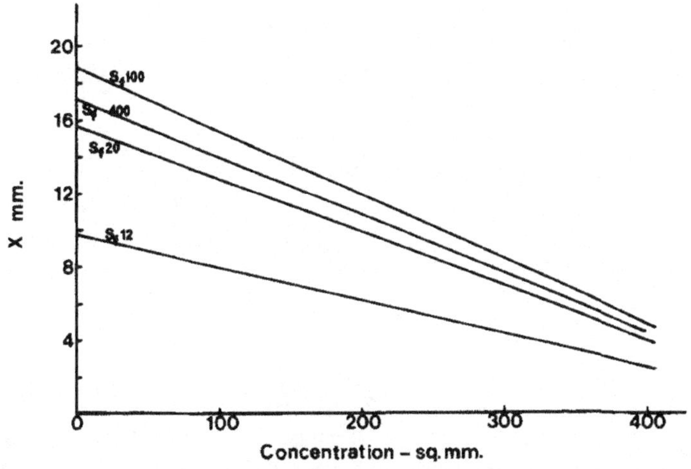

Fig. 7.1. A typical set of the standard curves that represent the function $S_f = S_f^o(1-K'C)$ for the S_f 12, 20, 100 and 400 boundaries. Points on the vertical axis represent the position of the boundary in the cell, at the time the photograph was taken, calculated at infinite dilution as described in Section 4.1.2. It is expedient to express the lipoprotein concentration as the area (in sq. mm) of the schlieren curve between the base of the cell and the boundary, therefore K' must also be expressed in these units, as explained on p. 340. Note that these curves are not linear at high concentrations (i.e. above about 350 sq. mm).

7.1). These standard curves are characteristic of each ultracentrifuge, provided that its calibration and the conditions of the experiment are kept constant. To determine the position of a boundary, estimate approximately where it will be and determine the area under the schlieren curve between that point and the base of the cell. Plot this point on the calibration curve. In general, it will lie to one side of the relevant line. Repeat this procedure for an estimated position that will lie on the opposite side of the line. Join the two points with a line that will cross the calibration curve at a point that marks the real position of the boundary and the area under that section of the curve.

The alternative to this procedure is to use a digital computer to find the position of the boundary. By contrast with the manual method, for which the boundaries must always be the same, the computer procedure allows them to be set at any convenient points, that may differ from one analysis to another. As before, the area under the schlieren curve is measured to each of these boundaries in turn. The apparent flotation rate at each boundary can be calculated from the equation

$$S_f = (\ln r_o - \ln r) \;/\; 4\pi^2 f^2 t$$

where r is the radial distance of the boundary from the centre of the rotor; r_o is the distance to the base of the cell; f is the speed of the rotor in rev/sec; t is the elapsed time at which the schlieren curve was recorded, including the acceleration time. (It is customary to use one third of the acceleration time as the equivalent at full speed.) This value of S_f can then be corrected to zero concentration by means of the equation

$$S_f^o = S_f \;/\; (1 - KC)$$

where the concentration (C) is expressed in terms of the area under the schlieren curve. It will be necessary to interpolate these corrected readings to obtain the areas at the required integer values of boundary flotation rate.

A second consequence of the dependence of flotation rate on concentration is the Johnston–Ogston effect (Johnston and Ogston, 1946) which requires that the apparent proportions of the lipoproteins as they are estimated from the schlieren pattern should be corrected for the accumulation of a slow-moving component behind a fast-moving boundary. The procedure to be used for a simple mixture of two different macro-molecules is well established and is expressed by the equation:

$$C_S^{obs} = C_S^F \frac{(S_F^{obs} - S_F^S)}{(S_F^{obs} - S_S^{obs})} \tag{1}$$

where S_S^{obs} is the observed sedimentation rate of the slow component; S_F^{obs} is the observed sedimentation rate of the fast component; S_S^F is the sedimentation rate of the slow component in the presence of the fast component; C_S^{obs} is the observed concentration of the slow component and C_F^S is the concentration of the slow component in the presence of the fast component (Schachman, 1959). The treatment to be used for such highly inhomogeneous mixture as the plasma lipoproteins is less certain. The method currently in use was developed at the Donner laboratory by Gofman and his colleagues and is based on the assumption that the lipoprotens between each delimiting boundary along the schlieren pattern can be regarded as a separate substance in the context of the Johnston–Ogston effect. The correction is then made by applying eqn. (1) to each of these 'lipoproteins' (e.g. S_f 0–12, S_f 12–20, etc.) at each boundary in turn. To do this, a value of flotation rate must be assigned to each 'lipoprotein'. This is equated to the S_f at the point that divides into two equal parts the area of the schlieren curve that corresponds to that 'lipoprotein'. Each of these median S_f rates is then corrected to zero concentration as explained above. As we shall describe in detail below, these corrected values can then be used to calculate the values of S_F^S that are inserted into eqn. (1). Since the number of times this equation is used when analysing a pattern is

$$n(n-2) - \sum_{x=2}^{x=n-1} x$$

where n is the number of delimiting boundaries over which the correction is to be made, it can easily be appreciated that the Johnston–Ogston correction is the most time-consuming computation of the whole process of data reduction and is where a digital computer can make its most valuable contribution.

For most 'normal' plasmas, the Johnston–Ogston correction can be reduced to negligible proportions by ensuring that the total concentration of lipoprotein in the centrifuge cell does not exceed 20–25 mg/ml (about $2C_o$). It is easy to exceed this limit if the subject is hypercholesterolaemic (Type II), but the concentration of the isolated lipoproteins can be checked by refractometry (Sections 6.1.1, 7.1) and the solution diluted if necessary. However, at this concentration of lipoproteins, there is some risk that at least a part of the VLD distribution may fall below the limits of detection. When the proportion of VLDL in the sample is high, as in Types I, IV and V, this way of eliminating the Johnston–Ogston effect is less satisfactory and some form of correction should be applied.

Both these corrections for the effects of concentration on flotation rate depend for their success on the assumption that a single value of K can be used for all lipoproteins over the whole range of concentration that is encountered. Surprisingly enough, there is little published evidence to support this. The earliest observations of K originated from the Donner laboratory, where a value of 0.0161 ml/mg was adopted for the flotation of low-density lipoproteins in a solvent of density 1.063 g/ml (deLalla and Gofman, 1954). The experimental basis for this value was not published, but deLalla et al. (1967) declare that the flotation rate of lipoproteins of S_f 6–8 is reduced by one half at a concentration of 32 ml/ml (i.e. $K=0.0156$ ml/mg) and that the S_f vs. C relationship is non-linear at concentrations above about 40 mg/ml. Latterly however, the Donner group appears, for unspecified reasons, to have adopted the value $K=0.0089$ ml/mg for use in their computerised film analysis (Ewing et al., 1965). The experiments of Lee and Alaupovic (1974a) confirm that, when LDL are centrifuged at a density of 1.063 g/ml, S_f and C are linearly related over the range of concentrations normally used but also suggest that, as would be expected from studies of other macromolecules, K is not the same for all lipoproteins. Unfortunately, the concentration in these experiments was not measured in absolute terms. In contrast with the centrifugation at 1.063 g/ml, del Gatto et al. (1959) found that K was at least an order of magnitude smaller for LDL in sodium bromide solution of density 1.20 g/ml. Moreover, whereas K for the smaller VLDL (approx. S_f 20–100) was 0.0064 ml/mg, the value for the larger VLDL (approx. S_f 100–400) was − 0.0056 ml/mg. These observations are partially confirmed by Fisher and Mauldin (1970), and by Fisher et al. (1971), who found that the flotation rate of LDL was almost independent of concentration in solutions of density 1.20 g/ml, at least in the range 2–7 mg/ml. There appear to be no published studies of the relationship between S_f and C at low concentrations, but observations in our laboratory have suggested that K may not only vary considerably from one preparation to another, but may also become less when the concentration is low. Given this degree of uncertainty in K, there is much to be said for performing the ultracentrifugal analysis at the lowest feasible concentration. As to the value of K to be adopted, a compromise of 0.01 ml/mg appears to be roughly correct for the analysis of low-density lipoproteins, and a value not exceeding 0.005 ml/mg for the high-density lipoproteins.

Apart from these considerations, it has been suggested that there may be association–dissociation phenomena to be taken into account when interpreting the hydrodynamic behaviour of lipoproteins such as LDL. In this context, it may be remarked that the smaller apo-lipoproteins can reversibly dissociate from a lipoprotein, with the formation of a protein-deficient particle. Thus two particles are formed that sediment in opposite directions. However, the consequence of this unusual situation, and of association phenomena in general, for the interpretation of the lipoprotein pattern has yet to be clarified and will be ignored in our discussion.

Data required. The following pieces of information are needed for an analysis.

(1) the calibration constant of the centrifuge, which relates area under the schlieren curve to concentration of lipoprotein.

(2) the distance of the base of the cell from the centre of the rotor (r_o).

(3) the time (in sec) for the rotor to accelerate to full speed under

the standard conditions used for the analysis (t_0).

(4) the time that elapses (in seconds) from the reaching of full speed to the taking of the photograph (t_s).

(5) the specific refractive increment of the lipoproteins in the solvent being used (Sections 6.1.1, 7.1).

(6) the value of the coefficient K that relates flotation rate with concentration.

(7) the magnification factor of the photographic enlarger (E).

(8) the area enclosed by the schlieren curve, the baseline and the ordinates erected at different points along the baseline.

(9) the factor by which the lipoproteins are concentrated during their isolation from the plasma (B).

(10) the density (d) and viscosity (η) of the solvent in which the lipoproteins are dissolved. This information is not essential but allows a correction to be made for the slight variations in the properties of the solvent that arise during the isolation process. Both d and η can be estimated by refractometry as explained below ('Preparation of sample'). If the solvent is essentially NaCl or NaBr solution, the required information can be had by reference to Tables 2.2. and 2.3. Otherwise, it will be necessary to prepare the appropriate standard curves for the salt mixture to be used.

Equipment

Preparative ultracentrifuge, rotor and accessories (cf. Chapter 2).

Analytical ultracentrifuge, rotor and accessories. In the majority of these instruments, a conventional schlieren optical system is used, in which the record is made photographically. In the M.S.E. Centriscan however, a scanning optical system is used and the schlieren curve is drawn by the pen of a chart recorder. This technique offers some useful practical advantages but also has the following significant handicap. In this instrument, the lipoprotein distribution and the solvent baseline are recorded in different cells, which it is difficult to make an exact match. Consequently, the superposition of the two curves is often inexact, and the precise area under the schlieren curve is difficult to determine. Optical analysers can also be attached to the

Beckman preparative centrifuge, thus giving this machine at least some of the capabilities of the analytical centrifuge. However, it appears that the performance of this system is, at present, inferior to that of the strictly analytical centrifuge and its place among the techniques of lipoprotein analysis is not yet clear.

In the context of the equipment, it is worth noting that the analysis of low-density lipoproteins requires that photographs be taken at intervals that are outside the scope of the usual sequence of automatic exposures. However, this inconvenience can easily be overcome by adding to the system an external stepping relay that will operate the camera at the required times. Alternatively, this function could be delegated to the micro-computer on which the computations will also be performed.

Centrifuge cells fitted with double-sector epoxy-resin centrepieces 12 mm thick. Two of these can be centrifuged simultaneously if one of them is fitted with a wedge window that will deflect the image towards the top of the camera. If a considerable number of analyses is in prospect, it will be convenient to have at least two sets of cells, but the component parts of these must never be interchanged. The epoxy centrepieces are not strictly designed for use at rotor speeds as high as the 50 000–52 000 rev/min that are customarily employed and the high stresses cause the resin to yield slightly, the centrepiece becoming wedge-shaped. This distortion makes the cell prone to leak and can cause a quartz window to break. If this happens during a run, the loss of fluid from one sector can lead to breakage of the centrepiece also. The distortion of a centrepiece can be monitored with an engineer's dial gauge and will also manifest itself through the need to seal the cell more tightly to prevent leaks. A new, lubricated Beckman cell should be essentially leak-free if tightened to a torque of 60–65 kg/cm but, as the cell ages, it may be necessary to increase this to 80–90 kg/cm. If a much-used cell shows persistent signs of leakage, the centrepiece should be discarded. In our experience, it is possible, in the case of Beckman cells, to postpone this writing-off quite considerably by using washers of greater thickness than usual between the quartz window and its housing. Washers cut from IBM punch cards have proved very satisfactory if they are changed regularly.

Syringes, 1 ml or 0.5 ml glass, for filling the cells. It is desirable to check the calibration of these syringes to ensure that exactly the same volume (0.45 ml) can be delivered into each compartment of the cell.

Volumetric flasks of 1 ml capacity. These take the form of a small test-tube of about 2.5 ml capacity, with calibrations at 0.5 and 1.0 ml. Stoppers are not needed.

A small bench centrifuge with an angle rotor capable of taking the 1 ml flasks.

Vials of about 2 ml capacity in which to store the isolated lipoproteins. These must have closely fitting caps with plastic liners.

Photographic film. Use the film that will give the best available combination of speed, contrast and fineness of grain. Users of the Beckman centrifuge may find it necessary to cut their own film from 20 × 25 cm sheets. Note that, if it is necessary to buy a substantial amount of film at one time, it can be kept almost indefinitely in a deep-freeze.

Dark-room facilities

A photographic enlarger. This will be needed unless the film is to be read on an electro-mechanical device that itself incorporates a projector.

Equipment to determine the area under the schlieren curve. A planimeter can be used but, in the current state of technology, it is probably easier to measure the ordinates of the curve at a number of points and to calculate the areas of the trapezia thus defined on a programmable calculator or a digital computer.

Rotor oven. Because it is at present the convention to run the lipoprotein analysis at 26 °C, it is convenient to keep the rotor at this temperature when it is not in use.

Reagents

Diluents of the appropriate density will be needed for the centrifugal isolation of the lipoproteins, as explained in Section 2.1.

'Baseline solutions'. In the relatively highly concentrated salt solutions that are used as solvents for the lipoproteins in this analysis, centrifugation brings about a significant re-distribution that is manifest as a marked curvature of the baseline that is recorded by the schlieren optics. If the area under the schlieren curve is to be estimated accurately, the lipoprotein curve and the baseline must be superimposed by centrifuging the lipoproteins and a matching 'baseline solution' in the same double sector cell. To ensure that the match is as perfect as possible, the baseline solution and the lipoprotein solvent must contain the same salts, in the same concentrations. For example, if HDL is prepared by adjusting plasma with a mixture of NaCl and NaBr, the same mixture, appropriately diluted with NaCl of density 1.0065 g/ml, must be used as the baseline solution. Note that, if the solvent contains NaBr, it may be necessary to allow for the fall in density that will result from the re-distribution in the preparative centrifuge.

Procedure

(*1*) *Auxiliary measurements*

Calibration of the centrifuge. Before carrying out this operation, it is essential that the optical system is cleaned and focussed in accordance with the manufacturers instructions (cf. also, Gropper, 1964; Richards et al., 1971; Rees et al., 1974). Any subsequent disturbance of the system will necessitate a re-calibration.

The concentration of solute in the centrifuge cell is related to the area under the schlieren curve by the equation:

$$C = \frac{A \tan \theta}{LhMm\,(\Delta n)}$$

where A is the area under the curve as measured on the film; θ is the phase plate angle; L is the optical lever (i.e. distance from the centre of the condensing lens to the phase plate); h is the height of the fluid column in the cell; M is the magnification of the camera lens; m is the magnification of the cylindrical lens and Δn is the relevant value for the specific refractive increment of the solute. *N.B. all linear measurements in this discussion will be in millimetres.*

All these parameters can easily be determined, although the estimation of L requires the use of interference optics. In practice however, it is usual to determine the value of tan $\theta/(LhMm)$ from a single experiment, either using a special calibration cell (Beckman Instruments Inc.), or by centrifuging a solution of serum albumin or sucrose of precisely known concentration. Since these techniques give a calibration factor at a fixed value of θ, the angle must be set at the time of the calibration and then left strictly undisturbed. To ensure that it cannot be inadvertently changed, it may be prudent to disconnect the phase plate from the adjusting mechanism. Of the two procedures, the calibration cell is the quickest and most reliable. Moreover, it also provides an estimate of M, which is needed when the flotation rate is calculated. If albumin is used as a standard, its ash and moisture content must be known. A solution of accurately known concentration (about 1%) in 0.15 M NaCl solution is then prepared by weighing, and is centrifuged with the phase plate set at the angle to be used for the lipoprotein analyses. The centrifugation can be carried out in either the standard cell, or in a synthetic boundary cell. The latter allows observations to be made more quickly, and also permits sucrose to be used instead of albumin. When the schlieren curve has become distinct, it is photographed and its area measured. The calibration factor can then be calculated from eqn. (2), making use of the value $\Delta n = 188 \times 10^{-6}$ ml/mg for the refractive increment of albumin ($\Delta n = 143 \times 10^{-6}$ ml/mg for sucrose in water).

Note. A phase plate angle of about 55° from the vertical should give satisfactory results, but the exact setting is not critical.

Magnification of the camera lens (M). This can be determined by taking a photograph of the calibration cell, or of the counterbalance, and measuring the distance between the vertical lines, or between the reference edges respectively (Chervenka, 1969). Alternative techniques are described by Gropper (1964), Richards et al. (1971), Rees et al. (1974) and in the manufacturer's handbooks. The constants of the Beckman calibration cell are stamped on the casing, while the distance between the reference edges is nominally 1.60 cm. The photographs should be measured with a comparator or a measuring microscope and, like the calibration factor, M must be re-determined whenever the optical system of the centrifuge is cleaned or otherwise disturbed.

Radial distance to the base of the cell (r_0). Run the rotor, with a cell and counterbalance, at full speed (usually about 50 000 rev/min) and take a photograph. Measure the distance from the bottom of the cell to the reference edge with a comparator. It is convenient to express the resulting value of r_0 in terms of the notional distance from the centre of the rotor as measured on the film. Thus, if M is approximately 2, r_0 will be about 142.75 mm in the Beckman centrifuge.

The 'up-to-speed' time (t_0). When the computations are performed entirely on the computer, it is feasible to measure the time taken to accelerate the rotor to top speed every time an analysis is made, and to include this as one of the items of data. Even under these circumstances, it may be desirable that t_0 should not vary too widely. If the computation is done manually, or has a large manual element, the rate of acceleration must be as nearly as possible identical for each run. On the Beckman centrifuge this is to say that the drive current must be kept constant at the highest practicable level (i.e. without blowing the fuse), until the run speed is reached. The value of t_0 can then be equated with one third of the acceleration time.

Magnification of the photographic enlarger (E). Perhaps the best standard for this purpose is a microscope eyepiece graticule that takes the form of a 1 cm square that is ruled into 100 1-mm squares. This will require to be supported on a slide that can be inserted into the enlarger. Alternatively, an acceptable standard can be made by scribing a circle of exactly 1 cm radius on an exposed photographic film with a pair of draughtsman's dividers. The film should then be mounted in the manner of a projection slide. Insert the standard into the enlarger and adjust to give a sharp image at the desired magnification, which can be determined with an engineer's precision scale. If the computations are to be done manually, it will be convenient to make E exactly equal to 5.

(2) Preparation of the sample

It is customary to isolate the lipoproteins by a sequential centrifugation, first in a solvent of density 1.063 g/ml and then after readjustment to 1.20 g/ml. For special purposes however, other densities may be adopted, for example, it may be convenient to use 1.0065 g/ml for the analysis of VLDL or, under some circumstances, to use 1.100 g/ml for the analysis of LDL (Mills et al., 1972). Unfortunately, because it is then necessary to adjust the computational procedure, this use of the non-traditional solvent densities may only be feasible if advantage can be taken of the flexibility that results from the use of a digital computer. The details of the isolation procedure are described in Section 2.1. A volume of 2 or 3 ml of plasma usually yields a lipoprotein sample of a concentration such that it can be analysed satisfactorily but in which the Johnston-Ogston effect will be minimised. If the plasma is either hyper- or hypo-lipoproteinaemic, the volume that is taken can be adjusted accordingly.

After the preparative centrifugation, the lipoproteins must be aspirated from the top of the centrifuge tube in exactly 1 ml (Section 2.1.1.7), with every precaution to ensure that recovery is essentially quantitative. In particular,

(1) after the main part of the lipoprotein solution has been taken off, material adhering to the walls of the tube must be recovered. If the tip of the pipette is held parallel to, and in contact with the wall of the tube, liquid will be drawn up by surface tension and a scraping motion will wash the lipoprotein residue into the body of the tube. This operation can be difficult if the concentration of VLDL is high, and it may the be necessary to recover the lipoproteins in a total volume of 2 ml. If this cannot be done, the preparative centrifugation must be repeated with a smaller volume of plasma.

(2) the lipoproteins adhering to the stem of the tube cap must be recovered with the pipette and the stem washed with a drop of the infranatant solution from just below the lipoprotein layer.

(3) the tube cap must be disassembled and any lipoprotein solution that has penetrated behind the gasket removed with appropriate washing.

If these precautions are scrupulously observed, the recovery of the lipoproteins should exceed 95%.

The isolated lipoproteins must be immediately transferred from the volumetric flask to their labelled vial, which is then kept tightly closed. The analytical centrifugation should be carried out with the least delay.

After the lipoproteins have been collected, the 1 ml of solution immediately below them in the centrifuge tube is carefully aspirated off and stored in a separate vial. The refractive index of this sample can be used

(1) to check that the density of the plasma was correctly adjusted before the preparative centrifugation.

(2) to check that there was no gross failure to recover the lipoproteins quantitatively in the top 1 ml of solution.

(3) to enable an estimate of the solvent density in the lipoprotein sample to be made by comparison with control samples of the relevant salt solution that have been centrifuged under the same conditions as the plasma (Sections 2.1, 2.1.2.5). Because the re-distribution of salt during the centrifugation is reproducible so long as the conditions under which it is carried are always the same, it can be assumed that the solvent density in the top 1 ml lipoprotein fraction is closely approximated by $d_2 - \delta$, were d_2 is the measured density of the second 1 ml fraction and δ is the difference between the densities of the first and second fractions in the control centrifugation. Examples of this redistribution have been described by Lindgren et al. (1964) and by Lee and Alaupovic (1974) but are applicable only to experiments done under the conditions used by these authors.

Clearly, if this estimate of solvent density is to be accurate, there must be no significant contamination of the second fraction with lipoproteins. A recovery of 95% of a sample of 'normal' plasma lipoproteins implies that the refractive index of fraction 2 may be spuriously raised by about 0.0001 units, a value that is approximately equivalent to a density of 0.0005 g/ml in a solution of NaBr. For most samples therefore, an acceptable estimate of solvent density can be made if the refractive index is measured to ± 0.0001. However, when the lipoprotein concentration is high, a recovery of 95% (which is then sometimes more difficult to achieve) implies that lipoproteins may make a detectable contribution to the refractive index of fraction 2.

(3) Analytical centrifugation

The following description will assume that a Beckman Model E centrifuge is used. Slight modifications of detail may be assumed if the instrument is of a different make.

Assemble a double-sector cell in accordance with the maker's instructions and tighten it to a torque of between 70 and 100 kg/cm, using the lowest value compatible with the prevention of leaks. Use a syringe fitted with a blunted 16 gauge needle to inject 0.45 ml of the sample *slowly* into one sector of the cell. Follow this with 0.45 ml of baseline solution into the other sector (N.B. it is prudent always to use the same sector for the sample.) Take great care to avoid scratching the centrepiece when assembling or filling the cell. Note too, that there will be a mismatch of the baselines if the two sectors of the cell contain significantly different volumes of solution. Lindgren et al. (1972a) have described a modification of the centrepiece which ensures that the menisci are the same distance from the centre of rotation but this result can be achieved to a sufficiently good approximation by the careful use of calibrated syringes. Lindgren also recommends that the sealed cell should be weighed to ±0.2 mg before and after the centrifugation to determine whether any leakage has occurred. Although large leaks can easily be detected by observing the position of the meniscus during the run, small ones, which can be a significant cause of inaccuracy in the estimation of flotation rate, may be more easily detected by weighing.

Insert the cell and a matched counterbalance (or a second, wedge-window cell if two samples are analysed simultaneously) into the rotor holes, with the filling holes towards the centre of the rotor. It is important that the sectors of the cells are carefully aligned with the radius of the rotor, or convective disturbances will develop during the centrifugation and will appear as spikes on the schlieren photograph. Even if a microscope is used to ensure an accurate setting, there is a danger that a loose-fitting cell may move while the rotor is being loaded into the centrifuge. This possibility can be minimised by smearing the cell with grease but, as Lindgren points out, it may then be difficult to weigh the cell accurately after the run.

Finally, load the rotor into the chamber, a film into the camera, and carry out the preliminaries that are required by the manufacturer's instruction manual. When these are complete, switch on the drive current and a stop-watch simultaneously. Accelerate the rotor as rapidly as possible, according to the pre-determined protocol, until the required speed, which we shall assume to be about 52 500 rev/min, is reached. At this instant, take the first photograph ($t = 0$) and re-set the stop-watch to zero. If it is low-density lipoproteins that are to be analysed, make two more exposures at $t = 6$ min and $t = 30$ min respectively. If the sample is an HDL, it will be sufficient to take only one photograph at $t = 60$ min. These exposure times have been found by experience to be satisfactory when the rotor speed is 52 640 rev/min but would need adjustment if a significantly different speed is used (see for example del Gatto et al. 1959). Moreover, this system may not be satisfactory if the lipoproteins to be analysed are those of an animal in which the LDL are more dense than those of man. Under these circumstances, a significant amount of LDL may be recovered in the fraction of density 1.063–1.20 g/ml. To estimate these lipoproteins, take an additional photograph at $t = 20$ min during the HDL analysis. However, if lipoproteins with this distribution are to be analysed frequently, it will be better to isolate and analyse the low-density lipoproteins at a density of 1.08 or 1.10 g/ml as mentioned earlier.

(4) Computation of results

The computation of an analysis of high-density lipoproteins is simpler than is the case for low-density lipoproteins insofar as all the data are available on one photograph. For this reason we shall begin with the procedure for HDL. Moreover, this same procedure can easily be adapted to the analysis of other lipoproteins that can be recorded on a single exposure, e.g. guinea pig LDL that has been isolated at $d = 1.100$ g/ml, or even LDL that has been isolated at 1.063 g/ml if the amount of associated VLDL is negligible. If the latter condition is not satisfied, the contribution of the VLDL to the over-all Johnston–Ogston effect cannot be ignored.

The first step is to make an enlarged tracing of the $t = 60$ min photograph on a piece of graph paper. Fix the paper on the base of the enlarger and align the image of base of the cell with one of the vertical scale lines near the right-hand margin. If the base of the cell is not clearly defined, measure off, from the chosen scale line, the known distance from the base of the cell to the reference trace (allowing for the magnification of the enlarger). Then set the reference trace at this point. Then trace the complete schlieren peak as shown in Fig. 7.2, using a hard pencil to draw along the centre of the

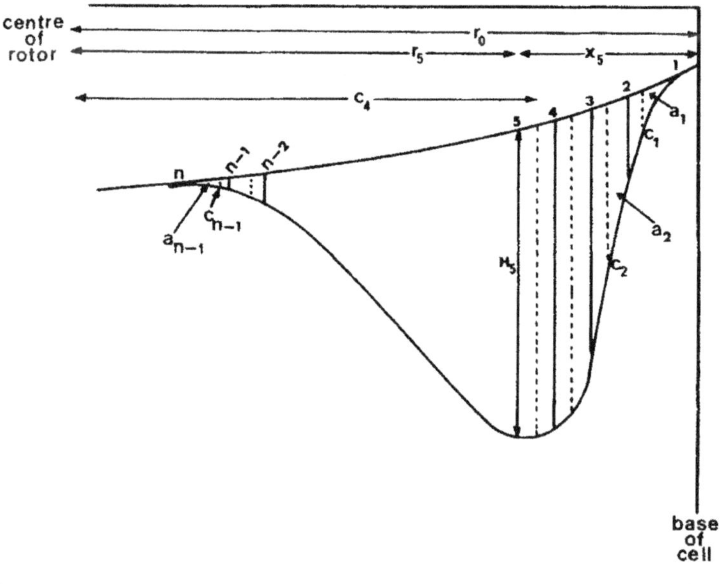

Fig. 7.2. Schematic summary of the measurements needed for the estimation of lipoprotein concentration from the ultracentrifuge schlieren curve. Notation as defined in the text.

TABLE 7.2

The calculation of a quantitative lipoprotein distribution from a single schlieren diagram: each column shows the results obtained at successive stages of the computation of a typical experiment in which the area of the peak was divided into 12 segments by 13 boundaries. The nomenclature is as defined in the text. In this example, the parameters have been set as follows:

Rotor speed = 52640 rev/min
$t_o + t = 3750$ sec
$r_o = 142.75$ mm (as measured on the photographic film)
$E = 6.13$
$K = 0.016$ ml/mg
$\Delta n = 0.00149$
Calibration of the centrifuge (w) = 0.145 mg/ml/sq. mm.
$B = 4$

n	X_n	H_n	r_n	h_n	C_n	A_n	F_n	F°_n	FC_n	FC°_n	L_n
1	1.04	0.00	142.58	0.00	141.97	2.81	0.11	0.11	0.48	0.48	9.50
2	6.31	39.85	141.72	6.50	141.25	7.54	0.64	0.64	0.93	0.94	24.08
3	11.34	73.38	140.90	11.97	140.47	10.98	1.15	1.17	1.41	1.47	34.04
4	16.31	95.87	140.09	15.64	139.70	11.93	1.65	1.74	1.90	2.02	38.11
5	21.15	92.99	139.30	15.17	138.71	16.35	2.15	2.33	2.52	2.79	55.53
6	29.06	66.51	138.01	10.85	137.63	7.63	2.96	3.35	3.21	3.66	26.97
7	34.08	51.12	137.19	8.34	136.82	5.51	3.48	4.02	3.72	4.33	20.43
8	38.99	36.11	136.39	5.89	135.98	4.29	4.00	4.69	4.26	5.02	16.17
9	44.26	28.26	135.53	4.61	135.13	3.35	4.55	5.40	4.82	5.74	12.84
10	49.53	21.88	134.67	3.57	134.21	2.74	5.12	6.12	5.42	6.51	10.68
11	56.64	8.95	133.51	1.46	132.84	1.62	5.87	7.08	6.32	7.63	6.06
12	66.94	3.74	131.83	0.61	130.55	1.24	6.98	8.46	7.84	9.52	4.27
13	93.79	0.00	127.45	0.00			9.95	12.10			

line projected by the enlarger.

From this drawing, measure and tabulate the values of X_n and H_n where H is the distance from the baseline to the lipoprotein curve at a point X mm from the base of the cell (Table 7.2). H can be measured directly from the printed graph scale or, if preferred, with an accurate rule. Next, calculate the notional distance of each boundary from the centre of the rotor as if it had been measured on the film itself (r_n) and the corresponding diminished values of the ordinates (h_n), by making use of the equations:

$$r_n = r_o - \frac{X_n}{E}$$

$$h_n = \frac{H_n}{E}$$

where r_o and E are defined in section (1) above.

If the area (a_n) that lies between a pair of ordinates (or boundaries) is considered as approximating to a trapezium, it can be determined from the equation:

$$a_n = 0.5(h + h_{h+1})\,(r_n - r_{n+1})$$

At this stage it is also convenient to apply the so-called Moring-Claesson correction for the time-dependency of the apparent concentration in the cell (Moring-Claesson, 1956). The area corrected for this effect (A_n) is given by:

$$A_n = a_n r_n / r_o$$

To proceed further with the analysis, it is necessary to find a value of r that can be used to characterise the average member of the population of lipoprotein particles that lies between each pair of boundaries. This value of r is defined as the point at which an ordinate will divide the trapezoidal area between the boundaries into two equal parts (the 'mid-area' point, Fig. 7.3). To obtain these values, first use the equation

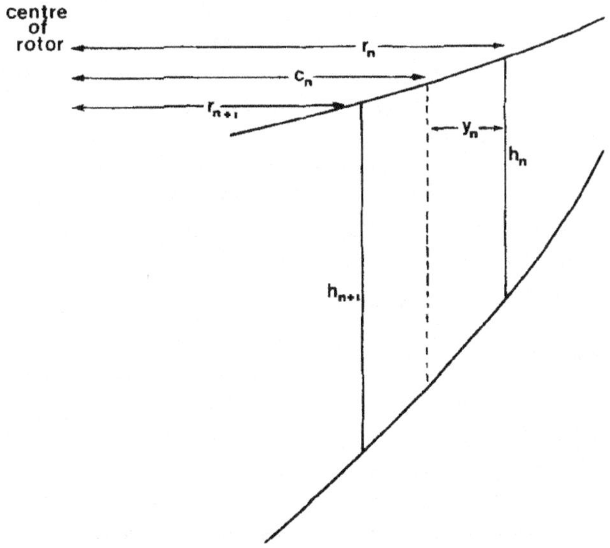

Fig. 7.3. One of the approximately trapezoidal areas into which the schlieren curve is divided, showing the bisection into two equal parts.

$$P_n = \frac{(0.5\,(h_n^2 + h_{n+1}^2))^{1/2} - h_n}{h_{n+1} - h_n}$$

to calculate P for each pair of boundaries, where P_n is the ratio $y_n/(r_n - r_{n+1})$ and y is defined as in Fig. 7.3. The distance of the required 'mid-area' point from the centre of the rotor (C_n) is then given by

$$C_n = r_n - P_n(r_n - r_{n+1})$$

Note that if h_{n+1} equals h_n, the equation for P becomes indeterminate and P must be set equal to 0.5. Remember that, if there are n boundaries altogether, there will be only $(n - 1)$ values for C and A.

It is at this stage that the apparent flotation rate corresponding to each value of r must be calculated from the expression

$$F'_n = \frac{\ln r_o - \ln r_n}{(2\pi f)^2 (t_o + t)\, 10^{-13}}$$

where f is the speed of the rotor in rev/sec and $(t_o + t)$ is also expressed in seconds. The same operation must be also be carried out for each of the 'mid-area' points, namely

$$FC'_n = \frac{\ln r_o - \ln C_n}{(2\pi f)^2 (t_o + t)\, 10^{-13}}$$

Note that the factor 10^{-13} is included in the denominator because it is convenient for the value of F to be determined in svedberg units.

The values of flotation rate that are obtained in this way are subject to the small errors that result from the inevitable slight variations in the salt composition of the lipoprotein solvent that were discussed earlier. These could be corrected by dialysis but this would lead not only to an uncertainty in the concentration, but to an unwelcome delay in the analysis. It is more convenient to calculate a correction from the relationship

$$F = F' \frac{(d_s - \sigma)\eta}{(d - \sigma)\eta_s}$$

where d_s and η_s are the putative density and viscosity of the solvent at the temperature of the analysis; d and η are the real values as estimated by Lindgren's refractometric technique (cf. 'Preparation of sample'), and σ is the hydrated density of the lipoprotein. Unfortunately, because σ is not a constant for all lipoprotein particles, the exact values that are required for this calculation are unknown. Nonetheless, there is a general correlation between flotation rate and particle density (Table 4.2) and Lindgren has made use of this to circumvent the difficulty in the following way. The measured value of F' is used to make a preliminary estimate of σ by interpolation of the tabulated data. This value is then used to calculate a roughly corrected F which is used in turn to find a revised estimate of σ. After several repetitions of this process, the corrected values of F will converge.

Several objections can be levelled at this procedure. Firstly, the iteration is tedious if carried out manually. Secondly, the published data that relate F and σ for high density lipoproteins are sparse. More data are available for the low-density lipoproteins, although even these are not remakable for their consistency. Thirdly, there may be some uncertainty about the significance of the correction. If d differs from d_s by 0.001 g/ml, the flotation rate will be in error by about 0.25 units for an LDL of S_f 12, and by about 1 unit for a particle of S_f 100. This will affect the results of the analysis in two ways, (1) there will be a small uncertainty in the identification of the boundaries and (2) there will be an error in the estimates of lipoprotein concentration. However, since the latter are obtained by difference, the change in the relative positions of the boundaries is probably negligible: only 0.6 svedbergs in the case of the S_f 20–100 lipoproteins in our example. The effect of this on the apparent mass distribution of the lipoproteins will depend on the shape of the distribution curve but, in a typical normal distribution, a change of 0.001 g/ml in d_s increased the S_f 12–20 class from 73.87 mg/dl to 75.61 mg/dl; an increment of 2.36%. Moreover, our observations suggest that $(d_s - d)$ is less than 0.001 mg/ml in 85 % of analyses and, if the value is large, it may be questioned whether the adjustment of the density or the recovery of the lipoprotein was at fault. It may therefore be said that, in most cases, this is a second order correction to a measurement that suffers from first order errors and those who lack computer facilities need have few qualms about omitting it.

For those who do have a computer, the procedure can best be illustrated by an example. Suppose a sample of low-density lipoproteins has been centrifuged in a NaCl solution that is estimated by refractometry to have a density of 1.062 g/ml and a viscosity of 1.0145 cp. Now consider a single boundary, for example one that has an apparent flotation rate of 8 units. By interpolation (Table 7.3), a lipoprotein of this S_f would have a hydrated density of about 1.0288 g/ml. Thus its flotation rate in NaCl of density 1.063 g/ml would be

$$8\,\frac{(1.063-1.0288)\ 1.0145}{(1.062-1.0288)\ 1.0173} = 8.22$$

TABLE 7.3

This table presents the coordinates from which the hydrated density (σ) of a lipoprotein of flotation rate S_f (in NaCl of density 1.063 g/ml at 26 °C) can be approximately estimated by linear interpolation of ln S_f and σ. These points were determined by fitting regression lines to the published data by the method of least squares.

S_f	ln S_f	σ (g/ml)
0.001	−7	1.063
4.0	1.386	1.049
164.0	5.1	0.939
2000.0	7.6	0.920

Interpolation using this value of the S_f gives a revised σ of 1.0277 g/ml. Use this to calculate a new approximation to the S_f i.e.

$$8 \frac{(1.063-1.0277)\ 1.0145}{(1.062-1.0277)\ 1.0173} = 8.21$$

A third interpolation from this value also gives σ = 1.0277 g/ml. S_f = 8.21 is therefore a close approximation to the corrected flotation rate. This whole procedure must be repeated for each value of F'_n and FC'_n.

It is at this stage in the computation that the corrections that derive from the concentration dependence of F can be introduced. The first, and the simplest of these to carry out, is the adjustment of each value of F_n to zero concentration by means of the expression

$$F^0_n = F_n \,/\, (1 - K' \sum_{1}^{n-1} A_i)$$

where F^0_n is the corrected flotation rate and K' is the coefficient that relates F to the lipoprotein concentration when the latter is expressed as area on the photographic film. Thus, if the calibration of the analytical centrifuge is such that 1 sq. mm is equivalent to w mg/ml of lipoprotein in the cell, $K' = K \cdot w$.

The second of these operations is the correction for the Johnston–Ogston effect, for which it is necessary to reduce the values of FC_n to zero concentration by means of the expression

$$FC_n^o = FC_n \,/\, (1 - K' \,(\sum_1^{n-1} A_i + 0.5\, A_n))$$

In describing the subsequent steps in the procedure it will be convenient to use the term Q_n to stand for the expression

$$(1 - K' \sum_1^{n} A_n)$$

The correction factor ($J_{j,i}$) that applies to the lipoproteins of the jth species at the ith boundary can then be expressed as

$$J_{j,i} = \frac{FC_i - (FC_j^{o})\ Q_{(i-1)}}{FC_i - (FC_j^{o})\ Q_i} \qquad (1)$$

where j takes all integer values from 1 to $(n-2)$ and, for each value of j, i takes all integer values from $(j+1)$ to $(n-1)$ in turn. The concentration of the jth lipoprotein species, corrected for the Johnston–Ogston effect at the ith delimiting boundary ($A'_{j,i}$) is then given by

$$A'_{j,i} = J_{j,i}\ (A_{j\ (i-1)})$$

and the corresponding effect on the ith species by

$$A'_i = A_i + (1 - J_{j,i})\ A_{j\ (i-1)}$$

In these equations, the values to be used for $A_{j\,(i-1)}$ and A_i are the partially corrected results that have been computed in the earlier steps of the chain of calculations.

To make this procedure clear, we shall illustrate it by the following example, which may be taken to represent an analysis of HDL. The data will be those given in Table 7.2 and, to ensure that the results of the correction are easily seen, we shall adopt the relatively high value of $K = 0{,}016$ ml/mg, which is equivalent to $K' = 0{,}0023\ \text{mm}^{-2}$.

(1) To correct the concentration of the first sub-species of lipoproteins (i.e. those corresponding to A_1) for the Johnston–Ogston effect at the boundary between the first and second sub-species (i.e. to obtain $A'_{1,2}$), calculate the relevant factor

$$J_{1,2} = \frac{0.928 - 0.481\ \ (0.993)}{0.928 - 0.481\ \ (0.976)} = 0.982$$

whence the corrected concentration of the first sub-species is

$$A'_{1,2} = 0.982\ A_1 = 2.81 \times 0.982 = 2.755 \text{ sq. mm.}$$

and that of the second,

$$A'_2 = A_2 + (2.81 - 2.755) = 7.595 \text{ sq. mm.}$$

(2) The first sub-species must now be corrected for the effect at the boundary between the second and third sub-species, i.e. at r_3. Here,

$$J_{1,3} = \frac{1.414 - 0.481\ (0.976)}{1.414 - 0.481\ (0.950)} = 0.987$$

whence the concentration of the first sub-species in the presence of the second,

$$A'_{1,3} = 2.755 \times 0.987 = 2.719 \text{ sq. mm.}$$

and

$$A'_3 = 10.98 + 0.036 = 11.016 \text{ sq. mm.}$$

(3) Continue this sequence of operations until the penultimate boundary is reached, when the factors to be applied to the first sub-species in the presence of the eleven of higher flotation rate is

$$J_{1,12} = \frac{7.84 - 0.481\ (0.825)}{7.84 - 0.481\ (0.823)} = 1.00$$

The eleven corrected concentrations of the first lipoprotein sub-species that result at this stage in the procedure are presented in the third column of Table 7.4.

(4) Next, treat the partially corrected value of A_2 from step (1) by the analogous sequence, beginning with

$$J_{2,3} = \frac{1.414 - 0.942\ (0.976)}{1.414 - 0.942\ (0.950)} = 0.953$$

whence

$$A'_{2,3} = 7.595 \times 0.953 = 7.238 \text{ sq. mm.}$$

and

$$A'_3 = 11.016 + 0.357 = 11.373 \text{ sq. mm.}$$

Note that this value of A'_3 is obtained from the partially corrected one that was calculated in step (2) above.

Continuing the correction over the next boundary,

TABLE 7.4

This table shows the successive stages of the Johnston–Ogston correction as applied to the example of Table 7.2. The value $K = 0.016$ ml/mg has been adopted to ensure that the effect is of a perceptible magnitude. The nomenclature is as defined in the text. The corrected areas (A^1) are those given at the foot of each column e.g. the corrected value of A_1 is that at the foot of the third column. Note that the operation has the effect of diminishing the values of A_n where n is small and of increasing them when n is large.

n	A_n	$A'_{1,i}$	$A'_{2,i}$	$A'_{3,i}$	$A'_{4,i}$	$A'_{5,i}$	$A'_{6,i}$	$A'_{7,i}$	$A'_{8,i}$	$A'_{9,i}$	$A'_{10,i}$	$A'_{11,i}$	$A'_{12,i}$
1	2.81	2.755											
2	7.54	2.719	7.241										
3	10.98	2.694	7.056	10.507									
4	11.93	2.671	6.905	10.024	11.622								
5	16.35	2.663	6.857	9.889	11.333	17.230							
6	7.63	2.658	6.828	9.813	11.185	16.770	8.534						
7	5.51	2.654	6.810	9.765	11.096	16.524	8.267	6.466					
8	4.29	2.652	6.797	9.734	11.040	16.379	8.132	6.283	5.109				
9	3.35	2.650	6.789	9.712	11.001	16.285	8.051	6.188	4.978	4.028			
10	2.74	2.650	6.784	9.702	10.983	16.242	8.018	6.151	4.934	3.973	3.358		
11	1.62	2.649	6.782	9.695	10.973	16.218	8.000	6.133	4.915	3.951	3.332	1.931	
12	1.24												1.409

$$J_{2,4} = \frac{1.895-0.942\ (0.950)}{1.895-0.942\ (0.922)} = 0.974$$

$$A'_{2,4} = 7.056 \text{ sq. mm.}$$
$$A'_{4} = 12.141 \text{ sq. mm.}$$

When this sequence has been completed to the 12th boundary, proceed to treat each successive value of A'_n in a similar way. This will eventually yield the matrix of data that is presented in Table 7.4, in which the finally corrected values of A_n are those at the foot of each column. If these are compared with the uncorrected values in the second column, it will be seen that the resultant of the correction procedure is to diminish the concentration of the slower moving components and to increase that of the faster ones by an equivalent amount.

When the Johnston–Ogston correction is completed, it remains only to convert each value of A'_n to the equivalent concentration of lipoprotein (L_n), in mg/dl of plasma, under the conditions at the start of the centrifugation, by means of the expression

$$L_n = \frac{100\ w\ A'_n}{B}\left(\frac{C_n}{r_o}\right)^2 \tag{2}$$

where w is the centrifuge calibration (mg/ml equivalent to 1 sq. mm) and B is the factor by which the lipoproteins were concentrated during their isolation. The term $(C_n/r_o)^2$ corrects for the concentration of the lipoproteins that occurs as they float up the sector-shaped cell. Each value of L then represents the effective concentration of lipoprotein to be found between two delimiting boundaries on the scale of flotation rate. These values of F will not generally be integral and it is expedient to sum the n values of L and interpolate to find the concentrations of lipoprotein between the conventional, or at least more convenient, limits.

It will be seen from this discussion that, while no single stage of the calculation is particularly difficult, the entire sequence is a considerable labour that requires a substantial 'memory' in which to store the intermediate data that are accumulated during the computation.

The number of readings of X that are taken must therefore be tailored to the capacity of the computing facilities that are available. A main-frame computer can easily cope with the load imposed when $n = 50$ or 100. With a sophisticated desk calculator, a value of 4 or 5 might be more realistic. However, the most practicable and economic facility is probably a micro-computer with an extended memory, which will allow an adequate resolution to be obtained from a single schlieren photograph.

Although the method of computing an analysis of low-density lipoproteins (S_f 0–400) is essentially the same as that just described, two complications are introduced by the fact that the data are recorded in three photographs that are taken at different times. In the first place, the three records must be overlapped to form an effectively continuous panorama of the lipoprotein distribution. Secondly, the technique for carrying out the Johnston–Ogston correction must take account of the progressive removal of the larger lipoproteins from the system during the centrifugation.

The essence of matching the records to form a continuum is the identification in one frame, of a value of r_n that is equivalent to the same S_f as the value r_m in the next later frame (Fig. 7.4). Traditionally, this has been done by determining the position of the S_f 100 boundary in the first and second photographs, and that of the S_f 20 boundary in the second and third. The simplest and in some ways the least objectionable procedure is to treat each photograph individually by the method described above. After the final interpolation of the three sets of results, the quantitative estimate of the S_f 100–400 lipoproteins can be taken from the analysis of the first photograph, that for the S_f 20–100 substances from the second and that for the S_f 0–20 from the third. One drawback to this procedure is the time-wasting transcription of all the data from each photograph, when only a part of it will be used. This objection can be minimised by taking only a few, widely spaced readings in the redundant parts of each frame, but there remains the more cogent reservation that it is difficult to make accurate measurements of area close to the base of the cell in the early photographs. In the alternative technique developed at the Donner

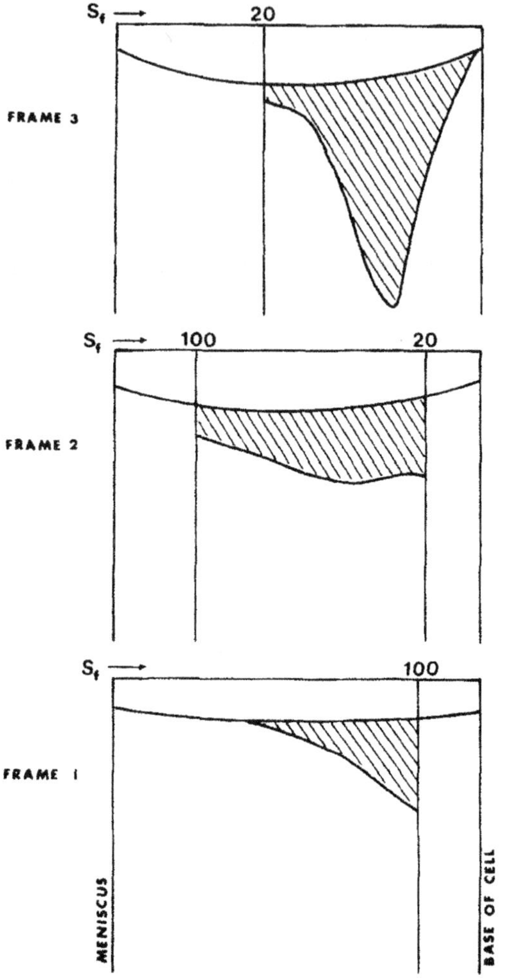

Fig. 7.4. The use of equivalent boundaries in matching the segments of the low-density lipoprotein distribution that are estimated in different ultracentrifuge photographs. The '100' and '20' boundaries mark the positions that particles of these S_f rates had reached at the time of the photograph. The hatched areas are those parts of the schlieren pattern that are actually used to make the estimation.

laboratory, redundant material is eliminated by first calculating the positions of the S_f 20 and S_f 100 boundaries at zero concentration. These can then be used to mark off the unwanted parts of each schlieren pattern. But, in reality, the boundaries are never at these positions and some of the lipoproteins that are ascribed to one photograph are actually measured in the adjacent one, where they are found in a different part of the cell (cf. Fig. 7.4). Compensation must then be made for the differences in concentration that arise from the sector shape of the cell. Alternatively, it can be assumed that the lipoproteins that have actually reached (for example) the apparent S_f 20 position in the second photograph will also have reached the equivalent nominal position in the third photograph. This premise will not be strictly true because the sectorial concentration of the lipoproteins will be slightly greater in the third photograph than in the second. An approximate correction can be made for the resulting error but, in practice, it is usually a second order effect that can be neglected.

The need to adapt the Johnston–Ogston correction arises when the calculation is done after the three patterns have been joined to form a continuum, and can be explained as follows. At the time of the first exposure, virtually all the low-density lipoproteins are still present in the cell and it is justifiable to make an estimate of the magnitude of the Johston–Ogston effect as it applies to the smallest lipoproteins in the presence of the largest. In later photographs however, the larger particles will have floated to the meniscus and will no longer contribute to the effect. Consequently, the procedure described for a single exposure must be adjusted to ensure that lipoproteins which are measured in the second or third frames are corrected only for the effects of those particles that are present in the cell at that time.

We shall illustrate the analysis of the low-density lipoproteins by an example. Firstly, calculate the distance of the S_f 20 boundary from the base of the cell, at zero concentration, in the second and the third exposures. Note that these distances depend on the speed of the rotor, the time of the exposure, the temperature of the centrifugation, the magnification of the optical system of the centrifuge, and the magnification of the enlarger. Consequently, if any of these variables is changed, the positions of the boundaries must be re-calculated. Next, determine the position of the S_f 100 boundary in the first and second frames in the same way. Mark these position on the sheet of graph paper and trace the appropriate portions of the enlarged schlieren patterns as indicated in Fig. 7.4. Also mark the positions of the meniscus in the second and third frames. If a large number of these analyses is to be done, it may be expedient to have the standard boundary lines accurately printed onto the sheets on which the enlargements will be traced.

Each of these tracings is now treated as in the single-frame analysis, beginning with the third exposure and ending with the first, but with the following modification. The cumulative lipoprotein concentration that is used to calculate F_n^0 in the second frame $(\sum A_i)_2$ must be set equal to

$$(\sum A)_3 + (\sum_{1}^{n-1} A_i)_2$$

where the first term is the total area under the peak from the base of the cell to the S_f 20 boundary in the third exposure, and the second term is the area from the S_f 20 to the $(n-1)$th boundary in the second frame. The cumulative area to be used in the first photograph is obtained in the analogous way.

For the Johnston–Ogston correction, the data from all three frames are treated in sequence, as though they originated from a single exposure. Thus, if there are n values of A in the third frame, the first of the m values measured in the second frame becomes A_{n+1} while the first of the p values measured in the first frame becomes A_{n+m+1}. The sequence of corrections to each value of A is therefore continued, as described for the single-frame analysis, until $i = n+m+p$. However, as we have said, it is necessary to distinguish the correction to A_j that is consequent upon the presence of faster moving lipoproteins in the third frame, from the correction to the even larger lipoproteins that results from the presence of the A_j species in frames 1 and 2. To do this, first calculate the flotation rates that are equivalent to the positions of the meniscus in the second and third exposures (M_2 and M_3). In the third frame, lipoproteins with a flotation rate greater than

M_3 (37.2 S in a typical case) have reached the meniscus and the only values of $J_{j,i}$ (eqn. 1) that apply to the S_f 0–20 lipoproteins are those which result when i is less than, or equal to M_3. Thus, provision must be made for recording the corrected values of the n readings of A that are measured in frame 3, at the point in the computation when i exceeds M_3. These are the values of A'_n that will be used in eqn. (2) to obtain the concentration of the S_f 0–20 lipoproteins. Similarly, when the second exposure was made, lipoproteins of flotation rate up to M_2 (i.e. about 155 S) were present in the cell. Accordingly, the m values of A that form the S_f 20–100 lipoproteins must be corrected up to the point where i becomes equal to M_2, at which stage they are recorded. Finally, the corrections to the remaining p values of A, that were measured in the first frame, are continued until the sequence is exhausted. Before leaving this description of the procedure, it should be emphasised that the Johnston–Ogston corrections to values of A that are measured in frames two and three do not cease at the menisci in those frames. In every case, the computation is continued across the full extent of the data from all three frames. The function of the procedure just described is to record the values assigned to A at points along the sequence of corrections that are fixed by the positions of the menisci.

When the calculation of the $(n+m+p)$ values of A' is completed, they can be converted to the concentration of lipoprotein in plasma (L) by means of eqn. (2). Note here that, since C_n is a measure of the distance of the area A_n from the centre of the rotor, its value returns towards r_0 at each transition from one frame to another. Care must therefore be taken to ensure that the values of C_n and A_n correspond. Finally, interpolate the values of L to obtain the total concentration of lipoprotein between the desired limits of flotation rate e.g. S_f 0–12, S_f 12–20, etc.

In a method of analysis as complex as the one we have been describing, there are many sources of error and variability. Some of these relate to the theoretical foundations of the method. For example, it is uncertain whether the model used to describe the hydrodynamic behaviour of the lipoproteins in the centrifuge cell accurately represents the properties of so heterogeneous a system. It is not known whether the procedure used for the Johnston–Ogston correction is an adequate approximation. Likewise, it has been suggested that at least a part of the system is a dissociating one, although this has not yet been verified.

As we have mentioned earlier, there is also uncertainty about the magnitude of some relevant properties of the system, for example, the refractive increment of the lipoproteins and the coefficient relating their flotation rate with concentration.

At a more practical level, it might be anticipated that the recovery of the lipoproteins from the top of the preparative centrifuge tube could be a significant source of error. This was confirmed by Milch et al. (1955) who found that the coefficient of variation for the recovery in replicate centrifugations (determined by weighing) was ± 4 %, although the replicates could differ by as much as 15 %. They also determined the precision of the analytical ultracentrifugation (including the computation stage) for replicate analyses of the S_f 0–12 and the S_f 20–400 lipoproteins. For the former, the overall range was 12.8 % of the mean, with a coefficient of variation of ± 4.2 % but, for the latter, the values were significantly greater, being 50 % and ± 19 % respectively. The relatively low precision of the estimates of VLDL is largely due to the uncertainty that is introduced by the finite thickness of the schlieren trace. Since the work of Milch et al., the magnitude of this effect has been somewhat reduced by the introduction of the phase plate in place of the diagonal wire schlieren diaphragm. Nonetheless, as the following data for duplicate analyses in our laboratory show, the estimations of VLDL are still less precise than those of LDL.

S_f	Mean (mg/dl)	Technical error
0–12	427	±11.3
12–20	66	± 4.9
20–100	134	± 7.3
100–400	98	±16.8

Here, the technical error is defined as

$$\sqrt{\frac{\sum d^2}{2N}}$$

where d is the difference between duplicates and N is the number of duplicate analyses.

7.2. *Nephelometry*

Because lipoprotein particles are of large diameter, they will scatter a significant proportion of an incident beam of light. Unfortunately, the intensity of the scattered light is proportional to the square of the particle weight and, in view of the inhomogeneity of lipoprotein preparations, this sets a severe limitation to the use of nephelometry as a precise method for the determination of absolute lipoprotein concentration. On the other hand, this property makes it possible to design a very simple instrument that will detect the light scattered by lipoproteins but not that scattered by the smaller plasma proteins. This concept was used by Thorp, Horsfall and Stone (1967) to develop an approximate method of estimating low-density lipoproteins for clinical purposes, which has the merit of simplicity, speed and cheapness, and does not require the lipoproteins to be isolated. However, it has the defect that the results are presented in arbitrary units and, since it is barely sensitive to the presence of LDL partcicles and cannot detect HDL at all, it is able to give direct information only about chylomicrons and VLDL. Nonetheless, Stone et al. (1970) showed that it is possible to convert the nephelometric measurement into an approximate estimate of the absolute level of VLDL. Moreover, they showed that, from this value and a total serum cholesterol measurement, it is possible to estimate the LDL level, thus enhancing the utility of the method.

Later designs of nephelometer that use a laser light source and more sophisticated detectors may offer a greater sensitivity than the Stone–Thorp design, but they are no less susceptible to the problems raised by lipoprotein heterogeneity.

It is also possible to use the nephelometer to detect and quantify the aggregation that results from an antibody–antigen reaction. This technique has been used to estimate apo-lipoproteins (Section 8.9) but there has been little report of attempts to apply it to the lipoproteins, even though it would seem to have a potential in the case of the HDL. However, it must be said that, when used in conjunction with the Stone–Thorp instrument, the method is not particularly economical in its consumption of antiserum.

Equipment

Nephelometer. Instruments based on the Stone–Thorp design are manufactured by Scientific Furnishings Ltd. The latest versions incorporate provisions for the automatic compensation of the readings for calibration which, in the older instruments had to be done manually.

Pressure filtration stand. This device, which is illustrated in Fig. 7.5, allows filtration to continue unattended under standard conditions. Its use is not essential. Plastic disposable syringes of 10 or 20 ml capacity. 25 mm stainless steel membrane-filter holders to fit the syringe. 25 mm membrane filters to cover the range of nominal pore sizes from 0.45 μm to 0.05 μm inclusive. Stone et al. (1970) recommend Sartorius filters.

Procedure. Because light scattering measurements are so sensitive to the presence of large particles, the solutions under observation must contain no dust particles, cells or air bubbles. Moreover, the nephelometer can easily detect the aggregation of lipoproteins that is the earliest sign of denaturation, and which may begin within a few hours of the collection of the blood. Measurements must therefore be made on fresh, unfrozen plasma. Note also that the intensity of the scattered light is proportional to the concentration of scattering particles only over a limited range. Above this both the incident and the scattered light are attenuated, leading to an apparent fall in the observed intensity. If therefore, the measured value approaches the upper limit of the range of the instrument, the sample must be diluted to ensure that the region of linear response is being used. Since the intensity of the scattered light is temperature dependent, the measure-

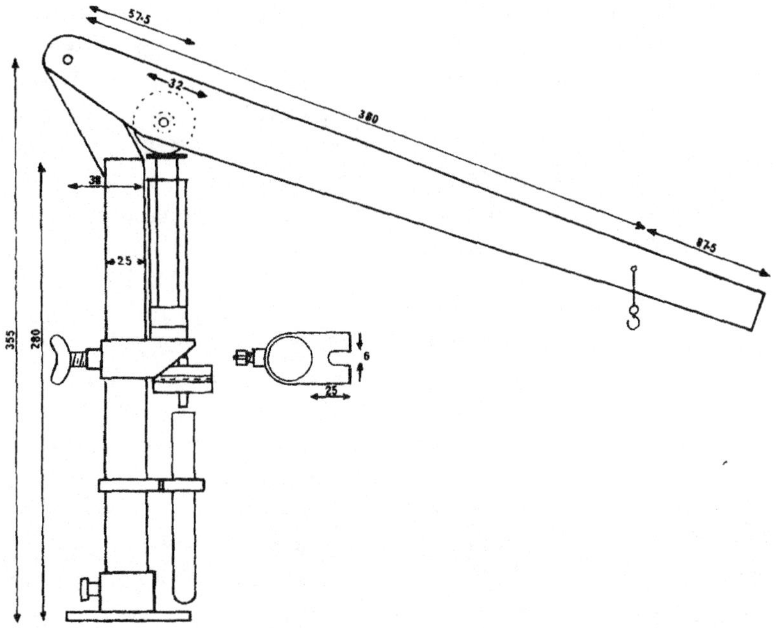

Fig. 7.5. Press for filtering diluted serum through membrane filters (after a design by Stone and Thorp: dimensions in mm). The sample is drawn into 10 ml syringe which is attached to a 25 mm filter holder. With coarse filters, an adequate flow is obtained under the weight of the handle alone. With a filter of 0.05 μm pore size, a 1 kg weight must be hung on the hook. Too great a pressure must be avoided since this leads to an apparent increase in porosity of the filter. The filtrate is collected directly into the nephelometer tube.

ments must also be made always at about the same temperature.

Calibration of the instrument. Because of the manufacturing tolerances on the electrical components, it is necessary to calibrate each instrument in terms of an arbitrary standard of light scattering that is supplied by the manufacturers. Details of the procedure are supplied with the equipment and it should be repeated at regular intervals, lest the calibration drift. Since the standards may likewise not be completely stable, they should also be replaced every year or two.

Preparation of serum. Because the nephelometer is highly sensitive to the presence of chylomicrons, there will be a substantial difference between the readings obtained with sera from fasting and non-fasting subjects. Blood should therefore always be taken after an overnight fast. Allow it to clot and then centrifuge for 20 min at 1500 *g*. Remove the serum with great care to exclude every trace of particulate material and then take the precaution of centrifuging it a second time. Proceed to the light-scattering measurements immediately.

Method of measurement. This is a subtractive technique, i.e. the intensity of the light scattered by the (diluted) whole serum is measured first. The sample is then filtered through a membrane that removes the largest particles (the 'chylomicrons'), and the light-scattering intensity (LSI) measured again. The difference between the two readings is a measure of the number of 'chylomicron' particles, while the second reading is an estimate of the concentration of 'VLDL'. Comparison with the analytical ultracentrifuge has shown that the fraction isolated by filtration through a membrane of pore size 0.05 μm approximates to the conventional VLDL and that the two measures are closely correlated (Stone et al., 1970). However, it is not certain that the published correlation will apply to all models of the Stone–Thorp nephelometer.

To avoid clogging the filter membrane and to bring the intensity of the light scattered by the chylomicrons within the range of the instrument, start by diluting the serum ten-fold with *clean* 0.9 % NaCl solution. If the serum is very lipaemic, it may be necessary to dilute it further but this tends to distort the apparent particle size distribution (Stone et al., 1970). It is therefore important to use the 1 in 10 dilution whenever possible and to exceed 1 in 40 only in the most extreme cases. Note that the LSI depends on the difference between the refractive index of the particle and that of the solvent. Diluted samples therefore cannot be compared directly with the original serum. However, this effect is a negligible source of error in samples that have been diluted by ten-fold or more.

To filter the diluted serum, fill a 10 ml syringe and attach it to a filter holder that is fitted with the appropriate grade of cellulose nitrate filter. Set the assembly vertically in a stand and apply a pressure of not more than 1.75 kg/sq. cm. of membrane. Discard the first six drops of filtrate and collect at least 3 ml of the subsequent flow in a dust-free nephelometer tube. Allow the filtrate to run down the side of the tube to limit the formation of air bubbles. Stand the filtrate for 15 min at room temperature to permit any micro-bubbles to disperse before reading the LSI.

N.B. Reject any nephelometer readings that are not stable. Fluctuations probably indicate the presence of large extraneous particles or bubbles, which must be removed.

The sequence of operations can be summarised as follows:

(a) If the ten-fold dilution of the serum has an LSI (I_t) of less than 100 units, as is usually the case, pass it through a filter of 0.05 μm pore size and read to LSI of the filtrate (I_f). The intensity of light scattered by the chylomicrons and by the VLDL is then given by:

$$\text{Chylomicron-LSI} = I_t - I_f \qquad (1)$$

$$\text{VLDL-LSI} = I_f \qquad (2)$$

(b) If the LSI of the diluted serum is between 100 and 500 units, filter it through a succession of membranes of decreasing pore size e.g. 0.15 μm followed by 0.05 μm. If the serum must be diluted more than 10-fold to reduce the LSI below 500 units, start the serial filtration with a membrane of 0.45 μm.

If the serum is diluted by more than 10-fold, the observations must be corrected to this dilution before they are inserted into eqns. (1) and (2). This can be done by means of the relation:

$$I_{0.1} = A(I_d) + B$$

where $I_{0.1}$ is the required intensity at 1 in 10 dilution, and I_d is the observed intensity at dilution d. The relevant values of the coefficients A and B are given in Table 7.5.

Note that the values of LSI measured after successive filtrations can be used to estimate the relative proportions of chylomicron particles of different sizes, subject to the restriction imposed by the square-law response mentioned above.

Estimation of lipoprotein concentration. The procedure just described provides an arbitrary measure of the serum level of chylomi-

TABLE 7.5
The coefficients of the equation $I_{0.1} = A(I_d) + B$

Material	Dilution (d)	A	B
Diluted, unfiltered serum	1:20	1.91	5
	1:40	3.89	5
	1:80[a]	1.68	29
After filtration through Sartorius 0.05 µm membrane	1:20	2.69	−8
	1:40	6.14	−5
	1:80[a]	2.57	−6

[a] The use of these coefficients corrects the LSI measured at 1:80 dilution to 1:40. Correct a second time to obtain the value at 1:10.

crons and VLDL. However, it is possible to convert the value of VLDL-LSI into the absolute level of VLDL with a modest degree of accuracy and, by making an additional measurement of total serum cholesterol, the level of LDL can be estimated as well.

Stone et al. (1970) found that the level of VLDL determined by analytical ultracentrifugation (S_f 20–400) was closely correlated ($r = 0.974$) with the LSI after filtration through a Sartorius 0.05 µm membrane (i.e. I_f) according to the equation:

$$\text{VLDL} = -0.276 + 0.1004 I_f - 0.000164(I_f)^2$$

where the concentration of VLDL is obtained in g/l serum. This equation can easily be handled on an electronic calculator but can be approximated, with some loss of accuracy, by the following two linear equations:

If $I_f < 70$ LSI units, $\text{VLDL} = 0.112 I_f - 0.58$ ($r = 0.96$)
If $I_f > 70$ LSI units, $\text{VLDL} = 0.31 I_f - 4.1$ ($r = 0.86$)

Note that these equations only apply to the filtrate from a 0.05 µm membrane. For details of the equations that relate to other conditions, consult Stone et al. (1970).

Since there is a close correlation between the total serum cholesterol level and the total level of low-density lipoproteins, it is possible to estimate the LDL concentration by combining a cholesterol measurement with the VLDL-LSI. Stone et al. (1970) found that these three variables were related ($r = 0.83$) by the relation:

$$LDL = 0.81(TC) - 0.19(\text{VLDL}) - 0.1 \qquad (3)$$

where *LDL* is the concentration of S_f 0–20 lipoproteins in g/l serum, *TC* is the total serum cholesterol level in mmoles/l and the VLDL level is that obtained from measurements of LSI as explained above. This equation can only be used in the form given if the estimates of cholesterol duplicate those made by Stone et al. (1970) using the method of Connerty et al. (1961). The resulting estimates of *LDL* are sufficiently accurate for many clinical purposes when the VLDL-LSI is less than 70 units. Above this, the method becomes progressively less reliable, largely because there is a change in the distribution of the VLD-lipoproteins in hyperlipaemic sera that leads to a disproportionate increase in the scattered light. In addition, eqn. (3) cannot be used when the serum is from a patient with Type III hyperlipoproteinaemia, in whom lipoproteins of abnormal composition are present. In these cases, the equation

$$LDL = 0.44\ (TC) - 0.14\ (\text{VLDL})$$

may be used (Stone et al., 1970). The correlation coefficient is only 0.77 however, and the estimates are consequently liable to a considerable uncertainty.

7.3. *The estimation of lipoproteins by chemical methods*

Although they are essentially estimations of lipids or protein, these methods form a part of this guide because they involve the fractionation and manipulation of lipoproteins, and we shall classify them in terms of the procedures that are used to resolve the lipoproteins. The best of these is preparative ultracentrifugation, by which the lipoproteins can be separated into well-defined classes according to their density. They can then be quantitatively estimated by the measurement of, for example, cholesterol. However, the centrifugal fractionation is both slow and expensive, and where the object is speed and economy, the resolution is more often brought about by precipitation or electropheresis. In the following sections we shall describe some typical examples of techniques that can be used in clinical studies where the speedy production of an approximate result is thought to be more important than accuracy or resolution.

7.3.1. Precipitation methods for 'HDL'-cholesterol

Because the 'low-density' lipoproteins can be cleanly and quickly precipitated from plasma with poly-anions, this procedure has been made the basis of an assay for 'HDL'-cholesterol. Simple though it is in concept, the method is less straight-forward than appears at first sight. In the first place, it is necessary to define what is meant by 'HDL'. As it is conventionally prepared by ultracentrifugation, HDL includes small amounts of lipoproteins that contain apo-lipoprotein B and which can be precipitated by poly-anions. In setting the conditions for the assay therefore, it has come to be an accepted criterion of an efficient fractionation that all the B-containing lipoproteins are precipitated, but none of those in which the major protein is apo-lipoprotein A. The complete removal of the low-density lipoproteins is particularly important because they far exceed the amount of the 'HDL' and are rich in cholesterol.

The general principles of the precipitation procedure have been discussed in Section 2.2.3. If the sodium phosphotungstate–Mn^{2+}, the dextran sulphate–Mg^{2+}, the heparin–Ca^{2+} and the heparin–Mn^{2+} systems are directly compared with each other, or with the fractionation by ultracentrifugation, the resulting estimates of HDL are found to be highly correlated (Bachorik et al., 1976; Ishikawa et al., 1977; Lopes-Virella et al., 1977; Srinivasan et al., 1978; McTaggart et al., 1978; Boni and Hendrikx, 1980; Seigler and Wu, 1981). Nonetheless, a careful study by Albers et al. (1978) showed that both phosphotungstate–Mg^{2+} and dextran sulphate–Mg^{2+} yield results for 'HDL'–cholesterol that are consistently some 3 mg/dl lower than those obtained with heparin–Mn^{2+}. By contrast, heparin–Ca^{2+} tended to give higher results. By immunochemical analysis it was also shown that this disagreement arose because phosphotungstate precipitated some of the HDL, whereas the removal of the B-containing lipoproteins was incomplete. In a comprehensive survey of the heparin–Mn^{2+} system which extended the earlier work of Bachorik et al. (1976), Ishikawa et al. (1977) found that the efficiency of the precipitation was unaffected by changes in the final heparin concentration between the limits 92–734 USP units/ml. Moreover, they observed that the concentration of Mn^{2+} could vary between 0.046 M and 0.23 M without affecting the amount of cholesterol found in the supernatant, but that a decrease of Mn^{2+} to 0.042 M led to a perceptible fall in the efficiency of the precipitation. Because the usually recommended concentration of 0.046 M Mn^{2+} was clearly very close to the lowest acceptable limit, Ishikawa et al. suggested that it would be prudent to use a higher concentration. Subsequently, as a result of a more critical examination, Warnick and Albers (1978) proposed that the standard Mn^{2+} concentration should be raised to 0.092 M because this improved the precipitation of the B-containing lipoproteins 'without excessive precipitation of high density lipoproteins from plasma'. This was later supported by Mao and Kottke (1980). Note however the reference to plasma. It appears that, whereas 0.046 M Mn^{2+} is adequate for the complete precipitation of the B-containing lipoproteins from serum, the presence of EDTA in plasma reduces the concentration of Mn^{2+} below the critical effective limit. However, there is an additional reason for using the higher level of Mn^{2+}. When the amount of VLDL in the plasma is large, the insoluble heparin complex is sometimes difficult to sediment completely. The proportion of samples in which this occurs is reduced to about 2.5 % when the precipitation is carried out in 0.092 M Mn^{2+}. The turbidity in this small number of samples can be avoided, or removed, in several ways:

(1) If the plasma is diluted with an equal volume of 0.15 M NaCl before the precipitation, the sedimentation is often made more effective. However, dilution tends to diminish the precision of the cholesterol measurements.

(2) The turbid supernatants can be re-centrifuged at high speed e.g. for 10 min at 12 000 *g*. The residual VLDL precipitate will then float and the clear solution can be carefully aspirated from beneath the pellicle.

(3) Warnick and Albers (1978b) recommend that these turbid supernatants are filtered through a 0.22 μm filter (25 mm diam.) which is protected by an AP15 and an AP20 prefilter in a Swinnex holder (Millipore Corp.).

Of these procedures, the last appears to be the most suitable for routine work.

As with other methods of lipoprotein analysis, it is important that the samples of plasma should be fresh. Warnick et al. (1979) found, for example, that estimates of HDL–cholesterol decreased in sera that were stored at 4 °C. Moreover, Bachorik et al. (1980) detected progressive changes in the apparent HDL–cholesterol after even a brief storage at −20 °C. They found this to be the result of changes in the precipitability of the lipoproteins, the low-density substances becoming gradually more difficult to precipitate, while the HDL were more easily precipitated. The final consequence was that, in samples with low HDL levels, the 'HDL'–cholesterol tended to increase with storage, whereas the opposite was the case in samples with a high HDL level. This had the effect of maintaining an approximate constancy of the group means, and underlines the insecurity of this statistical measure as an estimate of the reliability of an assay. These observations have since been confirmed by Gidez et al. (1982). Furthermore, Bachorik et al. (1982) have obtained qualitatively similar results with serum stored at −70 °C, although under these conditions the mean 'HDL'–cholesterol level decreased significantly after only one month. This is in contrast to the observations of Curb et al. (1980) who found the apparent HDL–cholesterol to increase during storage of up to 12 months at −80 °C but, despite a few such discordant observations, the weight of evidence strongly contradicts those who have maintained that satisfactory analyses can be made on stored, frozen serum (Miller et al., 1977; Lindgren et al., 1977).

The details of the precipitation procedure as modified by Warnick and Albers for use on plasma are as follows.

Reagents. Heparin solution containing 40000 USP units/ml. 1.06 M solution of $MnCl_2 \cdot 4H_2O$ (209.78 g/l) in water. Stock 0.2 M EDTA solution for the reconstitution of the reagent for enzymatic cholesterol determination: Disodium EDTA dihydrate, 4.27 g; tetrasodium EDTA dihydrate, 3.54 g; distilled water, 200 ml.

The reagents needed for the fractionation are also available in the form of commercially prepared kits.

Procedure. From the stock reagents, prepare a working solution by mixing 0.6 ml of heparin with 10 ml of 1.06 M $MnCl_2$.

Pipette 2 ml of fresh plasma into a conical centrifuge tube and add 0.2 ml of the working heparin reagent. Mix with a vortex mixer and incubate for 10 min at room temperature. Centrifuge at 1500 *g* for 30 min at 4 °C and carefully transfer the clear supernatant solution to a labelled, stoppered vial. (In our experience it may be better to perform this centrifugation at 12–15 °C. When setting up this analysis for the first time, it is important to check the efficiency of the fractionation by immunochemical analysis.)

The cholesterol content of the supernatant can be determined either by extraction and chemical estimation, e.g. with the Liebermann–Burchard reagent, or by enzymatic assay. However, some of the commercial enzyme kits contain phosphate buffer which reacts with Mn^{2+} ions to form a precipitate which interferes with the cholesterol estimation (Steele et al., 1976). Fortunately, it can be effectively suppressed by re-constituting the enzyme reagent in 0.004 M EDTA solution instead of water. This interference with the assay lead Kostner (1976b) to prefer the dextran sulphate–Mg^{2+} method for precipitating the low-density lipoproteins. But Henderson et al. (1980) found that this also could lead to low estimates of cholesterol, which they rectified by performing the incubation at 37 °C.

The interference of the heparin–Mn^{2+} reagent with the enzymatic cholesterol assay, coupled with its failure to precipitate hyperlipaemic plasma cleanly and its high cost, prompted Briggs et al. (1981) to propose polyethylene glycol as an alternative precipitant. In their hands, this method gave results that were highly correlated with those obtained by the heparin–Mn^{2+} procedure ($r = 0.975$) and the precipitated lipoproteins sedimented cleanly, even from the most lipaemic sera. However, although an electro-immunoassay of the fractions showed the efficiency of the technique to be high, its equality with the heparin–Mn^{2+} method has yet to be confirmed. It should also be noted that Briggs et al. worked with serum, not with plasma, and that their fractionations were carried out within 3 hours of the blood being drawn.

Reagent. Aqueous polyethylene glycol of molecular weight 6000 (PEG 6000; BDH Chemicals Ltd.): 200 g/l.

Procedure. To 0.5 ml of serum, add 0.5 ml of PEG 6000 solution, mix thoroughly and stand at room temperature for 10 min. Remove the precipitate by centrifugation at 4 °C for 30 min at 1500 *g*. Determine the cholesterol in a sample of the supernantant.

The accuracy with which either of these precipitation methods can reproduce the results obtained when the HDL is isolated by ultracentrifugation is significantly influenced by the amount of Lp(a) in the plasma, since this lipoprotein behaves like HDL during centrifugation but like LDL on precipitation. Thus, in plasma that was deficient in Lp(a), Bachorik et al. (1976) found the HDL–cholesterol in the heparin–Mn^{2+} supernatant to be an average of 1.2 mg/dl (2.7%) greater than that in the fraction isolated by centrifugation. By contrast, when the plasma was 'Lp(a) positive', the results of the centrifuge analysis were the greater by an average of 2.6 mg/dl (5.1%). Consequently, although the mean estimates of HDL–cholesterol in 129 randomly chosen samples differed by only 0.1 mg/dl plasma, the magnitude of the difference had a standard deviation of $\pm$3.8 ml/dl plasma. Ishikawa et al. (1977) also compared these two techniques and expressed their results in the form of the regression equation.

$$(\text{HDL–C})_H = 1.062\ (\text{HDL–C})_U - 1.65 \qquad r = 0.98$$

where the subscripts H and U refer to the heparin–Mn^{2+} and the ultracentrifuge techniques respectively. That the slope of this line is close to unity and the constant term is close to zero is further confirmation of the virtual equivalence of the two methods.

The polyethylene glycol precipitation has not been compared directly with the ultracentrifuge but Briggs et al. (1981) performed their analyses in parallel with the heparin–Mn^{2+} method and expressed their findings in the following equation:

$$(\text{HDL–C})_\text{P} = 1.05(\text{HDL–C})_\text{H} - 0.02 \qquad r = 0.975$$

which suggests that there is an adequate concordance between the two methods, although it also suggests that precipitation with polyethylene glycol may give results that are consistently slightly greater than those obtained with the heparin model.

In addition to these methods for the estimation of HDL–cholesterol, it is worth noting that Gidez et al. (1982) have described a method of precipitation with low molecular weight dextran sulphate, which they claim will yield two fractions that approximate closely to HDL_2 and HDL_3. Analyses made in parallel with analytical ultracentrifugation supported this contention in the case of HDL_2 but were less convincing for HDL_3. Nonetheless, the technique is of considerable interest. However, those who attempt to use it with dextran sulphate from a different source to that used by Gidez et al. must be sure to prove that the substitute will give the same performance.

7.3.2. The estimation of LDL– and VLDL–cholesterol

Although it has been claimed that 'LDL' and 'VLDL' can be separated by precipitation with polyanions, this technique has not been made the basis of an accepted and practical method of quantitative analysis. In an alternative procedure that was proposed by Friedewald et al. (1972), the VLDL–cholesterol is estimated indirectly from measurements of total plasma triglyceride. This depends on the validity of two assumptions, namely:

(1) That, in the majority of fasting plasma samples, chylomicrons are virtually absent and that VLDL make the largest contribution to total plasma triglyceride measurements.

(2) That the mass ratio of triglyceride to cholesterol is about 5 in the VLDL of almost all samples of plasma except for those from subjects suffering from the rare Fredrickson's Type III hyperlipoproteinaemia. An estimate of the amount of VLDL–cholesterol (in mg/dl) can therefore be made by dividing the total plasma triglyceride by 5. If HDL–cholesterol is also determined, using the precipitation

technique, it is then possible to estimate the LDL–cholesterol by difference.

The procedure can therefore be summarised as follows: (1) determine the total plasma cholesterol (C_T); (2) measure the total plasma triglyceride (G_T); (3) estimate the HDL–cholesterol as in Section 7.2.1 (HDL–C).

The approximate concentration of cholesterol in the low-density lipoprotein classes is then given by:

VLDL–cholesterol $= 0.2G_T$

LDL–cholesterol $= C_T - (\text{HDL–C}) - 0.2G_T$

Friedewald et al. (1972) found that the value of LDL–cholesterol obtained in this way correlated well with that determined from LDL isolated by preparative ultracentrifugation ($r = 0.98 - 0.99$), when the subjects were 'normal' or hypercholesterolaemic. Under these circumstances, the mean difference between the two estimates (expressed as a percentage of the value determined by centrifugation) amounted to 3–4%. Not surprisingly, in view of the assumptions on which the method is based, the correlation became progressively weaker in subjects with more VLDL. Thus, in a group of Type IV patients with total plasma triglycerides of less than 400 mg/dl, r was 0.94 and the mean difference between the estimates was 7% but, when subjects with a triglyceride level greater than 400 mg/dl were included, r fell to 0.85. It will be evident from these observations that this procedure may be useful for statistical studies of large groups but is less serviceable as a diagnostic method in individual cases, particularly in those where the level of VLDL is high.

A direct method of estimating VLDL–cholesterol that avoids the assumptions of the 'Friedewald approximation' was put forward by Ononogbu and Lewis (1976). This makes use of the observation that VLDL can be precipitated with sodium dodecyl sulphate (Burstein and Scholnick, 1972; Wilson and Spiger, 1973). Although it has not been widely adopted, the method has the merit that both cholesterol and triglyceride can be determined for each of the three main lipoprotein classes, which not only gives a more accurate and detailed estimate of the lipoprotein but allows the presence of abnormal lipoproteins to be inferred in, for example, Type III hyperlipoproteinaemia. Its authors found that the results of the method generally correlated well with those of the ultracentrifugal procedure. Thus, for the cholesterol estimates, the correlation coefficients were 0.98, 0.99 and 0.95 for VLDL, LDL and HDL respectively. For triglyceride, the coefficients, taken in the same order, were 0.99, 0.85 and 0.83. Moreover, the mean values of 35 parallel estimates of VLDL– and LDL–cholesterol made by the two methods agreed to within 2 mg/dl serum. Consequently, despite a somewhat larger discrepancy between the two measurements of HDL–cholesterol (average 5 mg/dl. i.e. 8.6%) this technique would seem to be clearly superior to the 'Friedewald approximation', albeit slightly more laborious.

In its original form, which we shall describe below, the HDL–cholesterol was estimated after the precipitation of the low-density lipoproteins with dextran sulphate. However, the heparin precipitation described in Section 7.3.1 could also be used.

Reagents.

10% sodium dodecyl sulphate (analytical grade) in 0.15 M NaCl.

5% dextran sulphate (mol. wt. 2×10^6; analytical grade) in water.

22.2% calcium chloride solution.

Procedure. Pipette 1 ml of serum into a small glass centrifuge tube and add 75 µl of the 10% SDS solution. Mix thoroughly and incubate the tube for 2 h in a water bath at 35 °C. Centrifuge off the precipitated VLDL at room temperature in a high speed machine (e.g. 10 min at 10 000 rev/min). The pellicle is easily disturbed, and the tube must be handled with care when it is removed from the centrifuge. Remove the clear solution with a pasteur pipette and wash the residue with 0.5 ml of 0.1% SDS in 0.15 M NaCl. The precipitate is finally dissolved in 1 ml of 1% SDS in 0.15 M NaCl by warming it for 30 min in the water bath at 35 °C. Reserve this solution for lipid analysis.

For the estimation of HDL–cholesterol, transfer 1 ml of serum to a 5 ml centrifuge tube and add 40 µl of 5% dextran sulphate and 50 µl of 22.2% calcium chloride solutions. Mix thoroughly and stand at 4 °C for 24 hours. Centrifuge off the precipitate and reserve the supernatant for lipid analysis.

Finally, estimate cholesterol and triglyceride in the VLDL and HDL fractions, and also in the original serum. The corresponding values for LDL can then be obtained by difference.

7.3.3. Derived estimates of lipoprotein concentration

If each of the three main classes of lipoprotein is assumed to contain a constant, characteristic proportion of cholesterol, measurements of lipoprotein cholesterol can easily be converted into estimates of the lipoprotein itself. However, the characteristic conversion factors are subject to significant (and in the case of VLDL large) uncertainties which become superimposed on those to which the estimates of lipoprotein cholesterol are already subject. Consequently, the derived estimates of lipoprotein concentration are likely to be of only modest accuracy. Unfortunately, there has been only one serious attempt to study this subject, namely that of Lindgren et al. (1977), who compared the estimates derived from the 'Friedewald' analysis with those obtained by analytical ultracentrifugation of the same sera. In these experiments, the characteristic cholesterol contents of VLDL, LDL and HDL were taken to be 14%, 33% and 17% respectively.

The results obtained from the analysis of 39 sera were expressed in the form of regression equations, as follows:

$$\text{VLDL} = 0.686\,(\text{VLDL–C}) - 51 \qquad r = 0.914$$
$$\text{LDL} = 0.920\,(\text{LDL–C}) - 21 \qquad r = 0.895$$
$$\text{HDL} = 0.785\,(\text{HDL–C}) + 78 \qquad r = 0.849$$

It is immediately evident that the correlation coefficients are worse than those quoted for the estimation of lipoprotein cholesterol in Sections 7.3.1 and 7.3.2. Moreover, the regression equations suggest that there were significant systematic differences between the two methods of analysis. For example, the estimation of VLDL–cholesterol indicated the presence of an average of 75 mg/dl of VLDL even in those samples in which virtually none was detected by analytical centrifugation. This over-estimation by the chemical analysis is evident in all the samples that were examined and may be partly ascribable to the uncertainties in the indirect estimation of VLDL–cholesterol. It is more surprising to find a substantial discordancy between the two methods of estimating HDL, since the HDL–cholesterol was determined directly. In this case, the chemical procedure apparently under-estimated the HDL concentration at low levels and over-estimated it at high levels. The fact that the disagreement between the two estimates of LDL was comparatively small is presumably due to the errors in the estimation of HDL– and VLDL–cholesterol roughly neutralising each other. On the face of it, these observations support our contention that it is better to measure lipoprotein by specific rather than by indirect methods. However, it must be emphasised that these observations are unconfirmed and need to be extended to a larger and more representative population than the group of young women from which they are derived.

7.3.4. Quantification of electrophoretic analyses

The prospect of measuring lipoproteins after they have been separated by electrophoresis has attracted many investigators, particularly before the development of precipitation methods. Even today, the micro-scale nature of the method is tempting but, despite this, none of the versions that have been proposed has come into widespread use.

In general, there are three ways of extracting quantitative data from the system after the lipoproteins have been resolved.

7.3.4.1. Elution of lipids The lipids can be eluted from the appropriate parts of the supporting medium e.g. a paper strip, and determined by a suitable micro-method. This technique was originally developed in the 1950's and in most examples, the electrophoretic analysis is run in duplicate. One of these strips is than stained and is used to mark the positions of the lipoproteins on the other. These portions are then cut out, the cholesterol extracted with solvent and determined (Anderson and Keys, 1956; Nury and Smith, 1957; Crawford, 1959). Alternatively, the staining process can be omitted and the whole of the electrophoretic strip cut into narrow segments, each of which is eluted and analysed for cholesterol. In the technique developed by Boyd (1954), each segment was put into a tube with 7 ml of acetone–ethanol (1:1 v/v) and heated in a water bath. The temperature of this bath was gradually raised from its initial 30 °C to 66 °C, where it was held for 10 min before being allowed to cool to 40 °C. The strips were then removed with forceps and rinsed into their tubes with acetone–ethanol. The solvent was then evaporated from the tubes at 60 °C under nitrogen and the residue finally dried at 110 °C for 30 min. The cholesterol content of the extracts was then determined by the Liebermann-Burchard method. The recovery in this procedure was said to be about 95%. Later, Searcy (1967) modified this procedure by using the cholesterol assay reagent to extract the cholesterol from the paper strips. Each segment was immersed in 6 ml of glacial acetic acid saturated with ferrous sulphate for 1 hour in a photometer cuvette. The strips were then removed and 2 ml of concentrated sulphuric acid added. After 15 min the optical density was read at 490 nm. One of the alleged advantages of this method was that, contrary to the observations of Langan et al. (1955), the electrophoretic strips could be stored without loss of cholesterol. In view of the well known susceptibility of cholesterol to oxidation, this may need confirmation. In a still more recently developed version of this technique, the cholesterol content of fractions isolated by electrophoresis in commercially available films of polyacrylamide gel is determined by gas phase chromatography (Gambert et al., 1980). This combines a high resolving power with a sensitivity that allows even the minor lipoproteins to be estimated in a sample of a few microlitres. However, its correlation with the ultracentrifugal analysis may leave something to be desired, being as low as 0.73 in the case of VLDL fractions. Nonetheless, this method may have some merit as a research tool.

7.3.4.2. Staining and photometric estimation This technique suffers from the serious handicap that there is no simple or reliable relationship between the amount of lipoprotein and the quantity of dye that it takes up. Moreover, the dyes themselves are mixtures of variable composition and properties, which adds to the difficulty of maintaining reproducibility between different laboratories. Much effort has been spent on attempts to overcome these difficulties and innumerable versions of the technique have been the result. The earliest of these necessarily made use of paper or cellulose acetate as the supporting medium. With their replacement by the essentially transparent film of agarose, the potential of the method was improved and the technique underwent an evolution that has been reviewed by Opplt (1980).

The first reasonably successful attempt to quantify the electrophoretically separated lipoproteins was due to Dyerberg and Hjørne (1970a). After resolution by a slight modification of Noble's technique, the stained portions of the dried agarose films were cut out, the dye (Sudan Black B) eluted with acetic acid: ethanol:water (11:5:4 by vol.) and its concentration determined photometrically at 590 nm. The amount of lipoprotein lipid in each fraction could then be estimated from a parallel gravimetric determination of total plasma lipid (T) by means of the relation:

$$L_F = T \cdot D_F / D_T$$

where D_F and D_T represent the amount of dye in the lipoprotein fraction and the sum total in all the fractions respectively. However, this procedure assumes that all the lipids absorb the dye equally well, which has long been known to be false (Jencks and Durrum, 1955). Accordingly, Dyerberg and Hjørne (1970b) devised a way of approximately correcting for the differences. From experiments with purified lipids, they determined that the relative absorption of dye by each class of lipid was:

cholesteryl ester (CE)	1.0
triglyceride (TG)	0.80
phospholipid (PL)	0.12
fatty acid (FA)	0.10
cholesterol (UC)	0.02

By combining these values with the corresponding proportions of each lipid in the lipoprotein, they then defined a series of factors (F) of the form:

$$F = \frac{1}{1.0\,CE + 0.80\,TG + 0.12\,PL + 0.10\,FA + 0.02\,UC}$$

where *CE*, *TG*, etc. are the mass fractions of each component of the lipid fraction of the relevant lipoprotein. For example, in the α-lipoprotein, $CE = 0.30$, $PL = 0.45$ etc. In this way, the factors for α-lipoprotein (F_α), β-lipoprotein (F_β), pre-β-lipoprotein (F_P) and chylomicrons (F_{CH}) were determined to be, relative to F_α,

$$F_\alpha = 1.0 \qquad F_\beta = 0.75 \qquad F_P = 0.75 \qquad F_{CH} = 0.63$$

The observed value for the photometric density of the stained lipoprotein eluted from the film (D_F) can then be corrected to a value approximating that which would be expected if dye absorption was uniform (D'_F) i.e. $D'_\alpha = D_\alpha \cdot F_\alpha$, etc. These values of D'_F can then be used to calculate corrected estimates of L_F which, in turn, can be used to estimate the amount of the relevant lipoprotein by making use of the known ratio of lipid to protein in each one.

Although the average results obtained in this way are plausible when compared with ultracentrifugal analyses made by other investigators, there appear to have been no concurrent comparisons of the two methods and the exact degree of correlation between them is unknown. By contrast, the technique developed by Lindgren and his collaborators over a period of some 10 years has been carefully compared with analytical ultracentrifugation. These investigators studied various methods of standardising the analysis, the most satisfactory of which requires the electrophoresis of the serum and also of at least one fraction that has been isolated by preparative ultracentrifugation (Lindgren et al., 1975). However, the use of the centrifuge tends to nullify the simplicity of the electrophoretic technique and it was suggested that, for all but a few hyperlipoproteinaemic samples, a simpler method which is standardised by reference to total serum cholesterol or triglyceride measurements would be adequate (Hatch et al., 1973). Nonetheless, it was considered that these procedures were too tedious for large-scale routine work (Wong et al., 1977) and, to overcome this objection, they were further developed into highly automated and computerised procedures which then became impracticable for the small, isolated laboratory (Wong et al., 1977; Lindgren et al., 1980).

In principle, the method of Hatch et al. (1973) is closely akin to that of Dyerberg and Hjørne but differs from it in 3 practical respects:

(1) The dried, stained agarose film is passed through a scanning photodensitometer and integrator, for example, the Beckman Analytrol. The amount of dye in each lipoprotein fraction (D_F) is therefore estimated as the integral of the relevant part of the absorption curve.

(2) The intensity of the stain is related to the amount of lipoprotein instead of to the quantity of lipid.

(3) The lipoproteins are stained with Fat Red 7B (Ciba) instead of Sudan Black.

The factors needed to compensate for the differential staining of the lipoproteins were determined by measuring the uptake of dye by fractions that were isolated by ultracentrifugation. This led to the values

$$F_\alpha = 1.75 \qquad F_\beta = 1.0 \qquad F_P = 0.82$$

in the nomenclature used above. In this system, chylomicrons were assumed to have the same staining capacity as the pre-β-lipoproteins.

To calibrate the system by reference to the total plasma cholesterol (TC), use is made of the relationship:

$$TC = 0.332(\beta) + 0.137(\text{pre-}\beta) + 0.168(\alpha) + 0.044(\text{chylos})$$

where the coefficients are the mean proportions by weight of cholesterol in the different lipoprotein classes. If it is assumed that the measured intensity of the stain is proportional to the weight of lipoprotein in the fractions, this can be approximated by:

$$TC = K_C(0.332D'_\beta + 0.137D'_P + 0.168D'_\alpha)$$

where K_C is a factor that is characteristic for each electrophoretic strip and depends on the amount of plasma applied to the gel, the efficiency of the staining process, etc. The value of K_C must therefore be determined for each analysis by means of this equation, when the amounts of the lipoproteins can be estimated from relations of the form: $L_{\beta} = K_C D'_{\beta}$ etc.

In an analogous way, the method can also be calibrated with measurements of total plasma triglyceride or total plasma lipid, by making use of the relations:

$$TG = K_G(0.072D'_{\beta} + 0.540D'_{P} + 0.039D'_{\alpha})$$

and

$$TL = K_L(0.77D'_{\alpha} + 0.91D'_{P} + 0.50D'_{\alpha})$$

If both cholesterol and triglyceride determinations are used for the calibration, a more reliable estimate of the lipoproteins may result. According to Hatch et al. (1973), it may then be advantageous to use a value for K that is the harmonic mean of K_C and K_G, i.e.

$$\frac{1}{K} = 0.5\left(\frac{1}{K_C} + \frac{1}{K_G}\right)$$

It will be evident that both this method and that of Dyerberg and Hjørne are critically dependent on the accuracy and constancy of the staining factors F_{α}, etc., and of the values adopted for the chemical composition of the lipoproteins. The effectiveness of the parameters used by Hatch et al. (1973) can be judged from the results of their parallel electrophoretic and ultracentrifugal analyses of the sera of 23 subjects, both normal and hyperlipidaemic. The correlation between the two sets of results could be expressed by the following regression equations:

VLDL = 0.743(pre-β) + 21.8	$r = 0.994$
LDL = 0.864(β) − 13.7	$r = 0.935$
HDL = 0.622(β) + 56.4	$r = 0.883$

which show that the electrophoretic analysis was subject to a substantial systematic error, the mean concentration of each lipoprotein class being over-estimated by an amount that ranged from 18 to 26%. One reason for this is doubtless the assumption that these classes can be regarded as being of constant, characteristic composition. The fallacy of this with respect to their content of triglyceride and cholesterol has been investigated by Myers et al. (1976), who found that the cholesterol:triglyceride ratio in LDL and HDL can be expressed by the equations:

$$\frac{\text{LDL-C}}{\text{LDL-TG}} = \frac{2180.25}{255 + X}$$

where X is the total serum triglyceride in mg/dl.

and

$$\frac{\text{HDL-C}}{\text{HDL-TG}} = 1.25 + 293/X$$

By making certain assumptions, they were then able to derive expressions for the proportions of these lipids in the lipoproteins. However, it seems that the refinement of the electrophoretic analysis by the incorporation of these observations has not been attempted.

Aside from this objection, it is uncertain whether the regression equations quoted by Hatch et al. (1973) would be valid if their method was to be reproduced in another laboratory (cf. Lindgren et al., 1975). To summarise, it appears to us that, contrary to the opinion of Lindgren et al. (1980), even the best available techniques for the quantitative estimation of lipoproteins by electrophoresis and staining are too unreliable, and probably too laborious, even for clinical use. Nonetheless, the refinements of technique that have been introduced in the attempt to improve reproducibility may well be studied even by the proponents of electrophoresis as a qualitative or semi-quantitative method of analysis. The most detailed discussion of these aspects of the methodology is that given by Lindgren et al. (1980), on which the following summary is based.

The procedure makes use of commercially prepared agarose gels (Bio-Gram A) in a Durrum type electrophoresis cell (Beckman Instruments). Albumin is not added to the electrophoresis buffer, but the slides are soaked in buffer which contains 0.05% human albumin for 24 hours. Do not pile the slides on top of each other during this operation, or distorted bands may result. Instead, arrange them in a rack in a covered dish. After the soaking, remove the slides and allow them to dry on the bench before applying the samples. This will take from 20 to 25 minutes under average laboratory conditions. If it is found that the sample floods the surface of the gel when it is applied, the gel is too wet and that run should be discarded.

Lindgren et al. (1980) use the sample applicators provided with the gel kits and, if care is taken to repeat each stage of the operation in exactly the same manner for each sample, the technique can be reproducible as well as speedy. A reference serum to which bromophenol blue has been added as a marker, should be run on the first and last slides of each set. Load the prepared slides, gel downwards into the electrophoresis tank, ensuring that there is good contact with the wick over a length of about 1 cm. Lindgren et al. (1980) recommend that a block of transparent plastic 2.5 cm thick and 4.5 cm wide should be put on top of the slides to make certain that good contact is maintained.

Run the electrophoresis at a constant potential of 150 volts for 15 to 25 minutes, i.e. until the trailing edge of the marker dye has moved about 15 mm from the origin. This distance can easily be measured if two lines are scribed in the appropriate positions on the under-side of the plastic weight. When the run is completed, transfer the slides to a rack and immerse them in a bath containing the following mixture:

benzene-free ethanol	570 ml
2-propanol	30 ml
glacial acetic acid	50 ml
water	350 ml

which is stirred with a magnetic stirrer. After 15 minutes, transfer the slides to a second, exactly similar fixing bath for the same period. (Discard the first bath after three or four uses and substitute it by the second, which is replaced by a fresh one.) Finally, dry the slides in a stream of warm air until the agarose film is completely transparent.

Of the many factors that contribute to inconsistent staining, Lindgren et al. (1980) claim to have identified two that are important when Fat Red 7B is used:

(1) The stock dye solution is mixed with sodium hydroxide and Triton just before use. If this mixture is not used immediately, the dye

will begin to precipitate before the slides are put into it.

(2) The temperature of the staining bath should be kept to within ±1 °C of room temperature during the process. This is made less easy by the fact that heat is liberated when the working stain is prepared and, for this reason, Lindgren et al. (1980) recommend the following procedure. At least 24 hours before use, dissolve 225 mg of stain in 1 litre of absolute methanol. Keep this stock solution at 4 °C, together with the 0.1 N NaOH that is used to prepare the working stain. When these cold solutions are mixed, the resulting solution has a temperature of about 20 °C.

The *procedure* for staining the slides is as follows. Load the dried slides into a staining rack. Measure 11 drops of Triton X-100 detergent into a staining dish that is set on a magnetic stirrer and add 184 ml of the cold stock stain. Stir well and continue to do so while adding 36 ml of cold 0.1 N NaOH. Measure the temperature after 1 minute, when it should be close to room temperature. Exactly 10 minutes after the addition of the NaOH, put the slides into the bath, where they must remain for exactly 15 minutes. Check that the temperature does not change by more than 1 °C during this time. Finally, wash the slides in running cold water for at least 1 minute and then dry at room temperature. Slides that are stained in this way must not be exposed to direct sunlight, but are said to undergo little or no deterioration when stored in the dark.

It is instructive to compare the results obtained by this rather sophisticated technique with those of Berends et al. (1972), who carried out parallel analyses by electrophoresis on Cellogel and by analytical ultracentrifugation. These electrophoretic analyses were quantified by reference to total lipid determinations but were not corrected for differential staining. With the results expressed as a percentage of the total measured plasma lipoproteins, the two methods could be correlated by the following regression equations:

$$\%\text{VLDL} = 0.544(\%\text{pre-}\beta) + 9.2 \qquad r = 0.70$$
$$\%\text{LDL} = 0.582(\%\beta) + 21.6 \qquad r = 0.506$$
$$\%\text{HDL} = 0.41(\%\alpha) + 9.8 \qquad r = 0.395$$

which strongly suggest that the refinements introduced by Lindgren and his collaborators do bring about a considerable improvement in the accuracy and precision of the electrophoretic analysis, even though its performance may still leave something to be desired. Much the same conclusion can be drawn from the work of Berenson et al. (1972) who used electrophoretic analysis to estimate β- and pre-β fractions with the aid of a turbidimetric estimate of their total level. When this was compared with the analytical ultracentrifuge, both the regression and the correlation coefficients were substantially closer to 1.0 than those obtained with the 'Berends' method, although they still did not equal those that can be obtained with the 'Lindgren' procedure.

7.3.4.3. The lipoproteins are precipitated in the agarose gel with polyanions In this technique, the lipoproteins are located in the electrophoresis gel by precipitation and are then estimated by measuring the optical density of the precipitate (Neubeck et al., 1977; Wieland and Seidel, 1978). This makes no assumptions about the composition of the lipoproteins, but requires that the method is standardised with respect to a reference method. In the version developed by Neubeck et al., the precipitating agent was dextran sulphate and standardisation was effected by means of gravimetrically analysed solutions of purified VLDL, LDL and HDL. Its originators allege that this is a highly reproducible method, the lipoprotein concentration being related to the absorbance by the expressions:

$$\log(\text{pre-}\beta) = 0.09A - 0.014 \qquad r = 0.999$$
$$\log(\beta) = 0.047A - 0.007 \qquad r = 0.999$$
$$\log(\alpha) = 0.056A - 0.0025 \qquad r = 0.999$$

where A is the optical absorbance expressed as a percentage of that of a standard filter. Unfortunately, even if identical equipment were to be used, there is nothing to suggest that these equations would be equally applicable in other laboratories. Moreover, the difficulty of transferring the small sample of serum accurately to the gel is a potentially important source of error. To overcome the latter problem, Bartholomé et al. (1980) proposed that the cholesterol in each lipoprotein fraction should be determined and the recovery of lipoprotein corrected by reference to a measurement of the total serum cholesterol. However, they also differ from Neubeck et al. (1977) in the use of sodium phosphotungstate as the precipitant.

Reagents. Electrophoresis gel: 10 g/l of agarose and 2 g/l of bovine serum albumin in pH 8.6 veronal buffer (50 mM).
Reagent A: 1.8 g of sodium phosphotungstate, 1.71 g of $MgCl_2$, 4.09 g of NaCl, water to 100 ml.
Reagent B: reagent A with the sodium chloride content increased to 9.53 g/dl.
Reagent kit for the enzymatic determination of cholesterol e.g. Merck.

Procedure. Mix the serum with an equal volume of albumin-free agarose solution and transfer 10 µl as accurately as possible to the prepared electrophoresis gel (Section 5.4). After electrophoresis for about 90 minutes at 10 V/cm, soak the slide at room temperature for 1 hour in reagent A and then for 2 hours in reagent B. To obtain the relative distribution of the lipoproteins, scan the slide (Neubeck et al., 1977; Wieland and Seidel, 1978) on an integrating densitometer such as the Integraph CH (Bender and Hobein) and refer to the calibration curves that have been prepared in advance. Then, soak the slide in distilled water for 45 minutes, cut out the bands of lipoprotein and transfer them to stoppered tubes containing 2 ml of the cholesterol colour reagent. Also cut control slices of the gel. Incubate these tubes at 37 °C for 12 hours, during which time the lipoproteins will dissolve. Then continue the determination of the cholesterol in accordance with the instructions given with the reagent kit but, before reading the absorbance, centrifuge the reaction mixture at 1500 *g* for 10 minutes to clear it of agarose. If a concurrent estimation of total serum cholesterol is made, it is then possible to correct the lipoprotein determinations by making use of the relationship:

$$TC = 0.33(\beta) + 0.14(\text{pre-}\beta) + 0.17(\alpha)$$

which was quoted in Section 7.3.4.2.

7.4. Estimation of LP–X

Because LP–X does not contain a specific apo-lipoprotein, its concentration in plasma cannot be satisfactorily determined by such methods as radial immunodiffusion. Micro-methods for its estimation therefore often take advantage of the ease with which it can be separated from other lipoproteins by electrophoresis in agar (Section 5.6.1). In the procedure developed by Neubeck and Seidel (1975), the LP–X is precipitated in the gel with polyanions and is estimated from the photometric density of this precipitate. This technique is analogous to the one described for normal lipoproteins in Section 7.3.4.3 and has the same disadvantages, notably the need to establish a standard curve by the analysis of solutions of purified LP–X of known concentration. Rittland (1975) has proposed a technique in which the plasma lipoproteins are first labelled by exchange with [^{14}C]cholesterol. The LP–X is then separated by electrophoresis and located by polyanion precipitation. The proportion of LP–X present can then be estimated from the relative radio-activity of this fraction. This method is sensitive but does not easily yield absolute values. Moreover the assumption of uniformity of labelling may be open to question. The immuno-electrophoretic method devised by Kostner et al. (1974) is also sensitive, but makes use of anti-serum to apo-lipoprotein C (or anti-LP–X) which is expensive and not readily available.

It is worth noting that the electrophoretic migration of all lipoproteins is more strongly anodic in post-heparin plasma, presumably because they adsorb the liberated free fatty acids. This effect can be sufficiently pronounced to abolish the cathodal migration of LP–X. Consequently, electrophoretic methods cannot be used to estimate LP–X in post-heparin plasma and may give false results in at least some post-prandial plasmas (Muckle et al., 1977; Sauar et al., 1978; Ras et al., 1978). Recently however, Bos et al. (1983) have described a precipitation technique that is claimed to be quick to execute, amenable to automation and can be used with post-prandial plasma. The lipoproteins that contain apo-lipoprotein B are first precipitated with an antiserum to this apolipoprotein and the LP–X then estimated turbidimetrically after precipitation with sodium dodecyl sulphate (cf. Section 7.3.2). At the time of writing, the reliability of this procedure is unconfirmed.

Of the methods available, the most solidly based and perhaps the most convenient for the occasional estimation, is that of Magnani and Alaupovic (1972). This is based on the hypothesis that the phospholipid content of LP–X is virtually constant at about 60% by weight, an assumption that seems to be acceptable in practice. However, since the LP–X is not made visible before it is cut from the gel, there may be difficulty in ensuring its complete recovery to the exclusion of all chylomicrons, particularly when the level of LP–X is high. It is therefore important to adjust the composition of the agar gel to achieve the best possible resolution of the lipoproteins (Section 5.6.1).

Procedure. Prepare agar gels for electrophoresis as in Section 5.6.1, and pipette 10 µl of plasma into each sample well, taking great care to ensure that the surface of the gel is not contaminated. When the electrophoresis is completed (about 80–90 minutes at 6 V/cm), examine the slides over a bright light to locate the position of chylomicrons that may have penetrated into the gel on the cathode side of the sample well. It may also be possible to see the LP–X as a transparent area in the gel. With a cutter 5–6 mm in diameter, punch out a disc of gel just to the cathode side of the well. This disc must include all the LP–X but must avoid any chylomicrons that may be present and must not break into the sample well. Control discs should also be cut from unoccupied parts of the gel. Place each disc in a clean test-tube that contains 0.7 ml of a mixture of equal volumes of 60% perchloric acid and concentrated sulphuric acid. Digest for 2.5 hours at 300 °C and estimate the phosphorus by any suitable method (Section 6.3.2.3; the originators used the method of Gerlach and Deuticke, 1963). The concentration of LP–X in the plasma is then given (in mg/dl) by: LP–X = 417.5(P), where P is the estimated phosphorus in µg. The method is said to give a linear response up to 840 mg LP–X/dl of plasma.

7.5. *Immunoassay of lipoproteins*

In the past, the immunoassays have occasionally been used to estimate intact lipoproteins but the methods are difficult to standardise and are of little value for the determination of VLDL, in which the protein content can vary widely. Moreover, because antisera to HDL are expensive and often of low titre, the technique must, in practice, be largely confined to the assay of LDL. At the present time however, immunoassay is almost entirely restricted to the quantitative analysis of apo-lipoproteins (Section 8.9) and we shall therefore give only a brief résumé of the techniques that have been used for lipoprotein determination. These can be classified under four main headings:

(*1*) *Turbidimetric assay.* In this technique, which apparently originated with Burstein and Samaille (1958), the lipoproteins are precipitated with the antiserum and are then estimated turbidimetrically or, better still, by nephelometry. It is a method that consumes relatively large quantities of antiserum which must be of high specificity, but this did not deter Kahan and Sundblad (1969) from developing it into an automated procedure for the estimation of total plasma low-density lipoproteins. However, in view of the difficulty of standardising these procedures in absolute terms and because there is little merit in relatively inaccurate estimates of the sum of LDL and VLDL, the value of the technique may be in the quantitative estimation of the lipoproteins (or other proteins) that contaminate a preparation e.g. the LDL in HDL. For this purpose, the high sensitivity of the nephelometer would be advantageous. The lipoprotein under test is first diluted to an appropriate extent and samples pipetted into a series of nephelometer tubes. Antiserum is then added, the amount being increased by 50–100 μl for each successive tube, and the light scattering intensity measured at intervals of 10 minutes until no further change takes place. The assay must be calibrated by carrying out of a similar titration with a lipoprotein solution of known concentration.

(*2*) *The 'immunocrit' method.* This method, devised by Heiskell et al. (1961) was later developed into a commercially available kit for the estimation of total plasma low-density lipoproteins (Hyland Laboratories). A drop of plasma is mixed with a drop of antiserum and is then drawn into a 50 μl capillary tube which is sealed in a flame. After 15 minutes, the tube is centrifuged in a haematocrit and the ratio of the length of the column of precipitate to that of the column of liquid is taken as a measure of the lipoprotein concentration. Like the turbidimetric method, this requires an antiserum of high specificity and reproducible titre.

(*3*) *Radial immuno-assay.* Lees (1970) used this method for the estimation of LDL that had been isolated by ultracentrifugation. However, its specificity appears to be its only recommendation as a general means of estimating lipoproteins that have been isolated by such a precise and expensive method. Nonetheless, Smith and Slater (1970) used it to effect in their early estimations of the small amounts of LDL that can be extracted from aortic tissue. In these experiments, the minced tissue was first extracted several times with cold 0.15 M NaCl and the LDL then isolated from this solution by ultracentrifugation. The lipoprotein was then assayed by an adaptation of the method of Fahey and McKelvey (1965): dissolve 0.7 g of agarose in 10 ml of 0.1 M phosphate buffer pH 7.2 at 100 °C. Cool the solution to 56 °C and add 0.15 ml of antiserum to human LDL immediately before pouring the plates. Use 3 ml per microscope slide. When the gel has set, punch 5 pairs of holes of 10 μl capacity along the length of the slide. Fill one row with doubling dilutions of a standard preparation of human plasma LDL and the other with one of the samples to be tested. Stand the plates for 18 hours at room temperature in a covered dish, and then measure the diameter of the rings with a magnifier. Plot a standard curve for dilution vs. diameter for each slide and use it to estimate the amount of LDL in the corresponding test solution. It is worth noting that Smith and Slater prepared a large volume of standard LDL solution, divided it into 0.5 ml portions, and froze these at −20 °C until needed. Unfortunately, they do not say for how long the samples may be stored under these unconventional conditions.

(*4*) *Electro-immunoassay.* This technique, which is widely used to estimate apo-lipoproteins (Section 8.9.3), has also been used to assay β-lipoproteins in plasma both by the 'Laurell' method (Kahan and Sundblad, 1969) and by the modification of Clarke and Freeman (Bradwell and Burnett, 1975). However, there is little to be said for it as a general method of lipoprotein assay except, perhaps, when the quantity available is very small. For example, Smith and Slater (1972) were able to adapt the technique to the estimation of β-lipoproteins in aortic tissue, thereby eliminating the preliminary extraction that is needed for their radial immunoassay which is described in paragraph (3). The procedure they used is as follows:

On to a 10 cm × 10 cm glass plate, pour 28 ml of a 1% solution of agarose in 50 mM veronal buffer of pH 8.6. When this has set, cut a window 9 cm × 6 cm in the upper part of the gel and fill it with 15 ml of agarose solution that contains antibody to human β-lipoproteins. Note that this solution must be no hotter than 56 °C and should contain sufficient antibody to keep the peak developed with 25 µl of serum within the bounds of the slide. Cut five slots, each 10 mm × 2 mm in the first gel, parallel with and 1.5 cm below the boundary with the antibody gel. Weigh samples of the finely minced tissue and wrap each one in a slip of porous paper of high wet-strength. Press these packets into three of the slots in the gel. Into the remaining two, insert a sliver of No. 17 Whatman filter paper and add 10 µl of control serum. Finally, add a few drops of molten agarose to each slot. Place the prepared slide upside-down in the electrophoresis tank and apply a potential of 4 V/cm for 40 hours. At the end of the run, wash and stain the slide as in Section 6.5. The amount of lipoprotein can then be estimated from the area under the precipitin curve by comparison with a range of samples of standard serum or lipoprotein solution.

All these methods have the following advantages in common; (1) they need only a small sample and (2) they are highly specific and the lipoproteins need not to be isolated before they are estimated. However, against these attractions must be set the need to prepare and maintain standards of known lipoprotein content but uncertain stability. Purified lipoprotein preparations of known cholesterol, protein or nitrogen content have all been used for this purpose and, in some cases, have apparently given satisfactory results even after a considerable period of storage. This suggests that at least some immunoassays may be insensitive to denaturation of the lipoprotein particle provided that its solubility and its reactivity towards the antiserum are unimpaired. In this context, the stability of lipoprotein in 20% sucrose solution may be significant, although it does not seem to have been investigated. However, under the present circumstances, a serum pool should be used as a secondary standard which should be re-calibrated against isolated lipoproteins at regular intervals. This standard serum should be frozen in small portions which are discarded after they have been thawed once. It may also be said that it is better to determine the primary standard in terms of lipoprotein, either by gravimetry or refractometry (Sections 6.1, 7.1), than in terms of attributes such as cholesterol or protein.

Isolation and assay of apo-lipoproteins

To isolate an apo-lipoprotein, it is only necessary to prepare a sample of the lipoprotein in which it is present in the highest proportion, to extract the lipids and to separate the mixture of proteins that remains. However, the 'best' strategies are as ill-defined and controversial as those of politics or economics. In the following discussion we shall outline the procedures that are currently available, but it will remain for the reader to decide which is the most suitable for his purpose. This is a choice that may be determined, for example, by the amount of protein that is required, the amount of blood that can be obtained (rat apo-lipoproteins are difficult to produce in large quantities) and by the laboratory facilities that are available. Notice too, that we shall only be concerned with the purification of the major apo-lipoproteins normally present in Man. Although the same principles will, in general, apply to the isolation of the analogous proteins of animals, small modifications may be needed on account of their slightly different properties.

The apo-lipoproteins are usually purified by a sequence of chromatographic separations, although electrophoretic techniques are of value in certain cases. Unfortunately, the apo-lipoproteins readily associate to form complexes, or become adsorbed to surfaces, and an ostensibly simple preparation is all too often wearisome and low-yielding. Contrary to the impression that is sometimes given by the literature, it is rarely possible to obtain a pure apo-lipoprotein from the raw mixture by a single pass through a chromatographic column. More often, each fraction must be analysed by polyacrylamide gel electrophoresis (Section 6.4.2) and the least contaminated set aside for further treatment.

Some of the difficulties that are experienced during these preparations may be the result of a partial degradation of some of the apo-lipoprotein during the early stages of its isolation. The subject of lipoprotein degradation is discussed in Chapter 2 where the appropriate preservative precautions are described. Not only must these be strictly observed during the preparation of an apo-lipoprotein, but each stage of the process must be performed without delay, thus excluding the use of 'out-dated' bank plasma as a source, and under conditions that will minimise oxidation. Like many other aspects of lipoprotein chemistry, the effects of oxidation are poorly understood. Nonetheless, there is sufficient evidence to suggest that it may be prudent to take preventative precautions. The most extensive work on this subject is due to Lee et al. (1981), who studied the degradative oxidation of apo-lipoprotein B and were led to propose that two different reactions may be involved:

(1) The peroxidation of lipids in the native particle results in the formation of malondialdehyde which can then condense with free amino groups in the protein molecules. If this hypothesis is correct, it would be relevant to the preparation of any apo-lipoprotein, suggesting that the conditions should always be made unfavourable to peroxidation. This can be done by working in an atmosphere of nitrogen, with de-oxygenated solutions and in the presence of antioxidants such as reduced glutathione. It is equally important, when the lipoproteins are extracted with mixtures that contain diethyl ether, that this should be freshly distilled and free of peroxide (Appendix 3). Lee et al. (1981) also add 0.02 % of butylated hydroxytoluene to the solvents used for the extraction.

(2) That there is a direct reaction between oxygen and a hypothetical site in the molecule of apo-lipoprotein B which results in cleavage of the polypeptide chain. If this is true, it would be necessary to exclude oxygen even from the purified apo-lipoprotein. In any event, there is no evidence for a reaction between oxygen and other apo-lipoproteins.

The scrupulously anaerobic technique used by Lee and her collaborators to counter these effects is summarised in Section 8.4.2.

It is not only degradation that complicates the preparative procedure. In order to dissociate the apo-lipoprotein complexes, they are often dissolved in concentrated solutions of urea or guanidine hydrochloride. The carbamylation of the proteins that occurs in urea solutions has already been mentioned (Section 6.4.2). To minimise this reaction, the urea solutions must be passed over a mixed bed ion-exchange resin (e.g. Amberlite MB-1 or Bio-Rex RG501-X8) to remove cyanate, and contact with the urea should be for the shortest possible time, at 4 °C. In addition the general grades of urea, guanidine–HCl and Tris buffer salt may all contain traces of substances that can interfere with the resolution of the apo-lipoproteins. For this reason, it is recommended that the 'ultra-pure' grades of these reagents should be used whenever this is feasible. However, ultra-pure reagents may be too costly for large-scale preparative work and it will then be necessary to re-crystallise the lower grade materials from de-ionised water in the laboratory. Concentrated solutions of laboratory grade guanidine HCl are also often discoloured, and absorb strongly at 280 nm. These impurities can be removed by passing the solution over activated charcoal.

It has also been said that dialysis tubing contains metal ions and possibly other substances that have a deleterious effect on the resolution of the apo-lipoproteins. Some workers therefore purify the tubing by, for example, immersing it in 1 mM EDTA solution which is slowly warmed to 80–90 °C, followed by a similar treatment with 0.1 % $NaHCO_3$ solution and then by thorough washing in de-ionised water.

Although the apo-lipoproteins are most often refined by chromatographic or electrophoretic methods, it is worth noting that differential solvent extraction is a simple method with considerable attractions for the preliminary processing of relatively large amounts of material. The discovery that apo-lipoprotein B can be selectively precipitated from a lipoprotein by 50 % aqueous tetramethylurea (Section 6.4.1.2), leaving the smaller apo-lipoproteins in solution, led to the first successful system of this kind. Unfortunately, this solvent is comparatively expensive and cannot be used to effect a differential extraction of the smaller apo-lipoproteins. Moreover, the large-scale separation of apo-lipoprotein B can be carried out rather more conveniently by other methods. TMU extraction has therefore rarely been used as a preparative procedure (cf. Weisgraber et al., 1978). In a search for more versatile extractants, Holmquist discovered that 50% aqueous trimethyl phosphate, dimethylsulphoxide or formamide will dissolve the C apo-lipoproteins more readily than the others (Holmquist and Carlson, 1977). He accordingly proposed the following scheme for the preparation of apo-lipoproteins C-I, C-II, C-III and E from VLDL.

(a) Add an equal volume of dimethylsulphoxide to a solution of VLDL in 0.195 molal NaCl that contains not more than 1 mg of protein/ml. Shake for 1 minute and centrifuge off the precipitate. The clear solution, which is enriched with respect to apo-lipoproteins C-II and C-III, is dialysed against water and then freeze-dried.

(b) Suspend the precipitate from the first step in the original volume of 0.195 molal NaCl and add an equal volume of iso-propanol. Centrifuge off the residue and retain the solution. Dialyse and freeze-dry this to obtain a preparation that is enriched in apo-lipoproteins C-I and E.

(c) Since both the dried products contain lipid, they must be washed with ethanol-ether and ether before the final treatment.

(d) The two protein concentrates are resolved by iso-electric focussing on cellulose acetate (Cellogel), to yield the purified apo-lipoproteins (the method of Section 8.1.2 should also be applicable). In this way, the three polymorphs of C-III can be obtained separately.

As it is described, this procedure is best adapted to small-scale preparations and, in practice, there is little independent evidence that it is effective. Nonetheless, it appears to be a technique that is open to further development, particularly insofar as it enables the smaller apo-lipoproteins to be separated from the B protein simultaneously with the (partial) extraction of the lipids. More commonly, the extraction of the lipid and the solution of the proteins are conducted as separate operations. The protein solvents used under these circumstances have ranged from acetic acid to dilute alkali, but extremes of pH are best avoided since they can bring about the hydrolysis of amide groups or of some sensitive peptide bonds. Unfortunately, neutral buffers are unsatisfactory because they do not sufficiently dissociate the apo-lipoproteins. It is therefore necessary to add a reagent such as urea, guanidine hydrochloride or an anionic detergent such as sodium decyl or dodecyl sulphate. Of these, only the detergent will dissolve apo-lipoprotein B completely. Moreover, detergents have the attraction that they are effective solvents at a low concentration, are relatively inexpensive and, unlike urea, carry no risk of a permanent modification of the protein. These advantages have led to their use, either alone or in conjunction with urea, for the complete solution of VLDL prior to the resolution of the individual apo-lipoproteins on a preparative scale. However, the detergent is tightly adsorbed to the protein molecules and can be difficult to remove completely, as we shall discuss in a later paragraph. In practice, concentrated solutions of guanidine hydrochloride or urea are probably the most widely used solvents for the crude apo-lipoprotein mixture, although 2 M acetic acid can be successfully used for apo-HDL. There is no convincing evidence that mixtures of urea with sodium dodecyl sulphate offer significant advantages, but a mixture of 3 M urea in 2 M acetic acid has the merit that it is not only a powerful solvent for the apo-lipoproteins but does not promote their carbamylation.

Once the proteins have been successfully dissolved, there are two general techniques that can be used to separate them:

(1) Chromatography. This is the most widely used and the most generally applicable technique, particularly when substantial quantities of purified apo-lipoprotein are required. The best results will usually be obtained when at least two different chromatographic systems are used in succession. Gel filtration is frequently used to effect a preliminary separation into fractions that contain mixtures of proteins of like molecular weight. However, there is much cross-

contamination of these fractions, which necessitates a re-chromatography on, for example, an ion-exchange column. In practice, nearly all the apo-lipoproteins have been purified by some form of this general two-stage process.

Recently, attempts to take advantage of the high resolving power of HPLC have been made, notably by Hancock et al. (1981) who used a reversed phase system, and by Wehr et al. (1982) who used a filtration column. By applying the samples of crude protein (apo-HDL) automatically, the latter were able to use the equipment continuously and thus to overcome the restriction imposed by the relatively small load (10 mg). Nevertheless, the apparatus is expensive and costly to run. Moreover, the peaks on the chromatogram did not correspond to pure apo-lipoproteins. Consequently, the need for judicious 'peak slicing' and re-chromatography was not eliminated. A similar use of the filtration technique for the isolation of apo-lipoprotein E has also been reported by Pfaffinger et al. (1983).

The reversed phase HPLC system may prove to be especially useful for the purification of the C apo-lipoproteins. This has hitherto been carried out by conventional ion-exchange chromatography (e.g. on DEAE-cellulose) but this system tends to give low and sometimes erratic recoveries. Moreover, the products are cross-contaminated. Ronan et al. (1982) have used a reversed phase system for the final purification of the apo-lipoprotein C-II obtained by chromatography on DEAE-cellulose, and were able to isolate a product that 'met several criteria of purity unmatched by the traditional protein isolation techniques'.

At the time of writing, the isolation of apo-lipoproteins by the technique of chromatofocussing has not been reported.

(2) Electrophoresis. The apparently good resolution of the apo-lipoproteins that can be obtained by electrophoresis on analytical polyacrylamide gels suggests that this could be developed into an effective way of purifying these proteins. Furthermore, the resolution of the method appears to be good enough to allow several proteins to be separated simultaneously, in the same gel. However, at least in the

case of the conventional urea gels, the results are disappointing because (a) there is substantial adsorption to the gel matrix, which leads to poor recoveries and (b) the bands of protein that can be isolated are not pure by the most exacting criteria. On the other hand, Hardman and Kane (1980) have claimed good recoveries of subfractions of apo-lipoprotein B by preparative electrophoresis in sodium dodecyl sulphate/polyacrylamide gel, using a modified form of the Shandon equipment (Shandon Scientific). A somewhat similar system has also been successfully used to isolate the apo-lipoproteins E and A-IV (Utermann, 1975; Weisgraber et al., 1978; Beisiegel and Utermann, 1979) but it is not clear whether it is as satisfactory for the preparation of other apo-lipoproteins. In any event, the removal of the detergent introduces an additional difficulty, as we have already pointed out.

Greater promise as a preparative method has been shown by the sister technique of iso-electric focussing (IEF). This was first successfully applied to the apo-lipoproteins in the form originally described by Vesterberg and Svensson (1966), but latterly attention has turned more towards the method of focussing in a flat bed of granular gel (Radola, 1973; Winter et al., 1975; Righetti and Drysdale, 1976). Sometimes, this technique has been used to complete the isolation of apo-lipoproteins that have been partially purified by other methods e.g. apo-lipoprotein E (Beisiegel, 1979), or C-II and C-III (Holmquist and Broström, 1979). In other laboratories, it has been used to isolate an apo-lipoprotein from a crude mixture in a single step e.g. the isolation of most of the apo-lipoproteins from VLDL (Marcel et al., 1979), or the preparation of apo-lipoprotein A-I from HDL (Nestruck et al., 1980; Forgez and Chapman, 1982). As in all methods for the preparation of apo-lipoproteins, the efficiency of flat-bed focussing is strongly affected by the extent to which the proteins are adsorbed on the support phase. In this context, Forgez and Chapman (1982) claim to have made a significant improvement by substituting polyacrylamide gel beads (Bio-Gel: Bio-Rad Laboratories) for the granular dextran used in earlier experiments. This change raised the total recovery of apo-HDL to 75 % and materially reduced the cross-contamination of the separated apo-lipoproteins. The speed and simplicity of this technique goes far to offset its comparatively small load factor (15–20 mg of protein) but the carrier ampholytes are difficult to remove completely (Section 8.1.2).

We have remarked earlier that bound detergents such as sodium dodecyl sulphate are also difficult to remove from proteins. Whereas free detergent in the solution can be removed by gel filtration, it may be necessary to dialyse for up to 5 days to reduce the proportion of bound dodecyl sulphate to 1 mole/mole of protein (Visser and Blout, 1971; Tuszynski and Warren, 1975). In this respect sodium decyl sulphate is preferable to the dodecyl salt because it is somewhat less tightly bound and forms more easily dialysable micelles. Nonetheless, apo-lipoproteins that are dissolved with the aid of detergent and are then purified by repeated gel filtration, or by electrophoresis in SDS-polyacrylamide gels, are likely to be contaminated. By contrast with dialysis, ion-exchange chromatography, either in the presence or the absence of urea, is an effective and speedy method of removing the bound detergent (Weber and Kuter, 1971; Lenard, 1971; Kapp and Vinogradov, 1978). Consequently, apo-lipoproteins that have undergone a refinement by chromatography on DEAE-cellulose may be almost un-contaminated, although the efficiency of this system seems not to have been quantitatively established. When an absolute freedom from detergent is an important consideration, it will therefore be prudent to avoid all contact with these substances. Alternatively the method of Amons and Schrier (1981) can be used to effect a final purification. In this procedure, up to 10 mg of protein, dissolved in a mixture of propionic acid: formic acid and water (2:1:2 by vol.) is run through a 10 cm column of Sephadex G-25 in the same solvent. The protein can be recovered from the eluate by low-temperature evaporation in vacuo, or by freeze-drying after a preliminary dialysis, first against water and then against 50 mM ammonium bicarbonate solution.

As the individual apo-lipoproteins become more highly purified, they also become more difficult to dissolve in mild solvents. Whether this is a result of their greater purity per se, or is merely a consequence of their repeatedly being dried, is not clear. That there is apparently some divergence in the experience of different laboratories suggests that the effect may be due, in part, to differences in manipulative procedures. Although the proteins can be dissolved in concentrated solutions of urea or guanidine hydrochloride (except of course for apo-lipoprotein B which requires the use of a detergent), these are unsuitable solvents in which to inject antigens. For this purpose, 0.1 M ammonium carbonate or bicarbonate are probably the most generally useful solvents, although 2 mM sodium decyl sulphate can also be used. However, apo-lipoprotein E can be particularly difficult to dissolve completely, requiring the use of detergent or 0.1 M ammonium hydroxide. Acetic acid (1.0 M) can also be a useful solvent in some circumstances and has been recommended for the solution of apo-lipoprotein C-I. If it is necessary to store the isolated apo-lipoproteins in solution, it is undesirable that they should be exposed to extremes of pH, or to the presence of urea. For this purpose, the best solvent is buffered guanidine hydrochloride, pH 7.5, at the lowest effective concentration. This may range from about 3 M in the case of the C apo-lipoproteins to 5–6 M for the more difficultly soluble apo-lipoprotein E.

It is an obvious stratagem to begin the isolation of an apo-lipoprotein by discarding those lipoproteins of which it is only a minor component. This is conventionally done by preparative centrifugation at judiciously chosen limiting densities (Chapter 2). The product must then be thoroughly washed by centrifugation to remove traces of the plasma proteins and especially of albumin. In this context it is worth noting that traces of albumin can be more quickly and economically stripped from a lipoprotein preparation on an immunoaffinity column (Holmquist and Carlson, 1978) than by centrifugation. Recently, attention has also turned to the use of immobilised antisera for the preliminary separation of the lipoprotein particles that contain the desired protein. This technique was used by McConathy and Alaupovic (1976) to prepare a lipoprotein fraction that was rich in apo-lipoprotein D, and has since been used to isolate fractions similarly enriched with apo-lipoprotein A or E (Kunitake et al., 1982; Gibson et al., 1982). Apart from its evident advantage when the desired apo-lipoprotein is present in only small amounts, the method is a technically simple and expeditious way of producing well-washed lipoprotein fractions. However, it suffers from the drawback that the adsorbents have a rather low capacity and require the preliminary production of a monospecific antiserum.

The final stage in the preparation of an apo-lipoprotein is the establishment of its purity. Any substantial contamination can be detected by electrophoresis in polyacrylamide gel (Section 6.4.2) and this is an important way of checking the progress of intermediate steps in the preparation. It is a comparatively insensitive assay however, and the final product must be subjected to immunochemical analysis (Section 6.5). The presence of plasma proteins can easily be detected with an antiserum raised against lipoprotein-free plasma and with commercially available anti-human serum albumin.

Although a judiciously prepared anti-apo-LDL is often essentially monospecific for apo-lipoprotein B, anti-sera to VLDL or HDL will generally react with more than one apo-lipoprotein. A carefully designed series of immunological analyses with these antisera will therefore often show whether the apo-lipoprotein under test is homogeneous and it is worth noting the power of two-dimensional immunoelectrophoresis in this context (Section 5.7). However, it will rarely be possible to identify the contaminating apo-lipoproteins that are detected in this way. Moreover, the reactivity of these antisera to lipoprotein fractions will not be uniform against all the potential antigens. Accordingly, although access to antisera against all the major apo-lipoproteins may be difficult, an effort should be made to obtain authentic, monospecific antisera to at least those apo-lipoproteins that are most likely to contaminate the protein which is being purified. In the case of apo-lipoprotein A-II for example, it would be desirable to test at least for the presence of A-I and D, the latter of which is highly immunogenic.

8.1. Isolation of apo-lipoprotein A-I

The richest source of this protein is HDL_3 and, if A-I is to be the only product, this is the best starting point. However, total HDL can also be used and this will be the better raw material if the object is to prepare both A-I and A-II. In any event, the lipoprotein must be freshly prepared from fresh plasma.

8.1.1. Chromatographic procedure

The first stage in this method is invariably a gel filtration through superfine Sephadex G-200 or (preferably) Sephacryl S-200 (or their equivalent). The choice of dissociating agent to be used in the eluent may require some consideration: 5 or 6 M guanidinium hydrochloride has the advantage that it does not degrade the proteins but is by far the most expensive reagent to use; 6 or 8 M urea is more economical but, during the course of a prolonged elution, there is inevitably some carbamylation of the proteins; 2 M acetic acid is the least expensive solvent available but there is some danger that prolonged exposure to the relatively low pH may hydrolyse some of the amide groups in the protein, although there is little evidence that this degradation is significant. Since the resolution that can be obtained is essentially the same in each solvent, the choice may be determined by financial considerations and, in this context, it must be remembered that only the best grades of urea or guanidine hydrochloride can be used.

Whatever solvent is used, the elution pattern will take the general form shown in Fig. 8.1a. Note, however, that the resolution of a freshly packed column may be less satisfactory than this but that it will generally improve with several cycles of use. As each protein is eluted, it should be collected in several fractions, each of which must be analysed by electrophoresis on polyacrylamide gel (Section 6.4.2.3). It will be found that there are fewer contaminants of the A-I under the leading limb of peak 2 than there are under the trailing limb. Moreover, the extent of this contamination (mainly with A-II) will be least when the column is only lightly loaded. Consequently, the best

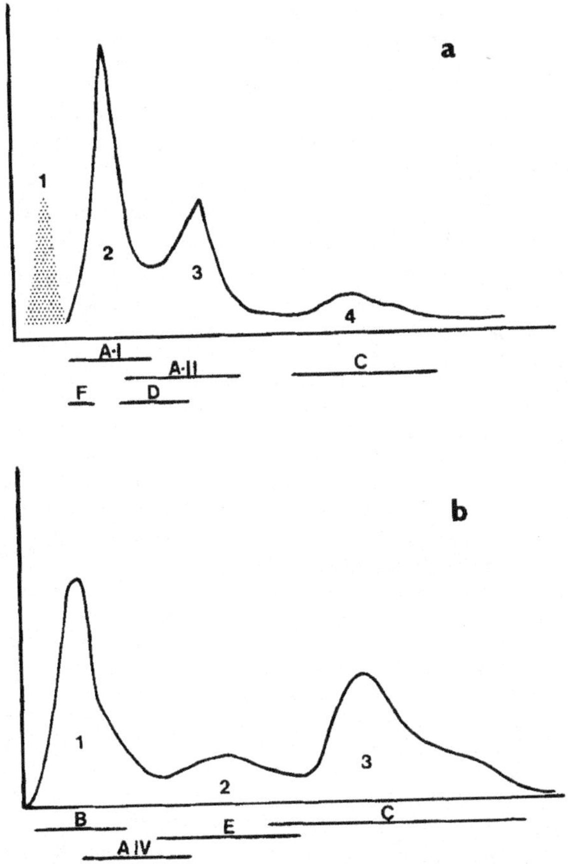

Fig. 8.1. Generalised gel filtration chromatograms for apo-lipoproteins. (a) apo-HDL. Peak 1 consists of aggregated apo-lipoproteins and can vary considerably in size according to the manipulative history of the sample and the way in which it is dissolved. It is disadvantageous to dry the apo-HDL before it is dissolved in the eluting solvent. (b) apo-VLDL. Note that the resolution of this system can be optimised by a judicious choice of gel and solvent. Thus, 2 mM SDS with a highly permeable gel will give a good resolution of peaks 1 and 2, whereas a gel of exclusion limit 10^5 can resolve the trailing limb of peak 3 if eluted with guanidine hydrochloride.

results will be obtained when the load is less than about 15 mg of protein per square cm of column cross-section, and the fractions are carefully selected before being pooled. Under these circumstances, a second cycle of chromatography on the same column may suffice to produce pure apo-lipoprotein A-I, albeit in a yield that is unlikely to exceed 25 % of the maximum. A somewhat better yield may result if the eluent used with the second column includes a reducing agent such as mercaptoethanol or thiodiglycol. This dissociates apo-lipoprotein A-II into its constituent peptide chains, which are more easily resolved from A-I than the native protein. If the A-II obtained from the first filtration column is to be discarded, it will be advantageous to run this too in the presence of reducing agent.

When relatively large quantities of A-I are needed, e.g. for physico-chemical studies, it may be a better strategy to load the gel filtration column rather more heavily, to collect a larger proportion of peak 2, and then to purify the A-I by chromatography on DEAE-cellulose. A gradient of increasing ionic strength elutes the apo-lipoproteins from this adsorbent in the sequence shown in Fig. 8.2. By collecting

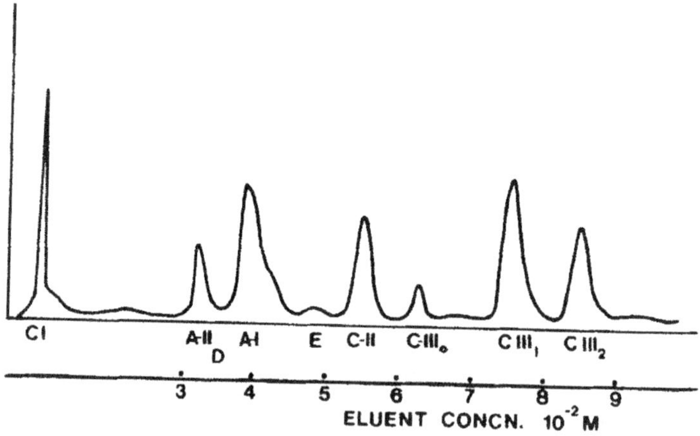

Fig. 8.2. The generalised sequence of apo-lipoproteins eluted from a DEAE-cellulose column. The elution is begun with a volume of 0.03 M Tris buffer/6 M urea (pH 8.0) equal to at least 2.5 times the volume of the column bed. It is then continued with a linear gradient of increasing ionic strength at pH 8.0.

peak 4 from this column in several fractions, each of which is analysed by electrophoresis on polyacrylamide gel, a pool of pure apo-lipoprotein A-I can be isolated. Note, however, that A-I may, under the best conditions, be eluted from this column as two or three partially resolved peaks.

Equipment. The following will be needed in addition to that required for the preparation of the lipoproteins.

(1) A glass column 2.5 cm × 150–200 cm, siliconised to reduce loss of apo-lipoprotein by adsorption, and packed (cf. Fischer, 1980) with superfine Sephacryl S-200 that has been thoroughly equilibrated with the chosen eluent (see below).

(2) A siliconised glass column 1 cm × 50 cm, packed with DE-52 microgranular DEAE-cellulose (Whatman) that has been equilibrated with 0.03 M Tris–HCl buffer, pH 8.0. Before it is used, wash the column with at least one column volume of the reagent B described below.

A fraction collector is required for each column and, if possible, equipment for continuously recording the photometric absorption of the eluate at 280 nm. A conductivity meter is a convenient way of checking the purity of the concentrated urea solutions, and of estimating the position of eluent fractions along the gradient used on the DE-52 column. The isolated apo-lipoprotein fractions can be concentrated either by ultrafiltration (e.g. in an Amicon cell using a UM-2 membrane filter), or by freeze-drying.

Reagents. Urea and guanidine hydrochloride must be of the highest available quality. If the 'ultra-pure' grade cannot be obtained, the analytical grade should be re-crystallised with the greatest care. All other reagents should be of analytical grade.

It is an almost uniform practice to buffer the guanidine or urea solutions used as eluents, with Tris–HCl of pH 8.0 However, the concentration of this buffer has variously ranged from 0.01 M (Scanu et al., 1969) to 0.2 M (Jackson and Gotto, 1972). There seems to be no published evidence to show that one of these is superior to the others and we have consequently chosen to use a concentration of 0.03 M.

The operator has a choice of three chromatographic eluents that are essentially equivalent, namely: (A) 0.03 M Tris–HCl buffer, pH 8.0, in 5 M guanidine hydrochloride solution. (B) 0.03 M Tris–HCl buffer, pH 8.0, in 6 M urea solution.

The urea solution used to prepare this solvent must be freshly made and be run over ion-exchange resin (Bio-Rex RG 501-X8 or equivalent) and should have a conductivity of less than 0.5 μ.mho. The eluent is best used at once, but may be briefly stored at 4 °C. Note that the concentration of urea used in different laboratories has ranged from 5.4 M (Jackson and Gotto, 1972) to 8 M (Scanu et al., 1969). There is however, no confirmed evidence that a high concentration gives a better result. (C) 2 M acetic acid. Despite their practical drawbacks, eluents of the type represented by B are the most widely used.

The gradient used to elute the DEAE-cellulose column (Lux et al., 1972) is made by mixing reagent B with the following limiting solution: (D) 0.07 M Tris–HCl buffer, pH 8.0, in 6 M urea solution. The preparation and use of this solution is subject to the conditions that apply to reagent B.

Procedure. Prepare HDL_3 (*d* 1.125–1.21 g/ml) by one of the methods described in Chapter 2 and wash it at least once by centrifugation at $d = 1.21$ g/ml. Dialyse the product against 1 mM EDTA solution in a closed vessel with minimal air space and then prepare the lipid-free protein as described in Appendix 3. Dissolve the product in sufficient of the solvent in which the Sephacryl column was packed to give a solution containing about 15–20 mg of protein/ml.

Transfer not more than 75 mg of apo-HDL to the column and elute with the chosen solvent (at room temperature with eluents A or C; at 4 °C with eluent B) at a flow-rate of 10–12 ml/h. If eluent B is used, purge the column with freshly made solution before the sample is applied, and use the same fresh solvent to dissolve the crude protein. Collect the eluate in fractions of 4–5 ml and dialyse, without delay, against 5 mM ammonium bicarbonate solution at 4 °C. Freeze-dry each of the fractions that correspond to peak 2 of Fig. 8.1a. Analyse each of these fractions by electrophoresis in both alkaline–urea and

SDS gels (Sections 6.4.2.3 and 6.4.2.4) and pool those that are the least contaminated.

Repeat this process on the pool of partially purified product, but with the addition of 0.1 % β-mercaptoethanol to the eluent. Retain those fractions that are shown, by electrophoretic analysis, to be uncontaminated. Confirm their purity by immuno-diffusion against antisera to plasma and HDL of the appropriate species.

If the alternative ion-exchange system is adopted in place of the second filtration chromatogram, transfer the partially purified A-I into 10 mM ammonium bicarbonate solution by dialysis, or by chromatography on a 30 cm column of Bio-Gel P-4. Lyophilise this solution and then re-dissolve the residue in a small volume of eluent B. Note that all stages of the procedure from this point must be carried out at 4 °C. Apply about 50 mg of the protein to a column of DE-52 that has been freshly purged with eluent B, and wash with 100 ml of the same solution before starting the gradient elution. This gradient should be linear in form and can be generated in the conventional way by the progressive mixture of equal volumes of eluents B and D in a two-chambered gradient maker (Peterson, 1981). To ensure a good resolution of A-I from A-II, the gradient must not be too steep. For the size of column proposed, 250 ml of each solution should give satisfactory results. Collect peak 3 (Fig. 8.2) in several fractions, dialyse them exhaustively against 10 mm ammonium bicarbonate and lyophilise. Finally, pool those fractions that are shown to be pure by electrophoresis in polyacrylamide gel, and confirm their purity by immuno-chemical analysis.

After the A-1 protein has emerged from the DE-52 column, the system should be purged with 100 ml of 0.5 M Tris–HCl, pH 8.0, in 6 M urea and then re-equilibrated with 0.03 M Tris–HCl, pH 8.0.

8.1.2. Isoelectric focussing

This method is well suited to the production of the small amounts of material needed for the production of antisera. The technique is thoroughly described by Righetti (1983) and much useful information is given by Winter et al. (1975). For the particular purpose of isolating apo-lipoprotein A-I we shall describe the procedure used by Forgez and Chapman (1982).

Equipment. The following details relate to the LKB 2117 Multiphor unit. With the appropriate modifications, it can be used with similar equipment from other makers. A power supply capable of operating in the 'constant power' mode is also required, e.g. the LKB 2103.

Reagents. Forgez and Chapman (1982) improved the resolution of their system by using a mixture of ampholytes that focussed to a pH gradient between the limits 4.6 and 5.9. This mixture was obtained by the methods of synthesis and fractionation described by Righetti et al. (1975) and Gianazza et al. (1975). Alternatively, a commercially available ampholyte mixture of pH range 4–6 (LKB-Produkter; Bio-Rad Laboratories) can be fractionated by pre-focussing in a dummy gel bed for 20 hours (cf. Procedure below). The section of gel that lies within the required limits of pH can then be cut out and used to make the bed in which the proteins are separated. In this procedure however, the gel in which the ampholytes are pre-focussed must not be concentrated by evaporation, as it is before the proteins are focussed.

Solution A: 36.0 g of urea
0.23 g of *N*-ethyl morpholine
Water to 100 ml, adjust to pH 8.6

Solution B: 27.0 g of urea
1.95 g of ampholyte
0.1 ml of β-mercaptoethanol
Water to 100 ml

Solution C: 360 g of urea
9.75 g of Tris
19.4 ml of 1.0 M HCl
Water to 1.0 litre, adjust to pH 8.6

The above urea solutions must be prepared with the precautions that have already been described in Section 8.1.1.

Solution D: 10 % trichloracetic acid in water

Solution E: 0.2 % Coomassie Blue R-250 in methanol–water–acetic acid (5:5:1 by vol.).

Anolyte: 0.1 M orthophosphoric acid solution

Catholyte: 0.1 M glycine solution

Procedure. Prepare apo-HDL_3 (Section 8.1.1) and dissolve it in solution A to a concentration of 5–10 mg of protein/ml.

To make up the gel bed, first suspend 6 g of Bio-Gel P-60 (50–100 mesh; Bio-Rad Laboratories) in 100 ml of solution B and de-aerate in vacuo. At each end of the gel tray, put a triple layer of the strips that are provided with the equipment. Each strip should be about 11 cm long and be soaked in solution B. Weigh the tray and fill it with a weighed amount of the well-stirred slurry of gel. Then warm the tray in an oven at 40 °C until 25 % of the water has evaporated.

Apply the solution of apo-HDL dropwise along a straight line across the gel, 7 cm from the cathode strip, until 15–20 mg of protein have been added. Saturate the electrode wicks with their respective electrolytes and apply the electrodes. Pass the current at a constant power of 10 W for 18–20 h, at 10 °C. During this time, the voltage will rise from the initial setting of 500 to about 1300 V.

To detect the position of the separated proteins, apply three strips of moist filter paper (e.g. Whatman No. 1, 1.5 cm wide) along the length of the gel bed, one at each edge and the third in the middle. Be sure not to trap air under the paper. When the strips are saturated, peel them carefully off the gel and dry at 110 °C for 10 min. Wash each print three times, for 15 min, in solution D. Then stain in solution E until the protein bands are clearly visible, de-staining in methanol–water–acetic acid (5:5:1 by vol.) if necessary. Finally, align these strips very carefully beneath the gel tray, to act as a guide to the positions of the proteins.

In this system, apo-lipoprotein A-I is resolved into the four 'polymorphs' that focus between pH 5.35 (A-I_4) and 5.6 (A-I_1) (Fig. 8.3). The latter, which forms the most intense of these bands, yields the purest product and can be recovered by carfully scooping out the gel with a small spatula. Transfer this gel to a short chromatography column (e.g. a plastic syringe barrel) and elute with solution C. Dialyse exhaustively against 5 mM ammonium bicarbonate solution at 4 °C and concentrate to suit requirements. Confirm the purity of the product by electrophoresis in polyacrylamide gel and by immunochemical analysis.

It may be observed that ampholytes are difficult to remove com-

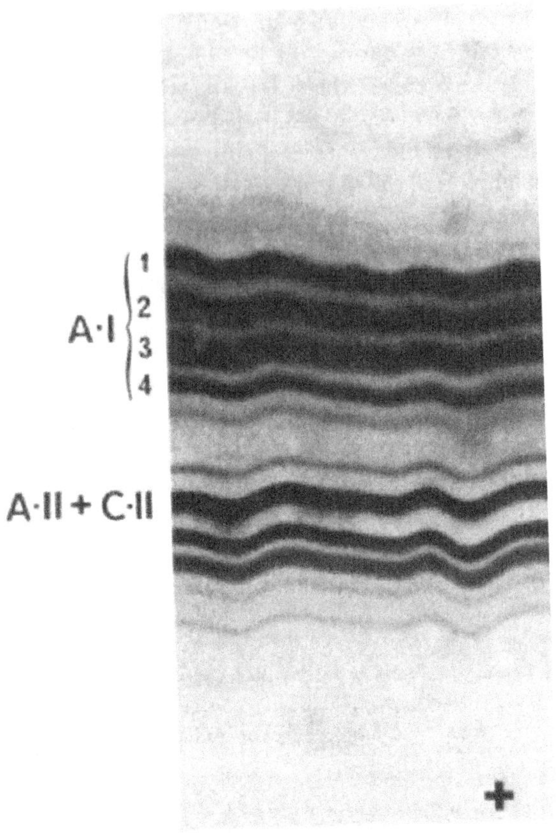

Fig. 8.3. Preparative isoelectric focussing of apo-HDL.

pletely from these protein preparations. Up to 5 days of conventional dialysis against repeated changes of 5 mM ammonium bicarbonate solution may be necessary. A quicker and more efficient separation can be achieved by electrodialysis (Marcel et al., 1979) which can easily be carried out in the ISCO electrophoretic concentrator (Instrumentation Specialities Co.) (Section 6.4.3). However, it will be found that, for many purpose, the presence of traces of ampholyte is of no consequence.

8.2. *Isolation of apo-lipoprotein A-II*

The strategy used to isolate this apo-lipoprotein by chromatography differs only in detail from that described for A-I in Section 8.1. Moreover, this is effectively the only procedure available, since isoelectric focussing will not satisfactorily resolve A-II from the A-IV, C-II or C-III_0 apo-lipoproteins.

In many laboratories, A-II is isolated from HDL (*d* 1.063–1.21 g/ml) but there may be some advantage in using a carefully purified preparation of HDL_2 which is a slightly richer source of this apo-lipoprotein. The raw protein is then usually refined by a stage of gel filtration, followed by one of ion-exchange chromatography, exactly as used for apo-lipoprotein A-I. However, Herbert et al. (1977) prefer to reverse this sequence, a preliminary purification on DEAE-cellulose being followed by gel filtration in guanidine hydrochloride. Alternatively, the gel filtration stage can be replaced by a second passage through the ion-exchange column. Indeed, according to Lux et al. (1972), it is possible to isolate pure A-II, in a yield of 75 %, from only a single stage of ion-exchange chromatography. Whichever plan is chosen, it is crucial that a large number of fractions is collected from the first column. Each of these must be analysed by electrophoresis, and the least contaminated pooled for a second stage of purification if necessary.

Procedure. The reagents, equipment and procedure are as described in Section 8.1.1. Collect peak 3 (Fig. 8.1a) from the gel filtration column in several fractions, and pool the purest of these for subsequent chromatography on DE-52 cellulose. Apo-lipoprotein A-II is the second protein to be eluted from this column (Fig. 8.2) and this peak must also be collected in several fractions. Concentrate and pool all those which consist of pure A-II, as shown by electrophoretic and immunological analysis.

8.3. Isolation of apo-lipoprotein A-IV

This apo-lipoprotein was first isolated from rat plasma (Swaney et al., 1977) but has since been obtained in small quantities from the VLDL of human thoracic duct lymph (Weisgraber et al., 1978) and from the plasma VLDL of non-fasting hypertriglyceridaemic subjects (Beisiegel and Utermann, 1979). With a molecular weight of about 46 000, this apo-lipoprotein runs slightly faster than apo-lipoprotein E on gel filtration chromatography (Fig. 8.1b) but an extra long column is needed to achieve a useful resolution. Moreover, if VLDL is the source material, the A-IV protein can be difficult to separate from the much larger amount of the dissolved apo-lipoprotein B. Weisgraber et al. (1978) minimised this problem by precipitating the B proteins with tetramethyl urea (Section 6.4.1.2) and then resolving the soluble apo-lipoproteins by gel filtration chromatography in buffered 4 M guanidine hydrochloride solution. By contrast, Swaney et al. (1977) effected a preliminary partial purification on a short (100 cm) filtration column in a mixture of 6 M urea with 2 mM sodium decyl sulphate. The appropriate fractions of this eluate were then concentrated, transfered into a solution of 6 M urea in 0.1 % dodecyl sulphate and re-chromatographed on a 270 cm gel column. A judicious fractionation of the eluate then yielded pure apo-lipoprotein A-IV. Both Weisgraber et al. (1978) and Swaney et al. (1977) also used the Stephens (1975) method of preparative electrophoresis in SDS-gel to refine their product. Beisiegel and Utermann (1979) likewise used SDS-gel electrophoresis, but in a continuous elution apparatus. Moreover, they apparently applied whole apo-VLDL to their gels without any prior purification. However, the simplicity and economy of this procedure must be set against the fact that the (unstated) yield of pure A-IV protein that was obtained from each 10 mg load must have been small.

According to Beisiegel and Utermann (1979), apo-lipoprotein A-IV is present in the VLDL of only some 15 % of non-fasting hypertriglyceridaemic subjects. It is therefore necessary to screen the potential donors before drawing blood.

Equipment. (1) a siliconised glass chromatographic column, 2.5 cm × 100 cm, packed with Sephadex G200 in solution A (see below). (2) an exactly similar column but of total length 300 cm. This can be conveniently arranged by using short lengths of fine-bore tubing to connect 3 of the 100 cm columns in tandem.

Reagents. Solution A: 0.1 M Tris–HCl, pH 8.0, in 5 M guanidine HCl solution.

Procedure. Isolate VLDL from the selected hypertriglyceridaemic plasma in the conventional way (Section 2.1) and wash at least twice by centrifugation at $d = 1.006$ g/ml. Prepare the lipid-free apo-VLDL (Appendix 3) and extract them thoroughly with 5–10 ml of solution A. Centrifuge off the undissolved residue and transfer 50–60 mg of the dissolved protein to the short chromatography column. Develop the chromatogram with solution A. Collect the trailing half of peak 1 together with peak 2, dialyse against 5 mM ammonium bicarbonate and lyophilise. Redissolve the protein in solution A and re-run on the 300 cm column in the same solvent. The proteins will appear as a diffuse, unresolved distribution which must be collected in a number of arbitrary fractions. Analyse these by electrophoresis in polyacrylamide gel (Sections 6.4.2.3 and 6.4.2.4) and select those which contain pure apo-lipoprotein A-IV. If the purification is inadequate, the product can be refined by one of the preparative methods of electrophoresis mentioned above.

8.4. *Isolation of apo-lipoprotein B*

When undertaking this preparation, it is important that all the phospholipids are extracted from the apo-LDL that is used as the raw material, otherwise an 'apo-lipoprotein' of spurious solubility will be obtained. Fully extracted apo-lipoprotein B is virtually insoluble even in 8 M urea and, in a number of early experiments, a more tractable product was obtained by chemical modification e.g. by succinylation or by maleylation of the protein (Scanu et al., 1968; Gotto et al., 1968; Kane et al., 1970). Here, however, we shall consider only methods for the isolation of the native apo-lipoprotein. The most successful of these make use of the fact that apo-lipoprotein B will dissolve moderately well in sodium decyl sulphate solutions and can be freed from accompanying apo-lipoproteins of lower molecular weight by gel filtration chromatography (Brown et al., 1969; Herbert et al., 1973). The protein isolated in this way can still be resolved into several components by electrophoresis in SDS-gel (Section 6.4.2.4), and Kane et al. (1980) have developed this into a preparative procedure. However, the chromatographically purified protein is satisfactory for the majority of purposes.

Lee et al. (1981) recently postulated that the insolubility of the isolated B protein may be the result of aerial oxidation during its isolation. By taking elaborate precautions to prevent this, they claimed that they could prepare apo-lipoprotein B in a form that was completely soluble in buffered guanidine hydrochloride solution in the presence of reducing agents. Moreover, in this process, the unwanted low-molecular weight apo-lipoproteins were separated from the B protein, not by chromatography, but by extraction into *N*-ethylmorpholine. If the advantages that are claimed for the method can be confirmed, it may be a valuable way of preparing relatively large amounts of apo-lipoprotein B. At the time of writing however, it is interesting mainly for its scrupulous precautions against oxidation.

The richest source of apo-lipoprotein B is the LDL fraction, of which it constitutes nearly 20 % by weight. However, it is an advantage to use a narrow cut from the centre of the LDL density range (e.g. 1.025–1.045 g/ml) since this contains the smallest proportion of the unwanted apo-lipoproteins.

8.4.1. Chromatographic procedure

In this system, the lipid-free apo-LDL is dissolved in 0.1 M sodium decyl sulphate solution which is adjusted to pH 8.0–8.5 with any convenient buffer, e.g. Tris–HCl, phosphate, or ammonium bicarbonate. The apo-lipoprotein B is then refined by filtration chromatography in buffered 2 mM SDeS. The columns used for this purpose have been reported to range in porosity from Sephadex G-150 to Bio-Gel A-1.5M, but no clear advantage is claimed for any one of these gels. In terms of practical convenience, the gels of lower porosity and compressibility may be preferable.

Equipment. A 2.5 cm × 100 cm siliconised glass column, packed with Sephadex G-150 in solution B (see reagents). Auxiliary equipment as in Section 8.1.1.

Reagents.

Solution A: 2.6 g of sodium decyl sulphate
10 mg of sodium azide
100 ml of 0.01 M Tris–HCl buffer, pH 8.0

Solution B: 0.52 g of sodium decyl sulphate
0.1 g of sodium azide
1.0 litre of 0.01 M Tris–HCl buffer, pH 8.0

Procedure. Extract the lipid from the LDL preparation, which must be freshly made from fresh plasma, and dissolve the protein in solution A to a concentration of 15–20 mg/ml (Appendix 3). Apply 50–60 mg of protein to the column and elute with solution B, at 20 °C, at a flow rate of about 50 ml/h (10 ml/h/cm^2). Collect the eluate in fractions of about 7.5 ml.

The apo-lipoprotein B will emerge from this column as an asymmetrical peak at the void volume, the trailing limb being contaminated

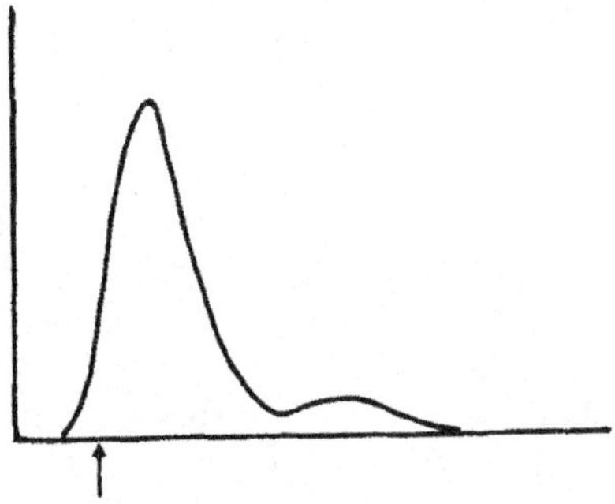

Fig. 8.4. The filtration chromatography of apo-LDL in 2 mM sodium decyl sulphate solution. In this generalised representation, the arrow indicates the void volume of the column and the smaller peak represents contaminants of low molecular weight, such as apo-lipoprotein C.

with what may be breakdown fragments of apo-lipoprotein B and with traces of other apo-lipoproteins. Collect the central one-third of this peak and freeze-dry it. Re-dissolve the product in 1–2 ml of water and chromatograph it a second time on the same column, Concentrate each of the fractions that compose the first peak to emerge and test their purity by electrophoresis in SDS-gel (Section 6.4.2.4) and by immunochemical analysis (Section 6.5) with antisera raised against lipoprotein-free plasma and against a TMU extract of plasma lipoproteins. Pool those fractions that show no contamination.

8.4.2. *Non-oxidative procedure*

The following is a summary of the procedure recommended by Lee et al. (1981).

Reagents. Solution A: 1.3% ε-aminocaproic acid, 1.0% EDTA and 0.5% glutathione in water, adjusted to pH 7.0.

Solution B: 5000 units/ml of penicillin-G, 500 µg/ml of streptomycin sulphate and 200 µg/ml of chloramphenicol in solution A.

Solution C: 0.05% thiodiglycol, 0.05 M *N*-ethylmorpholine acetate in a 10-fold dilution of solution A.

Solution D: 6 M ultra-pure guanidine hydrochloride–0.1 M Tris buffer (pH 8.3) in a 10-fold dilution of solution A which also contains 0.05% thiodiglycol, 0.1% dithiothreitol and 0.02% tryptamine.

Procedure. Lee et al. collected plasma by plasmapheresis into a plastic bag. If this facility is not available, the blood should be collected into a transfusion bag, the plasma separated and transferred to a clean bag with the least possible delay or exposure to air. Add 1 ml of solution B to each 9 ml of plasma and seal the bag with care to exclude all air. The isolation of the lipoproteins should preferably be started at once, although Lee et al. imply that the plasma may be kept for up to 3 days in the sealed bag at 4 °C.

To minimise contact with air during the isolation of the LDL, it is best to use a density gradient procedure in Beckman Quick-Seal tubes (Section 2.1). Lee et al. used this technique to prepare a fraction of density 1.032–1.043 g/ml by centrifugation at 45 000 rev/min in the Spinco 50.2 Ti rotor, for 26 hours at 5 °C. If another density fraction is to be isolated, or a different rotor must be used, the method can be easily adapted (Section 2.1). The solutions used to prepare the gradient should be de-aerated and contain 10 % of solution A. The Quick-Seal tubes should be filled as completely as possible and be sealed immediately. After the centrifugation, the required density fraction is collected with a hypodermic needle attached to a length of fine-bore plastic tubing. Thrust the needle through the wall of the centrifuge tube at the point that corresponds to the upper limiting density of the lipoprotein fraction (this must be determined from a control gradient), and allow the calculated volume of the contents to flow into an ice-cooled flask which is continuously purged with nitrogen. To each ml of this solution, add 0.3 ml of NaCl of density 1.19 g/ml (containing 10 % soln. A) and layer it in a Quick-Seal tube, under NaCl solution of density 1.063 g/ml (containing 10 % soln. A). Centrifuge for 20 hours at not less than 200 000 *g* and collect the floating layer of LDL as before.

Follow the isolation of the LDL fraction immediately by extracting the lipids with ethanol–diethyl ether at 5 °C. The ether used for this operation must be freed of peroxides, dried over anhydrous $CaCl_2$ and re-distilled immediately before use. The extraction procedure used by Lee et al. (1981) is as follows: add 36 ml of ice-cold absolute ethanol containing 0.02 % butylated hydroxytoluene (BHT) to a 50 ml tube and purge with nitrogen. While the gas is still passed through the alcohol, add 2.5 ml of the LDL solution, which should be equivalent to about 10 mg of protein. Stand in ice for 20 min, under nitrogen, before adding 12 ml of ice-cold ether which contains 0.02 % BHT. Mix thoroughly, purge with nitrogen and stopper the tube. Continue to agitate gently at 5 °C for 30 min before centrifuging down the precipitate and aspirating off the organic phase. Repeat this extraction four times with 0.02 % BHT in cold 3:1 (v/v) ethanol: ether, followed by a single wash with 0.02 % BHT in 5:1 (v/v) ethanol–ether. All these steps must be carried out in an atmosphere of nitrogen.

After the final extraction, suspend the residue in 20 ml of solution C and purge gently with nitrogen for 1–2 hours to remove the residual ether. Recover the precipitate by centrifugation and repeat the washing process with 5 ml of solution C. This extraction removes inorganic salts and is also said to extract most of the 'soluble' apo-lipoproteins that are present in LDL (A-I, C-II, C-III, D, E, F). After this treatment, Lee et al. claim that the apo-lipoprotein B is completely soluble, to a concentration of 5 mg/ml, in solution D.

8.5. *Isolation of the C apo-lipoproteins*

Since the largest proportion of these proteins is to be found in the largest lipoprotein particles, i.e. in chylomicrons or the high molecular weight VLDL, it is as well to start this preparation with plasma from hypertriglyceridaemic subjects. The isolation of the C apo-lipoproteins then takes a course similar to that used to prepare the A proteins (Section 8.1), namely the isolation of a crude mixture of the C proteins by gel filtration chromatography, followed by refinement on an ion-exchange column. However, there are several factors to be considered when designing the gel filtration system:

(a) Because apo-VLDL contains a substantial proportion of apo-lipoprotein B, it is not completely soluble in the solvents used for apo-HDL and must be dissolved in 0.1–0.2 M sodium decyl sulphate. This solution can then be put directly onto a gel filtration column and eluted with 2 mM SDeS (Section 8.4.1), or with alkaline 6 M urea solution. Unfortunately, SDeS has a low solubility in concentrated solutions of guanidine salts and its concentration must be reduced by dialysis or by gel filtration before the apo-VLDL can be fractionated in a guanidine hydrochloride solution.

As an alternative to the complete solution of the apo-VLDL, it can be extracted with a concentrated solution of urea or guanidine hydrochloride which is buffered to pH 8.0–8.5. In practice, this gives a good recovery of the C proteins, and the method has the advantage that the extract can be put directly onto a column that is equilibrated with guanidine hydrochloride.

(b) The resolution achieved can be determined, to some extent, by the nature of the gel filtration system that is used (Herbert et al.,

1977). Elution with 6 M urea or 2 mM SDeS from a column of Sephadex G200 will result in three peaks (Fig. 8.1b), the third of which will contain most of the C proteins, albeit they will be contaminated with other apo-lipoproteins (and plasma proteins). If a gel bed of greater porosity is used (e.g. Bio-Gel A-0.5M) the peaks eluted in SDeS become more diffuse but, by a judicious pooling of the collected fractions, it is possible to obtain C fractions that are not only less contaminated, but are also partially resolved. Because guanidine hydrochloride dissociates the apo-lipoproteins more completely than either SDeS or urea, it gives the better resolution of C-I from the C-II and C-III proteins, even on columns of lower porosity, e.g. Bio-Gel P-100. The high cost of ultra-pure guanidine hydrochloride or urea makes the SDeS system the more economical for large-scale use but, as we pointed out in the introduction to this chapter, the detergent is strongly adsorbed to the proteins and must be removed during the later stages of the purification.

The first crude mixture of C proteins is resolved by chromatography on DEAE-cellulose in a gradient of increasing ionic strength (Fig. 8.2). The system used is akin to that described in Section 8.1.1, for the purification of the A apo-lipoproteins but, since C-II and C-III are more strongly adsorbed than the A proteins, the final concentration of the eluent must be greater. Even after this treatment however, the separated C proteins are generally still contaminated and a careful selection must be made from the fractions collected across each peak if a pure product is to be obtained. In this context, it may be more profitable to collect a greater proportion of the peak and to subject the product to a third stage of purification.

Isoelectric focussing would be an attractively simple and accessible technique for this final refinement (Holmquist and Broström, 1979; Marcel et al., 1979), particularly if the technical modifications introduced by Forgez and Chapman (1982) are found to be effective in raising the recovery of the C proteins. Marcel et al. support their claim that the C proteins can be purified in this way by appeal to the evidence of electrophoretic and aminoacid analysis. However, it must be said that the apo-lipoproteins C-II, C-III$_0$, A-IV and A-II have very similar isoelectric points and are therefore difficult to resolve. Moreover, it is not yet certain whether complexes of these apo-lipoproteins are entirely dissociated during the focussing procedure. By contrast, because its p*I* is higher than that of the other major apo-lipoproteins, the C-I protein is probably a good subject for purification in this way, although it is worth noting that Herbert et al. (1977) recommend that this protein be purified by the simpler process of chromatography on Bio-Gel P-100 or Sephadex G-75, in 5 M guanidine hydrochloride. To those with access to equipment for HPLC, the reversed phase systems described by Hancock et al. (1981) and by Ronan et al. (1982) may also be attractive for the final purification of the C-II and C-III apo-lipoproteins. Of the two techniques however, iso-electric focussing seems to give the better resolution of the polymorphs of C-III. Moreover, focussing may be used to give at least a preliminary fractionation of the whole apo-VLDL mixture (Marcel et al., 1979).

Like all the apo-lipoproteins, the isolated C proteins tend to aggregate and to become progressively less easily soluble on repeated freeze-drying. When the dried protein is to be re-dissolved in preparation for a further stage of purification, and exposure to extreme pH can be brief, 0.1 M ammonium bicarbonate or 1–2 M acetic can be used, as appropriate. However, if the specimen is to be stored in solution, it is better to use 3 M guanidine hydrochloride, buffered to pH 7.5, at a temperature down to −15 °C.

8.5.1. *Chromatographic procedure*

Equipment. (1) A siliconised glass column 2.5 cm × 150 cm, packed with Bio-Gel P-100 (or Sephadex G100) in solution A (see below). (2) A siliconised glass column 2.5 cm × 50 cm, packed with DE-52 microgranular DEAE-cellulose (Whatman) in equilibrium with solution B (see below). Auxiliary equipment as in Section 8.1.1.

Reagents. Solution A: 477.7 g of guanidine hydrochloride (high grade)
3.95 g of ammonium bicarbonate
Water to 1.0 litre

Solution B: 0.01 M Tris–HCl buffer, pH 8.0, in 6 M urea
Solution C: 0.125 M Tris–HCl buffer, pH 8.0, in 6 M urea

Both solutions B and C must be prepared with deionised solutions of highly purified urea and should be stored for the shortest possible time at 4 °C before use (Section 8.1.1).

Procedure. Prepare VLDL ($d < 1.006$ g/ml) with the minimum of delay, from freshly collected plasma. Extract the lipids and suspend 75–100 mg of protein in solution A with gentle agitation overnight. Centrifuge off any insoluble material and transfer the supernatant to the gel filtration column. Elute with the solvent A, at 20 °C, at a flow rate of about 50 ml/h and collect fractions of 10–12 ml.

Analysis of the individual fractions from peak 3 (Fig. 8.1b) by electrophoresis on polyacrylamide gel (Section 6.4.2.3) will generally show that the leading limb consists of apo-lipoproteins C-II and C-III contaminated with larger apo-lipoproteins, whereas C-I is concentrated towards the trailing limb. Accordingly, a judicious selection of the fractions that are to be pooled may effect a useful preliminary separation of the C proteins. The combined fractions must then be dialysed exhaustively against 5 mM ammonium bicarbonate in tubing of molecular weight cut-off 3500 (Spectrapor 3; Spectrum Medical Industries Inc.) and freeze-dried.

To minimise the carbamylation of the proteins, all the following stages, including the running of the DEAE-cellulose column, must be carried out at 4 °C. Dissolve the crude preparation of C proteins in a small volume of freshly made solution B and dialyse (Spectrapor) against the same solvent. Purge the column of DE-52 with fresh solution B, apply up to 100 mg of the protein mixture and wash with 350 ml of solution B. Continue the elution with a linear gradient formed by the progressive mixture of 1.0 litre of solution B with 1.0 litre of solution C, until apo-lipoprotein C-III_2 has emerged (i.e. at a buffer concentration of about 0.1 M; conductivity approx. 2.5 mmho; Fig. 8.2). Collect the eluate from the column in fractions of 10 ml which should be dialysed without delay against 5 mM ammonium bicarbonate solution at 4 °C. Finally, wash the column with 250 ml of 0.5 M Tris–HCl, pH 8.0, in 6 M urea and re-equilibrate with 0.01 M Tris–HCl, pH 8.0. If the column is not to be immediately used again, it should be stored in this condition and purged with solution B when required.

Check the purity of each fraction from the peaks of C protein by electrophoresis (Sections 6.4.2.3 and 6.4.2.4) and select those that are the least contaminated. In this context, note that, according to Herbert et al. (1977), apo-lipoprotein C-III_2 is often contaminated with proteins of similar size and electrophoretic mobility. If further purification is necessary, use the method of isoelectric focussing described in Section 8.1.2.

The total yield of C protein that is obtained by this chromatographic procedure will not exceed 50% and, in practice, the yield of pure material will be still further reduced by the need to select fractions that are uncontaminated.

8.5.2. Isoelectric focussing

This technique was used by Holmquist and Broström (1979) to purify a mixture of apo-lipoproteins C-II and C-III that had been prepared by solvent extraction. These proteins were focussed on a gradient of pH 4.7–5.5 in a bed of Sephadex G-200, which was found to give a better resolution than Bio-Gel P-200. By contrast, Marcel et al. (1979) used the method on whole apo-VLDL and were able to isolate all the individual proteins in a single focussing on a gradient of pH 4–6. However, they obtained a better separation of C-II and C-III by a second focussing on a gradient of about pH 4.5–5.3. The first gradient is also intrinsically inappropriate for the preparation of apo-lipoprotein C-I, which has a p*I* of about 6.5. This protein was recovered in the experiments of Marcel et al., presumably because the sodium hydroxide that was used as catholyte modified the upper end of the gradient. Better results were obtained on a gradient from pH 5 to 7. In the following, we shall apply the method described in Section 8.1.2.

Equipment. As in Section 8.1.2.

Reagents. Ampholytes (LKB-Produkter; Bio-Rad Laboratories); for the separation of apo-VLDL, mixtures of pH range 4–6 and 5–7 are needed. For the final refinement

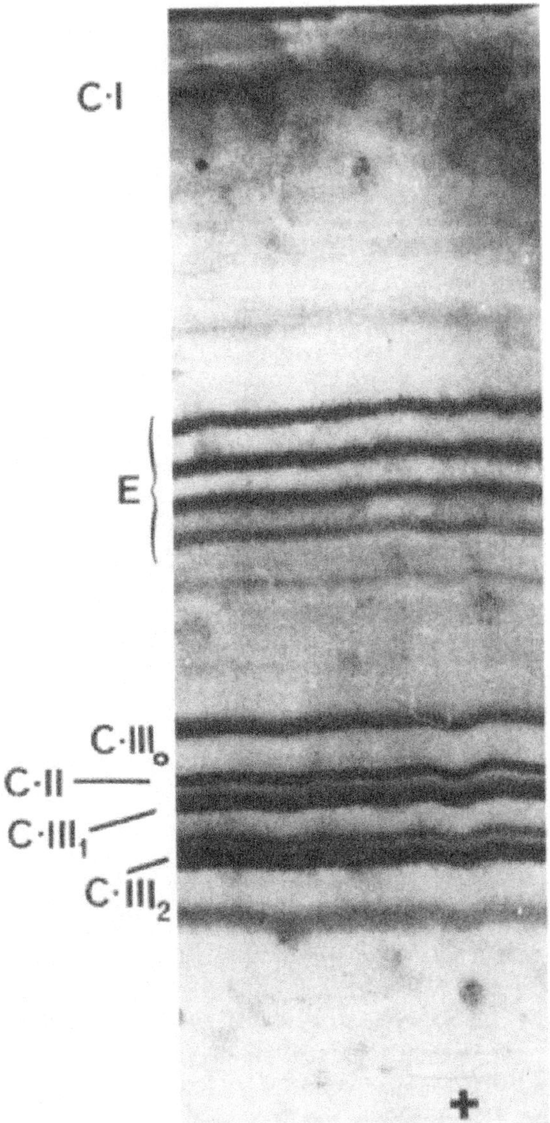

Fig. 8.5. Preparative isoelectric focussing of the soluble components of apo-VLDL.

of partially purified C-II and C-III, prepare a gradient of pH range 4.5–5.3 by pre-focussing, as described in Section 8.1.2. For the purification of C-I, use ampholyte of pH 5–7.

Solution B: 29.0 g of urea
1.95 g of ampholyte
Water to 100 ml

All other reagents as in Section 8.1.2

Procedure. Dissolve the purified apo-lipoprotein C in solution A to a concentration of 5–10 mg of protein/ml. (If the method is to be applied directly to apo-VLDL, extract this with the same solvent and centrifuge off the undissolved gel of apo-lipoprotein B.)

Fill the gel tray as described in Section 8.1.2, using the appropriate ampholyte. Apply the protein mixture and focus for 18–20 hours. The proteins form bands at the positions that correspond approximately to the following pH (Fig. 8.5):

Protein	pH
C-I	6.5
C-II	5.0
C-III_0	5.05
C-III_1	4.95
C-III_2	4.8

Recover the apo-lipoproteins, concentrate them and check their purity as before.

8.6. *Isolation of apo-lipoprotein D*

This is a difficult protein to purify, even in the small amounts required for the preparation of an antiserum. Indeed, Herbert et al. (1977) point out that it may be easier to raise an antiserum against crude apo-lipoprotein A-II and then to prepare a specific anti-apo-lipoprotein D from this by absorption. Once such an antiserum has been isolated, it can be used to prepare an immuno-adsorbent with which to separate the D-containing particles from the large excess of other lipoproteins. This technique has been used by McConathy and Alaupovic (1976), in combination with chromatography on hydroxyapatite (Section 2.3.3) to isolate lipoprotein particles which contain only apo-lipoprotein D ('LP-D'). However, the D protein was originally isolated from apo-HDL_3 by chromatography on hydroxyapatite (McConathy and Alaupovic, 1973). A more conventional procedure, closely akin to that used for the isolation of the A apo-lipoproteins, was used by Kostner (1974), namely a preliminary separation of apo-HDL_3 by chromatography in the acetic acid/Sephadex system, followed by chromatography on DEAE-cellulose. Since the overall yield from this process is only about 0.25%, it is clear that the preparation of apo-lipoprotein D should only be attempted by those to whom fresh plasma is freely available. However, because there has been so little published work on this subject, it is less clear whether the method of Kostner is superior to that of McConathy and Alaupovic. In the former, apo-lipoprotein D elutes from the ion-exchange column between apo-lipoproteins A-II and A-I, with both of which it is heavily contaminated. Kostner (1974) therefore found it necessary to collect the eluate in small fractions and to select those which were pure. The alternative method depends on the fact that apo-lipoprotein D is desorbed from hydroxyapatite at a lower ionic strength than other apo-lipoproteins. The protein fraction was therefore passed repeatedly over the adsorbent until only one band was seen on electrophoresis in urea gel (Section 6.4.2.3). A final stage of chromatography on Sephadex, in 2 M acetic acid, provided a pure product in an unstated yield.

In a laboratory where the A proteins are in regular production by the method described in Section 8.1.1, the appropriate D-containing fractions from the gel filtration column can be acculumated for later treatment. However, it is uncertain whether chromatography on hydroxyapatite or on DEAE-cellulose is the better choice for the final purification. Although at first sight, the hydroxyapatite procedure is the simplest to operate, it must be remembered that this adsorbent is notoriously variable in its properties and that it may be necessary to adjust the procedure if the hydroxyapatite is from a source different to that used by McConathy and Alaupovic.

Equipment. (1) A siliconised glass column 2.5 cm × 15 cm, packed with a mixture of 2 volumes of hydroxyapatite (Bio-Gel HT; Bio-Rad Laboratories) with 1 volume of microcrystalline cellulose (Baker TLC reagent; Baker Chemical Co.) in 0.001 M K_2HPO_4. These volumes are measured after the adsorbent has settled after suspension in the phosphate solution.

(2) A siliconised glass chromatography column, 2.5 cm × 100 cm, packed with Sephadex G100 in 2 M acetic acid.

Reagents.

Solution A: 8 M urea in 0.001 K_2HPO_4 solution
Solution B: 0.001 M K_2HPO_4 solution

Procedure. From 1 litre of fresh plasma, isolate the HDL fraction of density 1.12–1.27 g/ml by centrifugation and wash the product twice at d = 1.27 g/ml. Dialyse the HDL exhaustively against distilled water, freeze-dry and extract five times with chloroform:methanol, followed by two washes with diethyl ether. Dissolve the apo-HDL in 10 ml of solution A and dilute to 2 M with respect to urea by the addition of 30 ml of solution B.

Purge the hydroxyapatite column with 1.0 M K_2HPO_4 solution and re-equilibrate with solution B. Apply the apo-HDL solution to the column and wash with 50 ml of solution B. Repeat this whole process a second time and analyse the eluate by electrophoresis in basic urea-gel (Section 6.4.2.3). If more than one band is found, continue the purification until no contamination can be detected. To remove the urea and to effect a final refinement of this product, concentrate it and transfer it to the Sephadex G100 column. Elute with 2 M acetic acid and collect the principal protein peak. Confirm the identity of this by electrophoretic and immunochemical analysis.

8.7. Isolation of apo-lipoprotein E

This protein is difficult to isolate, in part because it is normally present in plasma at a relatively low concentration, and in part because it can only be dissolved in the most powerful solvents, i.e. 5–6 M guanidine hydrochloride, 0.1 M SDeS, or 0.1 M ammonium hydroxide. Conventionally, it can be prepared by the same sequence of gel filtration and ion-exchange chromatography that is used to isolate the C apo-lipoproteins (Section 8.1.1). The best of the gel filtration systems for this purpose is probably 5 M guanidine hydrochloride–Sepharose 6B (or Bio-Gel 1.5 M) and in some laboratories this procedure is repeated until the desired degree of purity is attained. Alternatively, the partially purified product can be refined by ion-exchange chromatography in the DEAE-cellulose/6 M urea system, from which apo-lipoprotein E is eluted by approximately 0.05 M Tris–HCl buffer, pH 8.2 (Shore and Shore, 1973; Shelburne and Quarfordt, 1974). Utermann (1975) has also successfully refined this protein by preparative electrophoresis in SDS–polyacrylamide gel. However, isoelectric focussing is, at present, the more versatile of the electrophoretic methods and has the advantage that it can resolve the polymorphs of apo-lipoprotein E.

Because filtration chromatography in ultra-pure guanidine or urea is expensive, some operators have used 2 M acetic or 2 mM sodium decyl sulphate instead. The crude preparations made in this way can then be further purified in the guanidine hydrochloride system. However, apo-lipoprotein E that has been dissolved in SDS will adsorb detergent that may not be completely removed by a gel filtration procedure. Protein that is contaminated in this way can be used to prepare antisera but may be unfit for other purposes. For critical experiments it may therefore be necessary to refine the apo-lipoprotein by chromatography on DEAE-cellulose, or to remove the detergent as described in the introduction to this chapter.

The richest source of apo-lipoprotein E is to be found in the VLDL fraction, and especially in that of patients who suffer from Fredrickson's Type III hyperlipoproteinaemia (primary dysbetalipoproteinaemia). However, this disorder is rare and is not generally a practicable source of the apo-lipoprotein, which must be isolated from the more common forms of hypertriglyceridaemia.

Equipment. (1) A siliconised glass column 2.5 cm × 250 cm, packed with Sepharose 6B-CL (or Sephadex G200) in solution A (see below). (2) Equipment for iso-electric focussing as in Section 8.1.2.

Auxiliary equipment as in Section 8.1.1.

Reagents. Solution A: 477.7 g of guanidine hydrochloride (high grade)
3.95 g of ammonium bicarbonate
Water to 1.0 litre

Also reagents for iso-electric focussing as in Section 8.1.2.

Procedure. Prepare washed VLDL from the fresh plasma of hypertriglyceridaemic subjects and extract the lipids (Appendix 3) with the minimum of delay. Suspend about 75 mg of the protein in 5–7 ml of solution A and agitate gently until dissolved, or overnight. Remove any insoluble material by centrifugation, transfer the supernatant to the gel filtration column and develop with solution A. The elution pattern will be in the form of three partially resolved peaks (Fig. 8.1b). Collect the second of these, concentrate it and re-chromatograph under the same conditions. Continue to re-run the central peak until it is possible to collect fractions from it that contain uncontaminated apo-lipoprotein E, as shown by electrophoresis in urea–polyacrylamide gel (Section 6.4.2.3).

The iso-electric focussing technique described in Section 8.1.2 is a quicker method of refinement that may also give better yields. If this is adopted, the partially purified E protein from the second stage of gel filtration chromatography must be concentrated and then thoroughly dialysed against 5 mM ammonium bicarbonate, followed by 20 mM *N*-ethylmorpholine in 6 M urea (solution A of Section 8.1.2). Focus this protein mixture on a pH gradient from about 4.6 to 5.9, as described for the preparation of apo-lipoprotein A-I. The polymorphs of apo-lipoprotein E will then be resolved into the five major bands at the alkaline end of the gradient (Fig. 8.5). The third and fourth bands of this group lie very close together and appear to be forms of E_2. Moreover, there is some cross-contamination of the other polymorphs and the most acidic of them (E_1) is usually slightly contaminated with apo-lipoprotein A-I. Elute the proteins separately and characterise them by electrophoresis (Sections 6.4.2.3 and 6.4.2.6) and by immunochemical analysis (Section 6.5). Finally, remove the ampholytes by exhaustive dialysis against 5 mM bicarbonate solution, or by electrodialysis, and concentrate as required.

8.8. *Isolation of apo-lipoprotein F*

This protein was isolated in minute yield from apo-HDL by Olofsson et al. (1978) and the subsequent preparation of an 'LP-F' (Koren et al., 1982) may lend some colour to the belief that F is a functional apo-lipoprotein. However, it is at present of purely academic interest. It is a protein of molecular weight about 30 000 and when dissolved in a highly dissociating solvent should run with apo-lipoprotein A-I on gel filtration chromatography (Fig. 8.1a). However, perhaps because of its acidic nature (p*I* approximately 4), it tends to form complexes when dissolved at high concentration in 2 M acetic acid. These complexes elute first from a column of Sephadex G100 (Section 8.1.1) and it was this material that Olofsson et al. (1978) used as their source of apo-lipoprotein F. This crude eluate was lyophilised, re-dissolved in a mixture of 2-butanol:acetic acid:water (4:1:5 by vol.) and chromatographed (Rudman et al., 1970) on a column of Sephadex LH-20, using the same solvent, to 'remove any residual lipid'. The necessity for this step was not explained. The lipid-free protein recovered from this column was dried, dissolved in an 8 M urea solution containing 0.001 M potassium dihydrogen phosphate, pH 3.5, and then run through a column of carboxymethyl cellulose (Cellex-CM; Bio-Rad Laboratories) in the same solvent. Apo-lipoprotein F is not retained under these conditions and, by repeating this step two or three times, it was possible to obtain a product that was free of contaminating apo-lipoproteins as judged by immuno-chemical analysis.

8.9. *The estimation of apo-lipoproteins*

At the outset, it must be said that there is little evidence that the available methods for the estimation of individual apo-lipoproteins are the products of exhaustive technical investigation. The different protocols used in different laboratories may indicate that variations in at least some of the experimental conditions can be tolerated.

TABLE 8.1
Mean apo-lipoprotein levels in normal human serum (Alaupovic et al., 1983)

	Apo-lipoprotein (mg/dl ± std. dev.)							
	A-I	A-II	B	C-I	C-II	C-III	D	E
Men	127 ± 19	65 ± 11	110 ± 32	8.5 ± 3	2.9 ± 0.7	8.8 ± 3.2	8.6 ± 2.6	10.7 ± 3.3
Women	146 ± 27	74 ± 16	94 ± 24	8.0 ± 3.8	3.0 ± 1.1	8.4 ± 2.0	8.0 ± 1.8	11.7 ± 4.4

However, the limits of this variation are not generally known and the procedures that are described below must be taken as a guide only. Each method must be exhaustively validated before it can be confidently brought into use. However, here the analyst may meet another obstacle. The specificity of immunoassays opens the prospect, especially attractive for routine clinical purposes, of estimating an apolipoprotein in a native lipoprotein, or in serum, without prior delipidation and isolation. Unfortunately, under these circumstances, there is no satisfactory reference method of analysis and each procedure must be confirmed by parallel analyses using *several* other methods.

Three general procedures have been used to quantify the apo-lipoproteins:

(1) Separation by chromatography, followed by conventional determination of the proteins. The principal disadvantages of this method are its low sensitivity and the poor resolution that is achieved. Added to this, it is first necessary to isolate a purified specimen of the lipoprotein, and the final recovery of protein is often substantially short of 100%. In general, the technique is too inaccurate for critical work and too cumbersome for clinical studies. Nevertheless, it has occasionally been used to obtain, for example, an approximate estimate of the amount of apo-lipoprotein C in VLDL. The value of the technique may be improved by the introduction of HPLC (Hancock et al., 1981) but the promise of this development has yet to be substantiated.

We do not propose to describe these methods in detail, but shall refer would-be users to the survey by Herbert et al. (1977).

(2) Separation by electrophoresis, followed by staining and photometric estimation. This technique is more sensitive than the chromatographic method but is subject to essentially the same disadvantages. Moreover, the proteins may not all be equally 'chromogenic' under the staining conditions that are used and, unless this can be decided with the aid of pure, authentic apo-lipoproteins, the estimates will be subject to a degree of uncertainty. In addition, the reproducibility of the staining and the linearity of the photometric response must not only be determined at the outset, but must be regularly monitored

while the method is in use. In short, if this method is to yield really creditable results, the most assiduous attention must be paid to the strict control of the working conditions.

Despite these drawbacks, electrophoretic analysis has attractions for those laboratories that lack facilities for quantitative immunoassay. In particular, it allows an estimate to be made of those apo-lipoproteins for which specific antisera are not readily available.

(3) Immunoassays. Four different immunochemical techniques have been used to determine plasma apo-lipoproteins. These have been reviewed in the context of apo-lipoproteins A-I and B by the Standardisation Committee of the International Union of Immunological Societies, whose summaries of the subject should be read by all who propose to estimate these proteins by immunoassay (Steinberg et al., 1983; Rosseneu et al., 1983). The three most widely used methods for the immunoassay of apo-lipoproteins have also been reviewed by Karlin and Rubenstein (1979). Practical descriptions of the general techniques are available in many monographs (e.g. Clausen, 1981; Chard, 1981; Walker, 1977).

These methods are all not only more sensitive and more specific than the chromatographic or electrophoretic procedures, but they do not require the preliminary isolation of the lipoproteins. Thus, although they are at least as technically exacting, they tend to be less time-consuming and are the better suited to clinical use. However, their requirement for monospecific antisera to the apo-lipoproteins is considerable and, in many laboratories, this is a serious limitation to their use. In addition, there are several practical difficulties that are common to most methods for the immunoassay of apo-lipoproteins and which can arise from the characteristic properties of antisera and lipoproteins (Karlin and Rubinstein, 1979; Albers et al., 1980; Cheung, 1981):

(a) In general, the antiserum to a given apo-lipoprotein will consist of a heterogeneous population of antibodies which are directed, with different affinities, to different antigenic sites. Consequently, different batches of antiserum may not yield the same results when used in the same assay system.

(b) When the apo-lipoprotein forms a part of the lipoprotein complex, some of its antigenic sites may be masked by the lipid moiety, with a consequent loss of reactivity towards an antiserum that has been raised against the purified apo-lipoprotein. This can lead to a serious error if the apo-lipoprotein in the intact lipoprotein is assayed by reference to the pure protein. To avoid this, it may be necessary to dissociate the lipid from the protein before the analysis, using the methods that we shall describe later.

(c) Because the lipoprotein particles are not all identical, their reactivities towards a given antiserum may also differ. This is particularly relevant in the case of apo-lipoprotein B, which must be assayed in the intact low-density lipoprotein particles. Because these are widely different in size and composition, their immuno-reactivity may not be uniform over the whole range, and the observed result will depend on the lipoprotein distribution.

(d) Free apo-lipoproteins tend to aggregate and to form a mixture of oligomers, the nature of which depends on the protein concentration and on the pH, ionic strength and temperature of the solution. Since these oligomers will differ in their physical properties (e.g. diffusivity) and possibly in their immunochemistry also, differences in the conditions under which the samples or standards are assayed can be a source of error. The apo-lipoproteins that are used as standards should be stored at low temperature and low concentration in 3 M guanidine hydrochloride solution.

(e) The immuno-reactivity of lipoproteins and of isolated apo-lipoproteins can decrease on storage. However, the rate at which the activity is 'lost' tends to be unpredictable and appears to depend on unidentified features of the storage conditions. According to Kahan and Sundblad (1969), serum apo-lipoprotein B is stable for 6 months at 5 °C but is slowly lost at −20 °C. By contrast, Karlin et al. (1978) found it to be stable for up to 12 months in frozen serum. More surprisingly perhaps, Albers et al. (1975) were able to recover 100% from lyophilised serum after 4 months and Alaupovic et al. (1979) found no loss of apo-lipoprotein B in serum that had been stored for 15 days at 37 °C. Under these conditions however, apo-lipoprotein A-I was markedly unstable, 40% being lost even when penicillin or 0.1% EDTA was present. The addition of 0.1% Thimerosal improved the stability slightly, cutting the loss after 15 days to 28%. The importance of temperature is reflected in the fact that 13% of the A-I in native serum was lost after 15 days at 4 °C, while there was no loss after the same period at −5 °C. In these experiments, storage at 37 °C had no effect on the recovery of apo-lipoproteins A-II or C-III.

By contrast with the apparent stability of apo-lipoprotein B in serum, its immunological integrity in isolated LDL falls by about 1% per day during storage at 4 °C. Clearly, this phenomenon is significant for the maintenance of reliable standards, and for the preservation of specimens prior to their analysis. It is one that should be investigated under local conditions.

(*1*) *Radial immuno-diffusion*. This method (Mancini et al., 1965) has the merit that it requires the simplest of apparatus and is economical to run. On the other hand, it consumes relatively large quantities of antiserum and is comparatively insensitive, having a lower limit of about 10 μg/ml, although intensification techniques will lower this to about 1 μg/ml (Simmons, 1971; Sieber and Becker, 1974; Wieja and Smith, 1976). It is also comparatively slow because the time that must elapse before the plates can be read is usually considerably more than 24 hours.

In a critical discussion of the technique, Albers et al. (1980) pointed out that, not only must the antibody be specific for the antigen but that the latter must be able to migrate freely into the agar gel. Moreover, the protein under assay should have about the same specific immunoreactivity as the standard and also have a comparable molecular weight or molecular weight distribution. That these conditions may not always be satisfied can be seen from paragraphs (b) to (d) above. In particular, it is impossible to obtain generally reliable estimates of total plasma apo-lipoprotein B by RID, since the larger of the low-density lipoprotein particles may not enter the gel at all. Moreover, it is important to note that apo-lipoprotein A-I is largely inaccessible to antiserum when it is bound to the lipoprotein. This obstacle can be overcome by adding tetramethyl urea to the serum and then diluting it with Tris/8 M urea before the assay. Denaturation is apparently unnecessary for the estimation of other apo-lipoproteins, or for Lp(a). Although Kushwaha et al. (1977) added TMU to their assays of apo-lipoprotein E, this was apparently a precautionary measure.

Despite its simplicity, RID requires the most careful attention to technical detail if accuracy is to be maintained. For example, errors can result if the gel is of uneven thickness or is not thoroughly mixed with the antiserum, if the test wells are too close together or are clumsily filled, if the samples are inaccurately diluted, or the temperature varies while the plates are incubated.

Methods have been published for the estimation of the following apo-lipoproteins by RID:

A-I: Albers et al. (1976); Reman and Vermond (1978)
A-II: Cheung and Albers (1977)
B: Lees (1970); Albers et al. (1975); Sniderman et al. (1975); Curry et al. (1978); Havekes et al. (1981)
D: Curry et al. (1977); Albers et al. (1981)
E: Curry et al. (1976b); Kushwaha et al. (1977)
and also for Lp(a): Albers et al. (1975a).

(2) Electro-immunoassay (EIA). The technique can be more sensitive than RID, uses less antiserum and is considerably quicker. However, it is no less demanding in its standard of expertise and requires the equipment for electrophoresis in agarose gels. All the major apo-lipoproteins have been assayed by this method, with what appear to be satisfactory results, despite some unresolved technical reservations. For example, Laurell (1972) considered that EIA was not satisfactory for the estimation of substances with a molecular weight much in excess of 40 000, and cited β-lipoprotein as an example which would fail, probably because the antigen-antibody precipitate would clog the pores of the gel. Nonetheless, EIA has been used to obtain estimates of the apo-lipoprotein B in both LDL and VLDL, and also of the apo-lipoprotein A (i.e. A-I + A-II) in HDL, which were in acceptable agreement with the results of gravimetric analysis (Curry et al., 1976a,

1978). Moreover, the estimates of apo-lipoprotein B in LDL were confirmed by parallel analyses made by RID and by radio-immunoassay (RIA) (Curry et al., 1978). However, a similar comparison of the analyses of isolated VLDL showed that the amount of apo-lipoprotein B measured by EIA was about 65% greater than was estimated by either RID or RIA. Calvert et al. (1979) have also found the same difference between EIA and RIA, in experiments in which the EIA was confirmed by a colorimetric TMU extraction method. The difference between EIA and RID may be at least partly explained by the restricted diffusion of the large VLDL particles in agarose gel. It is less easy to account for the disagreement with RIA, in which VLDL is said to react like LDL. Calvert et al. (1979) have suggested that the concordance between the EIA and gravimetric analyses of VLDL is fortuitous, low immuno-reactivity in these particles being counter-balanced by their greater electrophoretic mobility, but this hypothesis is unconfirmed. While it is clear that EIA can be used to make acceptable estimates of apo-lipoprotein B in normal sera, the circumstances in which it can be used to obtain valid results with hypertriglyceridaemic sera deserve further investigation.

A second uncertainty arises from the fact that the reactivity of apo-lipoprotein A-I towards its antiserum in the EIA is apparently the same when it is bound to the lipoprotein particle as when it is free (Curry et al., 1976a). This is in contrast to the other immunoassays, in which it has been found necessary to add dissociating agents to plasma or lipoproteins to ensure that all the antigenic sites on the A-I protein are available. In practice, the addition of urea, tetramethyl urea, Triton X-100 or sodium dodecyl sulphate lowered the recovery of apo-lipoproteins A-I and A-II in the electro-immunoassays reported by Alaupovic et al. (1979). Even the removal of lipid by solvent extraction did not expose any additional determinants to the antisera to these two proteins (Curry et al., 1976a).

Some doubt has also been expressed as to exactly what is measured if the antiserum used in the EIA is polyspecific, e.g. anti-apo-HDL or anti-apo-lipoprotein C. In practice however, it appears that acceptable results can be obtained with antisera of this kind, which are often the only ones that are commercially available.

As it is generally operated, the sensitivity of the EIA is such that the lowest measurable concentration of apo-lipoprotein is approximately 2 μg/ml (i.e. 20 ng/10 μl sample). To attain and maintain a satisfactory performance at this sensitivity requires careful attention to the experimental conditions under which the assay is run. These are complex (Laurell, 1972; Weeke, 1973; Verbruggen, 1975) and must be determined by an exhaustive study for each apo-lipoprotein. This is a laborious operation which makes EIA less suitable for casual analyses than RID. On the other hand, the greater sensitivity of EIA makes it more easily possible to measure apo-lipoproteins such as D and E directly in plasma despite their relatively low concentration. Unfortunately, EIA is less well adapted to large-scale routine clinical use than RIA, even when the precipitin 'rockets' are read directly into a computer (Weech, 1981), and it is probably significant that essentially only one laboratory has adopted this procedure on a large scale.

The intra-assay coefficient of variation for EIA has been variously reported to be between about 1.5 and 8% with an average of about 4.5%. Between assays, the CV has ranged from about 3% to about 10%, with a mean of about 6.5%.

Detailed procedures have been published for the electro-immunoassay of the following apo-lipoproteins.

A-I and A-II:	Curry et al. (1976a); Hosler (1980); Miller et al. (1980); Mordasini and Riesen (1980); Fruchart et al. (1982)
B:	Curry et al. (1978); Reardon et al. (1981); Fruchart et al. (1981, 1982); Havekes et al. (1981)
C-I and C-II:	Curry et al. (1981); Jauhiainen et al. (1983)
C-III:	Curry et al. (1980)
D:	Curry et al. (1977)
E:	Curry et al. (1976b)

(3) Radio-immunoassay (RIA). Of the three immunoassays, this is the most sensitive and economical of antiserum, but it requires the most elaborate equipment. High sensitivity is not generally needed for the determination of the apo-lipoproteins in human serum but it makes RIA the method of choice for metabolic studies, where the amount of protein is often small, e.g. in small animals, perfused livers, or in

tissue cultures. On the other hand, RIA is a well established technique, which is widely documented (e.g. Chard, 1981). Its use for the routine estimation of the apo-lipoproteins in plasma may therefore be attractive to laboratories in which other radio-immunoassays are already in use. This application of RIA has been reviewed by Karlin and Rubenstein (1979), and follows established practice, although there are special considerations that arise from the nature of the lipoproteins:

(a) As usual, the antiserum should be monospecific and of high titre. However, if it is prepared from a pure apo-lipoprotein, the antibodies may be directed towards determinants that are not fully exposed in the native lipoprotein. It is then necessary to unmask these sites, as is well established for the case of apo-lipoprotein A-I. This can be done by extracting the lipids (Schonfeld and Pfleger, 1974), but it is a laborious procedure when many samples must be treated. Karlin et al. (1976) simply diluted the plasma (or HDL) and heated it at 52 °C for 3 hours and found this to give recoveries as good as those obtained after extraction with solvents. However, the same result can be achieved even more easily by performing the assay in the presence of either 0.05 M sodium decyl sulphate (Fainaru et al., 1976) or 0.32% of Tween 20 (Mao and Kottke, 1980a).

The apo-lipoprotein B component of LDL appears to be completely available to antisera to either the isolated protein, or to whole LDL, but there is some uncertainty about the accessibility of the apo-lipoprotein B in VLDL. According to Albers et al. (1975), the RIA will detect only 65% of the apo-lipoprotein B in isolated VLDL but will measure all of it when the VLDL is present in serum. By contrast, Curry et al. (1978) found low recoveries of apo-lipoprotein B in both isolated VLDL and in plasma. Schonfeld et al. (1974) differed from both of these in finding substantial agreement between RIA and a chromatographic assay of the B protein in isolated VLDL. Unfortunately, as we have already pointed out, immunoassays are difficult to check by reference chemical analysis and, despite the optimistic reports from some laboratories, further studies are needed to determine the conditions under which satisfactory analyses of VLDL are possible.

It is generally agreed that no denaturation is necessary for the estimation of apo-lipoproteins A-II and C-III, but the position is less clear in the case of the C-II and E proteins. According to Kashyap et al. (1977), bound apo-lipoprotein C-II is completely accessible and requires no pretreatment of serum. By contrast, Barr et al. (1981) found that some of the antigenic sites on this protein are masked but could be made accessible by adding 0.06% of Tween-20 to the assay medium. Likewise, Fainaru et al. (1977a) found pretreatment to be unnecessary for the estimation of apo-lipoprotein E, whereas Falko et al. (1980) incubated plasma in 7.2 M urea for 2 hours at 37 °C. The reason for these disagreements is not certain but may be the result of differences between the antisera used in the experiments. In any event, it appears that it may always be prudent to check the necessity for pretreatment of serum samples under the conditions that prevail in each laboratory.

(b) Labelling of the tracer antigen. This should be of a high specific activity but must be immunochemically indistinguishable from the unlabelled antigen. This presents little problem with the apo-lipoproteins which, with the exception of apo-lipoprotein C-I, can be satisfactorily labelled with radioactive iodine by any of the conventional methods, e.g. using chloramine-T (Greenwood et al., 1963; Hunter, 1970), iodine monochloride (McFarlane, 1958) or lactoperoxidase (Marchalonis, 1969). The labelling of intact lipoproteins can be more difficult however, because the relatively fragile particles are easily degraded and the presence of many unsaturated lipids leads to the incorporation of label into the lipid moiety. The latter can be a particularly difficult problem if the lipoprotein is not human, because the proportion of unsaturated fatty acids in animals if often greater than in man. For the technique of labelling, see Appendix 6. The labelled tracer is open to degradation by the radiolytic fission products of the water in which it is dissolved, a process that can be minimised by the addition of an excess (e.g. 1–3 g/dl) of a scavenger protein such as bovine serum albumin. In the case of apo-lipoprotein E however, Fainaru et al. (1977a) found that both the labelled and the unlabelled protein were 'unstable' in the absence of sodium decyl sulphate. They therefore dissolved the labelled E protein in a solution of 0.1 M sodium decyl sulphate in 0.05 M phosphate buffer, pH 7.2, which contained 1% of bovine serum albumin. Under these conditions, the tracer could be stored for up to 4 weeks at 4 °C. Despite the presence of albumin however, all the tracer proteins undergo some degradation during storage at 4 °C and it is a good practice to remove the breakdown products by gel chromatography, not less than 24 hours before use.

Like the EIA, radio-immunoassay is laborious to set up and to maintain, and is therefore unsuitable for casual analyses. The useful limit of detection can extend down to about 3–5 ng of apo-lipoprotein, with a discrimination of about 1 ng within the working range. However, since this sensitivity is rarely needed for the estimation of apo-lipoprotein in human plasma, it may be convenient to adjust the conditions of the assay to give a lower response. This also lessens the need for an extensive dilution of the plasma, which can be a source of error. The average coefficient of variation of the published procedures is about 6% within assays (3.5–10.5) and about 8% between assays (4–11).

Since apo-lipoprotein C-I contains no tyrosine, it cannot be labelled satisfactorily with radioactive iodine.

Most of the procedures that have been published for the RIA of the other apo-lipoproteins are of the double-antibody type:

A-I: Schonfeld and Pfleger (1974); Fainaru et al. (1975, 1976); Karlin et al. (1976, 1979); Schonfeld et al. (1976); Assman et al. (1977)

A-II: Mao et al. (1975); Assman et al. (1977); Schonfeld et al. (1977); Goldberg et al. (1980)

B: Eaton and Kipnis (1969); Gotto et al. (1973); Schonfeld et al. (1974); Albers et al. (1975); Bautovich et al. (1975); Bedford et al. (1976); Karlin et al. (1978)

C-II: Kashyap et al. (1977); Schonfeld et al. (1979); Barr et al. (1981)

C-III: Schonfeld et al. (1979); Mao et al. (1980); Kashyap et al. (1981)

E: Falko et al. (1980); Blum et al. (1980).

In addition, a solid-phase assay for apo-lipoprotein B has been described by Thompson et al. (1976). For the assay of Lp(a), see Albers et al. (1977).

(*4*) *Immuno-nephelometry*. This comparatively new technique has been successfully used to determine serum proteins and could probably be used equally well to estimate isolated apo-lipoproteins. In practice however, interest has centred entirely on its use for the measurement of apo-lipoproteins A-I and B in native serum. Unfortunately, the presence of the lipoproteins themselves is a serious obstacle insofar as they contribute significantly to the light scattering potential of the serum, especially in cases of hypertriglyceridaemia. The effects of this have been considered is some detail by Heuck et al. (1983), who considered that satisfactory estimates of apo-lipoproteins A-I could not be made in solutions of intact lipoproteins. They therefore extracted the serum with organic solvents before the assay. Other workers have minimised the problem by restricting their assays to normolipaemic sera, and still others have ignored it. This last approach has led, at least in some cases, to results that are clearly unsatisfactory. Regrettably it must be said that, at the time of writing, the immunonephelometric assay of apo-lipoproteins is at an early and tentative stage of development and we shall not describe it in detail. To those who wish to study the potential of the method, we recommend, in addition to the paper by Heuck et al. (1983), the useful practical survey by Whicher and Blow (1980) and the description of the rate nephelometer by Sternberg (1977).

Preparation and storage of reference standards. A significant difficulty that is common to all micro-methods of apo-lipoprotein assay is that of standardisation. Although reference solutions of the more readily soluble apo-lipoproteins can be prepared from pure, dry protein in the conventional way, this can make demands on the supply of protein that it is not feasible to satisfy. Moreover, as we have already remarked, the isolated apo-lipoproteins are not completely stable in solution, their immuno-reactivity declining slowly during storage. It is therefore usually necessary to set up secondary standards of serum in which the relevant apo-lipoproteins have been accurately determined by reference to a freshly prepared primary standard. Since these secondary standards are stable only when frozen and must not be repeatedly frozen and thawed, a large batch must be divided into many separate small samples which can be used for a brief period and then discarded.

Because of its insolubility, primary standards of apo-lipoprotein B are somewhat more difficult to prepare and maintain than those of other apo-lipoproteins. Purified apo-lipoprotein B is the best reference, and is relatively easy to prepare in large amounts. However, it is often dissolved in alkaline solvents that are not conducive to prolonged stability. LP-B is soluble in neutral buffers but is tedious to prepare and is no more stable than any other isolated lipoprotein. In fact, it is common practice to use a narrow fraction of LDL, or the protein extract thereof, on the assumption that the error introduced by the impurity of such a standard is comparable with that intrinsic to the assay procedure. For example, Karlin et al. (1978) dissolved apo-LDL (from LDL of d = 1.030–1.040 g/ml) in 0.02 M glycine buffer of pH 10 and stored this at 4 °C, but they quote no effective life-time for the solution. The same authors also used a standard solution of LDL which was said to be usable for up to 4 weeks if stored under nitrogen at 4 °C, an experience that confirmed the observations of Thompson et al. (1976). It is possible that a longer period of stability would result if the lipoprotein is dissolved in 20 % sucrose solution, but this has apparently not been investigated.

The protein content of the primary standard, whether it is prepared from lipoprotein or apo-lipoprotein, must be confirmed by chemical estimation:

(1) The method of Lowry et al. (1951) is the most widely used for this purpose, appropriate adjustments being made for differences between the chromogenicity of the apo-lipoproteins and of the albumin that is commonly used as the standard (Section 6.4.1).

(2) Quantitative aminoacid analysis has also frequently been used. This has the advantage that it will allow the identity and purity of an apo-lipoprotein that is characterised by the lack of an aminoacid(s) to be confirmed.

(3) Kjeldahl nitrogen determination (Williams and Chase, 1968). The factor that is needed to convert the estimate of nitrogen into weight of apo-lipoprotein can be determined from the published aminoacid analyses in each case.

Secondary standards, which contain more than one apo-lipoprotein, are generally calibrated by an immunoassay in which the reference has been standardised by one or more of the above methods. However, Weech et al. (1980) have pointed out that it is also possible to do this successfully by isotope dilution methods.

The method used by Albers et al. (1976; 1981) to prepare secondary standards is simple and effective, but it is not clear for how long the storage can be extended. A preservative solution containing 5.0 g of sodium azide, 0.1 g of chloramphenicol and 0.05 g of gentamycin per litre is added to a normal human serum at the rate of 1 part per 100 by volume. Small samples of this mixture (0.25–0.5 ml) are sealed into ampoules, quick frozen on dry-ice and stored at −20 °C until needed.

In the procedure described by Curry et al. (1976a), which has been used to standardise the immuno-assay of several apo-lipoproteins, a normal human serum pool is diluted to the appropriate degree with a solution that contains 85.5 g of sucrose, 9 g of NaCl and 1.0 g of sodium azide per litre. The diluted serum is then dialysed against the same diluent, after which the concentration of the relevant apo-lipoprotein is carefully determined and the solution sub-divided into volumes of 0.5 ml, in 2 ml ampoules. These are rapidly frozen at −70 °C and lyophilised, following which the ampoules are purged with nitrogen and sealed. Under these conditions, the serum is said to be stable at −30 °C for up to three years and can be re-dissolved in distilled water when required.

8.9.1. Apo-lipoprotein assay by electrophoresis

The following procedure is that of Kane et al. (1975), in which the apo-lipoproteins are resolved by electrophoresis in alkaline–urea gel. In some cases however, iso-electric focussing may give a better resolution, and Catapano et al. (1978) have described a similar method that was designed to analyse the C apo-lipoproteins in this way.

The lipoprotein is treated with tetramethyl urea (Kane, 1973) and the soluble apo-lipoproteins separated by electrophoresis, essentially as described in Section 6.4.2.3.

Equipment. As in Section 6.4.2.3. Kane et al. (1975) also recommend the use of a continuous-flow de-staining apparatus (e.g. Hoefer Scientific Instruments) to ensure the rapid and reproducible de-staining of the gels. A photometric densitometer is also needed. This must be capable of scanning a 6 mm gel of up to 120 mm length.

Reagents. In addition to those given in Section 6.4.2.3, the following solutions are needed.

Reducing reagent:	0.1 ml of thiodiglycol 0.25 ml of mercaptoacetic acid 0.2 g of tetrasodium ethylenediamine tetra-acetate 10 ml of cathode buffer adjusted to pH 8.5
pH 9.7 buffer:	the Tris–glycine catholyte of Section 6.4.2.3 is titrated to pH 9.7 with NaOH
Stain:	1 % Amidoschwarz 10B (w/v) in 7 % acetic acid (v/v)

Apo-lipoprotein standards: dissolve the pure protein in 0.05 M ammonium bicarbonate solution and determine the concentration by the Lowry method (Section 6.4.1.1). Dilute with pH 8.91 cathode buffer (Section 6.4.2.3) to give a series of six standards with concentrations in the range 0.2–5.0 mg/ml.

Procedure. The gels are prepared as described in Section 6.4.2.3, except that the separating gel is increased in length to an extent that is consistent with the size of the sample that is to be applied (a sample of 0.25 ml is the largest that can be accommodated on a gel 115 mm long).

When the method is first set up, it is necessary to establish the dose–response curve for each apo-lipoprotein under the conditions of the analysis. This must be done by running pure apo-lipoprotein preparations at several different known concentrations, which are determined by conventional protein estimation (Kane et al., 1975, used aminoacid analysis). A plot of protein concentration against the corresponding area under the densitometric response curve should be a straight line of characteristic slope. Under the conditions used by Kane et al., the values for the following apo-lipoproteins were

C-I	1.36 ± 0.11 O.D. units/μg of protein
C-II	1.59 ± 0.1 O.D. units/μg of protein
C-III	1.09 ± 0.06 O.D. units/μg of protein
E	1.06 ± 0.02 O.D. units/μg of protein

Note that it is important to ascertain that the dose–response curve obtained when different dilutions of lipoprotein are analysed is parallel to that given by the pure apo-lipoprotein.

The range over which the response is linear may extend to as much as 400 μg of protein. In practice however, the upper limit is likely to be set by the proportion of the apo-lipoprotein in the material under analysis. For example, because apo-lipoproteins C-I and C-II are normally present in only small amounts, it may be necessary to apply as much as 200 μg of TMU soluble protein to the gel in order to obtain reliable analyses. But a load of this magnitude may lead to an inadequate resolution of other, more prominent apo-lipoproteins which must be estimated from gels run with a lower load factor. Moreover, Kane et al. (1975) found that, although the response curve obtained with pure apo-lipoprotein E is linear up to 150 μg, there is a progressive fall in the recovery from VLDL when the total protein in the sample exceeded 350 μg. This accordingly sets a practical limit to the load when this apo-lipoprotein is estimated.

Each lipoprotein sample should therefore be run in duplicate, at several different loads of between 50 and 250 μg of TMU-soluble protein, in a volume of at least 50 μl of buffer of ionic strength not less than 0.05, and of pH between 6 and 9. Replicate standard mixtures must also be run concurrently, under the same conditions. Pipette each sample onto the prepared gel and immediately add an equal volume of pure TMU. Mix with the tip of a Pasteur pipette and add the reducing reagent and 80 % sucrose solution, both in amount equal to 10 % of the sample volume. After a final mix, fill the gel tube and the cathode chamber with the pH 8.91 electrolyte as described in Section 6.4.2.3. Apply a current of 1.25 mA/gel at once. When the tracking dye has entered the separating gel, remove the cathode buffer and replace it with the pH 9.7 buffer. Continue the electrophoresis at a current of 2.5 mA/gel until 30 min after the tracking dye has passed out of the bottom of the tube.

Take the tube from the apparatus and wash the precipitate of lipid and apo-lipoprotein B off the top of the gel with a fine jet of water. Next remove the gel from the tube as in Section 5.5.1 and immerse it in a test-tube full of the Amidoschwarz stain for 18 hours. De-stain the gel in 7 % acetic acid in the continuous-flow apparatus for 5 hours. Then re-charge the apparatus with fresh 7 % acetic acid and continue to de-stain for a further 18 hours. Finally, discard this extract and replace it by a solution of 0.5 mg of Amidoschwarz/litre of 7 % acetic acid, in which the gel is allowed to equilibrate for 24 hours.

Scan each gel in the photo-densitometer and determine the area under each peak. The dose-response curves can then be plotted for each apo-lipoprotein standard and used to estimate the amount present in the test samples. This operation can easily and advantageously be carried out on a micro-computer.

8.9.2. Radial immuno-diffusion

The following general method for the assay of apo-lipoproteins is based on the procedures of Albers and Hazzard (1974) and Albers et al. (1981). In each case however, the details of the method must be established by comprehensive trials before it is put to use.

Equipment. It is convenient and economical to cast the gels in rectangular plastic dishes (e.g. 125 × 125 × 12 mm) which must be rinsed with a 1 % solution of silicone before use. Provision must also be made for incubating the prepared plates at a constant temperature, in a moist atmosphere.

The diameter of the precipitin rings must be measured to an accuracy of 0.1 mm, either with a magnifying lens that incorporates a micrometer scale (e.g. Bausch and Lomb Inc.) or with equipment specifically designed for the purpose (e.g. Transidyne General Corp.).

Reagents.

Agarose as in Section 5.4
Tetramethylurea (Section 6.4.1.2)
8 M urea in half-strength buffer A.
Buffer A: 2.42 g of Tris(hydroxymethyl)aminomethane
30 ml of 0.1 M HCl
8.78 g of sodium chloride
0.372 g of disodium EDTA
0.5 g of sodium azide
Water to 1.0 litre (adjust pH to 8.0)
Buffer B: 1 g of bovine serum albumin in 100 ml of buffer A

Apo-lipoprotein standards: dissolve the pure protein in 0.05 M ammonium bicarbonate solution and determine the concentration by the Lowry method (Section 6.4.1.1). Dilute with buffer A, to which 0.1 % Tween 20 has been added, to give a series of six standards with concentrations in the range 10–450 µg/ml.

Procedure. The antiserum must first be diluted with buffer B to an extent that must be determined by preliminary experiments and may range from 1:10 to 1:100. A sufficient quantity of the dilute antiserum is then mixed with an equal volume of 2 % agarose in buffer A to provide enough mixture to fill the dishes to a depth of 1.5 mm (i.e. 15 ml for each 100 sq. cm. of area).

Prepare the 2 % agarose solution by heating on a water-bath as in Section 5.4. Meanwhile warm the appropriate volume of diluted antiserum and the siliconed dishes to 55 °C. When the agarose is completely dissolved, cool it to 55 °C and mix thoroughly with the warm antiserum (avoid introducing air bubbles). Set the warm dishes on a level surface and immediately pour in the calculated volume of gel mixture. Allow to set for 30 minutes.

The number of wells that can be punched in the gel will depend on their size and on the concentration of the antiserum. As a guide, 4 µl wells (about 1.8 mm diam.) can be 11–12 mm apart from centre to centre, whereas 5 µl wells (about 2.2 mm diam.) should be 14–15 mm apart. Set the cold gel over an appropriately ruled grid and punch the required array of holes.

Preparation of samples. Samples of lipoprotein or serum that are to be assayed for the A-I and A-II apo-lipoproteins must first be treated with tetramethyl urea and urea as follows. To 50 µl of sample, add 50 µl of TMU and mix. Next add 400 µl of 8 M urea in 0.01 M Tris buffer and stand for 30 minutes before transferring an aliquot to the gel.

Because RID is unreliable for the estimation of apo-lipoprotein B in the VLDL, these should be removed from plasma by a preliminary centrifugation before samples, suitably diluted with buffer A, are applied to the gel.

Samples that are to be assayed for apo-lipoproteins D and E need only to be appropriately diluted with buffer A before analysis.

Notice that the dilution and volume of the sample should be such that, at the concentration of antiserum used, the precipitin rings are from about 3.5 to 10 mm in diameter.

Run each sample in duplicate, in wells that are different quadrants of the gel. Likewise, distribute each of the six different standards, in duplicate, at random across the gel. As soon as the gel is loaded, transfer it to a level surface in the humid chamber and incubate at 37 °C until the precipitin rings reach their equilibrium positions. This may take from 24 to 172 hours. Finally, measure the diameter of the rings, in at least two directions, to 0.1 mm.

Computation. Calculate the mean area of the precipitin ring for each sample and standard. A plot of area vs. concentration for the six standards should fit closely to a straight line ($r > 0.99$) over the working range of the assay, although the latter, like the slope of the line, will be different for each apo-lipoprotein. Note that, when the assay is first set up, it is important to determine whether dilutions of a single plasma give a straight line of the same slope as that obtained with standards of pure apo-lipoprotein. If this is not the case, as when the A apo-lipoproteins are assayed without the preliminary treatment with urea–TMU, the conditions of the assay must be adjusted.

8.9.3. Electro-immunoassay

The following procedures were developed at the Oklahoma Medical Research Foundation by Alaupovic and his colleagues. Because of variations in the properties of different batches of agarose, in the titre of antiserum and the geometry of the electrophoresis equipment, it may be necessary to adjust the details slightly to maintain a satisfactory performance. In this context, it is important to ensure that the immuno-precipitate is formed under stable, equilibrium conditions i.e. that the tip of the 'rocket' is pointed rather than rounded.

Equipment. The gels should be of uniform thickness and are best cast between two glass plates that are separated by a U-shaped spacer 1.5 mm thick. A sheet of Cronar film (type P-47, clear base, thickness

0.007 inch; duPont de Nemours and Co.) placed between the spacer and one of the plates, provides a support for the gel when the mould is dismantled. The details that follow refer to a gel that is 20.5 cm × 11.0 cm.

The electrophoresis chamber and auxiliary equipment are as used for conventional electrophoresis in agarose gel (Section 5.4). The power supply should be capable of providing at least 150 mA at 250 V. The reproducibility of the assay is improved if the plate on which the gels lie during the electrophoresis is cooled by a flow of water at a constant 15 °C.

Reagents.

Agarose (the original procedures were developed with Indubiose, obtained from the Industrie Biologique Française)
Agar: special agar-Noble (Difco Laboratory)
Dextran T-10 (Pharmacia Fine Chemicals)
Buffer A: 8.6 g of sodium diethylbarbiturate
1.66 g of diethylbarbituric acid
0.13 g of sodium azide
0.08 g of Thimerosal
Water to 1.0 litre (pH 8.5)
Buffer B: 48.4 g of tris (hydroxymethyl)aminomethane
9 ml of concentrated HCl
0.13 g of sodium azide
0.08 g of Thimerosal
Water to 1.0 litre (adjust pH to 8.5)
Buffer C: Mix equal volumes of buffers A and B
Buffer D: Mix buffers A and B with water in the proportions 2:0.75:1.25 by volume
Stock stain solution: 0.6 g of Coomassie Brilliant Blue R250/litre of water
Working stain solution: 40 ml of stock stain
20 ml of acetic acid
10 ml of 95 % ethanol

Primary standards: standards of the pure apo-lipoproteins should cover the range 5–50 μg/ml in buffer A. Note that the ionic strength of the standard should not differ significantly from that of the samples to be assayed.

Procedure. The optimum conditions are slightly different for the electroimmuno assay of each apo-lipoprotein, those summarised in Table 8.2 being typical. Notice however, that the amount of agar to be added to the gel used for the estimation of the C apo-lipoproteins will depend on the agaropectin content of the agarose and may require adjustment.

TABLE 8.2
Conditions for electroimmunoassays

Assay variable	Apo-lipoprotein							
	A-I	A-II	B	C-I	C-II	C-III	D	E
Buffer	A	A	A	C	C	D	A	C
Agarose concn. (g/dl)	2	2	2.5	3	3	3	2	2
Dextran concn. (g/dl)	5	5	0	5	5	5	5	5
Agar concn. (g/dl)	0	0	0	0.4	0.4	0.4	0	0
Antiserum concn. (ml/dl)	1	1	0.4	4	4	4	1	2
Sample dilution	1/150	1/150	1/20	1/6	1/6	1/6	1/8	1/6
Potential gradient (V/cm)	10	10	10	6	6	10	10	10
Time (h)	4.5	3.5	3	5.5	5.5	3.5	3	3.5

Dissolve the required weight of agarose in the appropriate buffer (Table 8.2) by heating on a boiling water bath, with continuous stirring. Add the necessary amount of dextran and agar to this mixture and continue to stir on the water-bath until they are completely dissolved. Cool the solution to 50–55 °C and mix in the monospecific antiserum to the apo-lipoprotein that is to be determined. The amount of this antiserum that will give the best results must be determined by preliminary trials; the dilutions given in the table may be taken as a guide. Pour the warm mixture into the gel mould and allow it to congeal. For the maximum reproducibility, mature the gels in a moist atmosphere for 24 hours before use.

A slab of gel 20 cm long will accommodate 18 wells 4 mm in diameter, spaced 10 mm from centre to centre, or 35 wells, 2.5 mm in diameter at 5 mm spacing. Punch the chosen size of well along a straight line parallel to, and 2 cm away from the long edge of the gel. Before starting to load the gel, prepare the appropriate dilutions of the samples (Table 8.2) and standard, taking into account that the standard must be run in duplicate at three different concentrations on each gel and that each test sample must also be run in duplicate at each of two (or better, three) concentrations. Note that it is not necessary to dissociate the lipoproteins unless the concentration of chylomicrons is very high, i.e. the plasma triglyceride exceeds about 650 mg/dl (7.5 mmoles/l). Under these conditions the recovery of the C and E apo-lipoproteins will be low unless the sample is subjected to a preliminary extraction with *n*-butanol/di-isopropyl ether (see final section of Appendix 3). Use a micro-dispenser (Drummond) to transfer, as accurately as possible, a known volume of diluted sample or standard into the wells: put 10 μl in a 4 mm well and 5 μl into the 2.5 mm size.

As soon as the loading is complete, transfer the gel to the electrophoresis cell and apply the potential gradient specified in Table 8.2 (Note that the electrode compartments are filled with the same buffer that is used to prepare the gel.) During the run, maintain the temperature of the plates at 15°C with circulating water.

When the electrophoresis has continued for the appropriate time

(Table 8.2), transfer the gel to a bath of 0.9% NaCl solution for 2 hours, then blot off the surplus liquid with filter paper and dry, either at room temperature overnight, or in a stream of warm air. Finally, immerse the dry gel in the working stain for 10 minutes and then de-stain in acetic acid/95% ethanol/water (1:3:5 by vol.) until the background is colourless.

When setting up the analysis, it is important to establish two characteristics of the system.

(1) Whether it is necessary to measure the area of the rocket, or whether measurements of height alone will suffice. The latter, easier procedure can only be used when the precipitin curve develops in such a way that its area is proportional to its height. This has been found to be the case with apo-lipoproteins A-I, A-II and B. With apo-lipoproteins C-I, C-II, C-III and E however, it may be necessary to measure the area within the curve to obtain a linear response.

Although the dimensions of the 'rocket' can be measured to 0.5 mm with an ordinary rule, it is better to use a simple magnifying comparator with which estimates can be made to 0.1 mm (Section 8.9.2). Measure the height of the peak (h) from the centre of the sample well, and its width (w) at the half-height position. The area of the curve is then approximated by the product hw.

(2) Determine the slope and linearity of the calibration curve. This is done by running replicate analyses of 6–8 dilutions of the apo-lipoprotein standard on at least three different gels. Fit a straight line to these readings by the method of least squares. If an adequate fit does not result (i.e. $r<0.97$), some adjustment to the system may be needed. Even so, it may sometimes be difficult to obtain a linear response over an adequate range (cf. the estimation of apo-lipoprotein D; Curry et al., 1977). Under these circumstances, the points can be fitted to an exponential or a polynomial function. However, it is then less easy to compare the dose–response curve given by the standards that are run during a routine analysis with the calibration curve.

Note. A re-calibration should be carried out whenever a new batch of antiserum is used, and periodic checks should also be made against the primary standards to confirm the stability of the working standards and of the general performance of the assay. Unfortunately, no recommendations on this aspect of the method have been published.

Before calculating the results of an assay, plot a graph of the readings obtained for the standards on the relevant gel. If these do not fall on a straight line, or if the slope of the line differs significantly from that of the calibration curve, the assay is suspect and should be repeated. A small, constant difference between the working and the primary calibration curves can be corrected by an appropriate factor.

8.9.4. Radio-immunoassay

The following procedure is typical of those used for the RIA of apo-lipoproteins but may need modification to suit it to particular requirements or circumstances. Before attempting to set it up, an inexperienced operator should study either a comprehensive manual on the technique of RIA, or at least the concise introduction by Walker (1977).

Equipment. Test-tubes, 10 mm × 75 mm. These may be of glass, coated with silicone (e.g. Siliclad; Clay Adams Inc.) to minimise non-specific adsorption, or of plastic, e.g. polypropylene. The latter have the advantage that they transmit more of the weak radiation from ^{125}I. Do not mix tubes of different batches or manufacture, since they can vary slightly in thickness or composition and this will diminish the precision of the assay.

Piston pipettes can be used for diluting the samples and for dispensing the reagents but, if the work load is heavy, automatic pipetting equipment may be needed. To maintain the highest precision, use the same pipette for both the standards and the samples.

A centrifuge capable of handling large numbers of the 10 mm × 75 mm tubes at about 2500 *g* is required.

Since a large number of tubes must be counted, it is desirable that the gamma-spectrometer has a large capacity and is capable of prolonged automatic operation.

Reagents.

Assay buffer: 9.0 of sodium diethyl barbiturate
64.5 ml of 0.1 M HCl

0.29 g of EDTA
3 g of bovine serum albumin (Fraction V)
Water to 1.0 litre (adjust to pH 8.6)

Radio-iodinated tracer: label with ^{125}I as in Appendix 6, to a specific activity of about 10 mCi/mg protein. Use LDL of density 1.025–1.045 g/ml as the tracer for the assay of apo-lipoprotein B. For the estimation of other apo-lipoproteins, label the purified protein. Stock solutions of tracer can be stored briefly in the assay buffer, as follows:

Labelled LDL: 2–3 weeks at 4 °C.

Labelled apo-lipoprotein: up to 4 weeks at −20 °C (but also see above for apo-lipoprotein E).

Before preparing the working tracer solution, purify the stock by passing it through a 50 cm gel filtration column in the assay buffer. Use a column of Sephadex G-150 for the LDL tracer and one of G-75 for the apo-lipoprotein preparations. Determine the activity of this filtered product and dilute it to give a working tracer with an activity of about 10000 cpm/100 μl.

Rabbit anti-apo-lipoprotein serum: this must be mono-specific for the apo-lipoprotein to be determined. The native anti-serum must be diluted with assay buffer to such an extent that it will bind 60–70 % of the labelled tracer in the absence of unlabelled standard. Preliminary experiments will establish the magnitude of this dilution, which will vary with the batch of antiserum from about 1:1000 to 1:50000.

Non-immune rabbit serum: dilute with assay buffer to about 1:100.

Anti-serum to rabbit γ-globulin: this is produced commercially in goat or sheep and must be diluted to a concentration that will bring about maximum precipitation of the apo-lipoprotein–tracer complex when 100 μl is added to the assay system. This dilution will typically be between 1:5 and 1:50.

Apo-lipoprotein standard: prepare a stock solution as in Section 8.9.1, and dilute with assay buffer to give working standards at the concentrations defined below.

Procedure. Each assay should consist of the following series of tubes, all of them in duplicate:

(1) Control tubes, containing only tracer and buffer, to check the level of adsorption to the tubes.

(2) Control for non-specific precipitation, in which no anti-apo-lipoprotein is added.

(3) The standards. The first of these is the zero level, i.e. no unlabelled apo-lipoprotein is added. This is followed by 5 or 6 standards which are distributed over the range of the assay. Of these, the first should be slightly above the lower limit of the assay. If the estimates are to be made for diagnostic purposes, it is prudent to have a standard near the upper and lower limits of the normal range.

(4) Quality control samples. These should be in the low, middle and upper registers of the assay.

(5) The unknown samples. Run each sample at two dilutions. If experience shows that the assay is sufficiently reliable, these dilutions need not be run in duplicate. Insert a mid-range quality control at every tenth position in the series.

(6) Repeat the quality controls. Prepare these tubes as follows:

Control 1:	400 μl of assay buffer
	100 μl of labelled tracer
Control 2:	250 μl of assay buffer
	100 μl of labelled tracer
	50 μl of non-immune rabbit serum
	100 μl of standard
Standards:	150 μl of assay buffer
	100 μl of labelled tracer
	50 μl of non-immune rabbit serum
	100 μl of apo-lipoprotein standard
	100 μl of anti-apo-lipoprotein
Assay tubes:	make these up in the same way as the standards, but with the latter replaced by 100 μl of the diluted sample

Mix the prepared tubes thoroughly and stand them for 36–48 hours at 4 °C. Then add 100 μl of anti-rabbit γ-globulin, mix and stand for a further 24 hours at 4 °C. Centrifuge the tubes for 30 minutes at 2500 *g*, aspirate off the supernatant and wash the precipitate twice by centrifugation in 2 ml of ice-cold assay buffer. Finally, count the radio-activity in the tubes to an accuracy of 1.5 %.

Computation. Calculate the following variables (Rodbard et al., 1969).

B_o = (mean counts in zero standard) − (mean counts in control 2)
B = (mean counts in sample) − (mean counts in control 2)

$$\text{logit } B/B_o = \ln \frac{B/B_o}{1 - B/B_o}$$

A plot of logit B/B_o (ordinate) against log (dose of protein standard) should be linear over a substantial range and can be used to estimate the amount of apo-lipoprotein in the test samples. Fit a straight line to the observed points, preferably by a weighted least-squares procedure (Rodbard and Lewald, 1970; Dell et al., 1973).

As with the other immunoassays, it is important that the slope of the line produced by different dilutions of the material being assayed does not differ significantly from that of the standard curve. Although this complication is not often reported in apo-lipoprotein assays, Goldberg et al. (1980) found difficulty in satisfying this condition when estimating serum apo-lipoprotein A-II. This they were able to overcome by a gel-filtration of the iodinated tracer A-II on Sephadex G-100. When the fraction that emerged at the void volume was used as the tracer, satisfactory assays resulted.

The collection and preservation of blood plasma

The different lipoproteins are present in the plasma in amounts that are the steady state resultant of many biochemical reactions, some of which are influenced by environmental factors such as diet or exercise. It is therefore important to be sure, so far as is possible, that subjects whose lipoproteins are to be examined are in a stable and reproducible condition. They should have been consuming their normal diet for several days before blood samples are taken and, to ensure the most reproducible profiles of the VLD-lipoproteins, should have fasted overnight. This is as true for laboratory animals as it is for Man, and it is important to maintain all animals that are to be used for lipoprotein studies in the very best of health, under strictly defined conditions. For example, changes in the amount or nature of the diet, or in the ambient temperature, cause marked changes in the lipoprotein profile of guinea pigs. We have also found that rabbits of the same breed, that are maintained in different environments, may differ widely in the way their lipoproteins are distributed.

According to Tan et al. (1973) the concentration of plasma lipids can be influenced by the posture of the patient when the blood sample is drawn. After 30 minutes in a sitting position, the levels of cholesterol and triglyceride may have fallen by 5%, and after 30 minutes lying flat the difference may amount to 10%. In practice, it is difficult to control this variable and the majority of subjects are probably bled while seated because this is convenient. As the N.I.H. Manual of Laboratory Operations (1975) points out, many subjects will have spent a considerable time in a waiting room at a clinic, but it is rarely possible to standardise this situation.

There is also some evidence that the prolonged application of a tourniquet (i.e. 5–10 min) can significantly raise the concentration of plasma lipoproteins (Koerselman et al., 1961; Page and Moinuddin, 1962). It is prudent therefore to release the tourniquet as soon as possible, although in some cases it may be necessary to maintain the constriction to obtain a good flow of blood. But even then, it should be possible to draw the sample in a time substantially less than that just quoted.

As a general rule, plasma should be used as soon as possible after preparation. However, limitations imposed by equipment or personnel may enforce the storage of some specimens and the following paragraphs will summarise the conditions under which this may be done, and the consequences to be expected.

It is well established that plasma contains enzymes that can, at least under some conditions, modify the composition and structure of the lipoproteins. the best characterised being lipoprotein lipase and lecithin:cholesterol acyl transferase (LCAT). However, there are probably others that have not yet been described. To minimise the effects of these enzymes, it is essential to keep the plasma cold (cf. below). It is also desirable to add Ellman's reagent (5,5′-dithiobis(2-nitrobenzoic acid)) to a concentration of 1.5 mM to inhibit LCAT, and phenylmethyl-sulphonyl fluoride or diisopropyl fluorophosphonate (2 mM) to inhibit proteolytic enzymes. If the plasma is kept at 0–4 °C, a bactericide such as Thimerosal (also known as Thiomersal) or sodium azide should also be added. According to Lee (1976), Thimerosal may have the additional merit of inhibiting lipoprotein lipase.

It has been known for many years (Ray et al., 1954) that lipoproteins are susceptible to oxidative degradation that is catalysed by heavy metals. Schuh et al. (1978) found this reaction to be stimulated by the presence of azide. At the time of writing, this observation has not been confirmed, but it may be prudent, when designing an experiment, to consider whether such an azide-promoted oxidation could be accepted. Oxidation can be greatly reduced by the addition to the plasma of 1.0 mM disodium ethylenediamine tetraacetate (EDTA) to sequester the heavy metal catalysts.

According to Lee (1980), malondialdehyde is liberated during the storage of plasma, and this can be used as an index of the extent of lipid oxidation. In this way, she was able to confirm that EDTA is effective in reducing the oxidation but also found that a mixture of EDTA (0.05%) and glutathione (0.02%) was four times better. In later studies on LDL, Lee et al. (1981) found that the apo-lipoprotein B was also susceptible to degradation that could be greatly reduced by the addition of EDTA and glutathione to the plasma, which was then stored for the minimum of time, in the absence of air.

Without the fore-mentioned precautions, the lipoproteins undergo a steady change in properties and plasma should not be stored at 4 °C for more than 2–3 days. The earliest signs of this aging process result from changes in the surface of the lipoprotein particle. If plasma is kept for only 18 hours at 4 °C there can be a loss of resolution on electrophoresis as exemplified in Fig. A1.1. In the fresh sample, there are two bands of pre-β-lipoprotein which are no longer resolved if the analysis is delayed until the following day. Loss of integrity of the larger lipoproteins can also be detected at an early stage by an increase in the amount of light scattered by the plasma, i.e. by nephelometry (Section 7.2). As the storage time is prolonged, the increase in the electrophoretic mobility of the lipoproteins becomes more evident

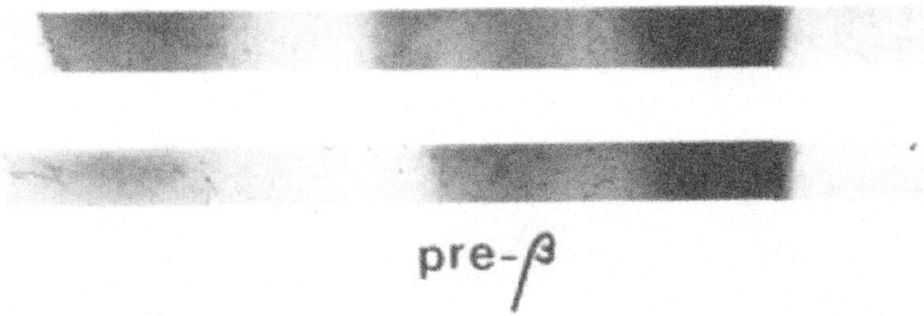

Fig. A1.1. This is an example of the loss of integrity of VLDL that can result from the storage of plasma, as detected by electrophoresis in agarose gel. The upper diagram shows the pattern obtained when the plasma was analysed as soon after its preparation as was practicable. The lower diagram shows the pattern obtained from the same plasma after 18 hours at 4 °C.

TABLE A1.1

The weight percentage composition of LDL isolated by ultracentrifugation from plasma that was stored at 4 °C. The insoluble aggregates were isolated and washed by centrifugation at $d = 1.006$ g/ml

	I	II	III	IV
Cholesteryl ester	39.3	37.0	32.0	13.7
Cholesterol	9.2	3.7	7.4	9.5
Triglyceride	7.4	17.6	16.4	35.8
Phospholipid	21.0	13.0	22.2	14.7
Protein	23.1	28.7	22.0	26.4

I, LDL from fresh native plasma.

II, LDL from the same plasma after storage for 21 days at 4 °C.

III, LDL from the same plasma after storage for 21 days at 4 °C in the presence of preservatives.

IV, The insoluble aggregates formed after prolonged storage at 4 °C.

TABLE A1.2

The distribution of phospholipids in LDL isolated from plasma that was stored at 4 °C for 21 days. The analyses show the phospholipid phosphorus as a percentage of the total phosphorus.

	Fresh	Stored
Lecithin	67.4	43.5
Lysolecithin	2.6	9.2
Sphingomyelin	24.6	40.7
Phosphatidylserine } Phosphatidylinositol }	2.2	2.9
Phosphatidylethanolamine	3.2	3.0

TABLE A1.3

The distribution of low-density lipoproteins in plasma before and after storage for 21 days at 4 °C. The values quoted are mg lipoprotein/dl plasma.

S_f	Fresh	Stored
0–12	248	229
12–20	61	38
20–100	65	5
100–400	23	1

and, at the same time, they begin to bind albumin. Both these effects can be seen in Fig. A1.2, where the presence of albumin is demonstrated by means of anti-serum to human serum albumin. Concurrent with these effects there is a progressive change in the chemical composition of the lipoprotein that is illustrated by the analyses of LDL that are summarised in Table A1.1. The first detectable change is a decrease in the proportion of cholesterol and phospholipid, and an increase in triglyceride. At the same time, there is a progressive fall in the proportion of lecithin and a rise in that of lysolecithin (Table A1.2). Since there is an increase in the proportion of sphingomyelin, it seems likely that the net loss of phospholipid that is shown in Table A1.1 results from the hydrolysis of lecithin, followed by diffusion of the lysolecithin into the surrounding tissues. This loss of phospholipid from the particles would make an important contribution to the

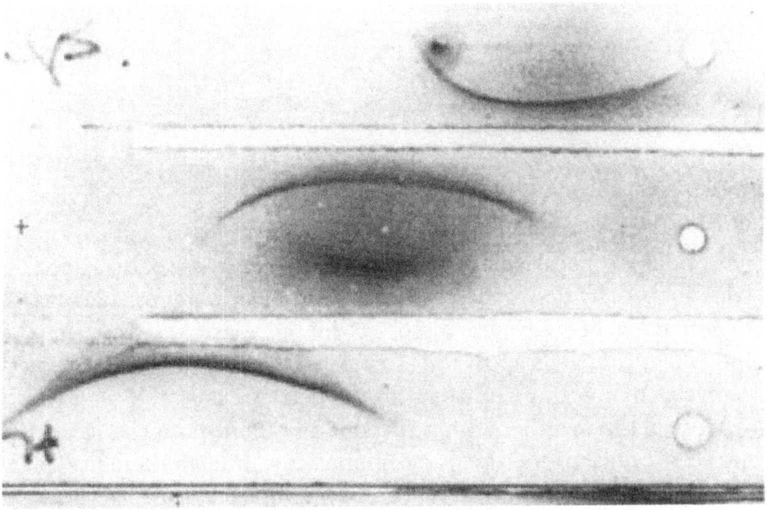

Fig. A1.2. The immuno-electrophoretic demonstration of the change in properties of LDL when it is isolated from stored plasma. The troughs contained the following antisera: upper, anti-LDL; lower, anti-albumin. The contents of the wells were as follows: top, LDL prepared from fresh plasma; middle, LDL prepared from the same plasma after storage; bottom, human plasma albumin.

aggregation that is the final result of long-continued storage. These aggregates have a very low content of cholesteryl ester and a high triglyceride (Table A1.1). The effect of these changes in chemical composition is to shift the ultracentrifugal profile towards the higher densities, so that the level of low-density lipoproteins falls, as is shown in Table A1.3.

The addition of the recommended enzyme inhibitors will not entirely eliminate the degradation of the plasma lipoproteins (Table A1.1). For example phospholipolysis can be almost completely inhibited but this does not prevent a slow loss of unesterified cholesterol. Moreover, the inhibitors do not stop the rise in the proportion of triglyceride, nor do they prevent the adsorption of albumin by the lipoproteins. It follows therefore that, if plasma must be stored, it should be for the briefest possible time at a temperature near 0 °C, in the presence of the following additives:

Ellman's reagent	0.6 mg/ml
Phenylmethylsulphonyl fluoride	0.35 mg/ml
Thimerosal	0.08 mg/ml
Sodium azide	0.13 mg/ml
EDTA	0.37 mg/ml

Note. Some workers prefer to include 0.005% chloramphenicol with this mixture (cf. also Lee et al., 1981).

If circumstances make a storage period of more than 2–3 days unavoidable, the plasma must be frozen. Whereas isolated lipoproteins are usually destroyed by freezing, they are more resistant to this treatment when they are dissolved in the plasma. However, the following comments apply only to plasma that is frozen once; repeated cycles of freezing and thawing are deleterious.

Most plasma can be frozen without any immediate change in the ultracentrifugal properties of the lipoproteins, though 'jaundice' plasma is an exception, and some other hyperlipoproteinaemic plasmas may also be affected. By contrast, the electrical properties of the lipoproteins are altered immediately, and it is important that specimens intended for electrophoretic analysis should not be frozen. The changes in lipoprotein composition that take place during storage at

4 °C do not cease when the plasma is frozen, even at −50 °C. However, they proceed at a much reduced rate and their consequences are not usually detectable in the ultracentrifuge until after 3 months at −20 °C. The progressive transformation of low-density lipoproteins into substances of higher density then starts to become evident and, after 12 months, aggregation and disproportionation begin (Mills and Wilkinson, 1962).

It will be clear from these observations that it may be difficult to maintain the strict integrity of the lipoproteins when plasma must be shipped from one laboratory to another. If this is unavoidable, the recommended preservatives should all be added and, if possible, the plasma should be shipped frozen and sealed under nitrogen. Note however, that the plasma must then be kept frozen until it is analysed. If this cannot be assured, it should be shipped in ice at 0 °C. If arrangements are made to deliver the samples to, and collect them from an airport, it is rarely necessary for them to spend more than a few hours in transit.

Reagents

Solution A. Dissolve 0.75 g of Thimerosal and 1.3 g of sodium azide in 50 ml of water and adjust to pH 8 with 5% sodium bicarbonate solution. Dilute with water to 100 ml and filter.

Solution B. This must be freshly prepared. Dissolve 0.595 g of 5,5′-dithio-bis(2-nitrobenzoic acid) in 10 ml of 0.2 M sodium bicarbonate solution.

Solution C. Dissolve 0.35 g of phenylmethylsulphonyl fluoride in 10 ml of 2-propanol.

Equipment

Blood may be collected with the conventional syringe and needle, or with the aid of pre-evacuated tubes such as those marketed by Becton-Dickinson and Co. under the Vacutainer trade mark. In the latter system, the blood is drawn directly into a 15 ml glass tube that contains 16 mg of solid EDTA. After mixing the contents, the tube can be centrifuged to separate the plasma. However, if this system is used, it is advisable to specify tubes with stoppers that are lubricated with silicone, rather than with glycerine which may interfere with the measurement of plasma triglyceride, e.g.

Vacutainer tubes Cat. No. 3218 XF 282
$1\frac{1}{2}''$ 20 gauge multiple sample needle Cat. No. 5749

Storage vials: there is a wide variety of glass or plastic tubes with push-in or screw-on plastic caps, that can be used for the short-term storage of plasma. However, caps that have cardboard liners should not be used. Lindgren (1975) advocates the use of screw-caps with Teflon liners, but cone-seal caps can also be used. The N.I.H. Manual (1975) recommends 20 ml liquid scintillation vials with cone-seal cap liners (e.g. Packard Instrument Co.), but these may leave an undesirably large air-space when the volume of the plasma sample is small. In all cases, care must be taken to leave an adequate space for expansion if the sample is frozen.

If plasma is to be shipped to another laboratory, vials with push-on caps must not be used. Under these circumstances, it is best to seal the samples into stout glass ampoules, but a tube with a strong cap, firmly screwed down is almost as reliable.

Procedures

Drawing blood. This operation should only be performed by qualified personnel. When the subject is an infant, the operator should be a physician. Detailed instructions are given in the N.I.H. Manual (1975), on which the following description is based. *N.B.* Exercise great care when handling blood and plasma. Do not pipette by mouth and avoid any contact with the skin.

Before drawing the blood, ascertain: (a) the subjects's age; (b) when and what the subject last ate. In the case of an infant, determine the kind of milk, or formula diet that was used; (c) what drugs are being taken; (d) if the subject is a woman, is she pregnant, or taking oral contraceptives?

Where possible, the subject should be seated and the blood drawn from a convenient arm vein, usually an antecubital vein. Use a tourniquet but release it as soon as is practicable, even before the blood is drawn, if this is possible. If a syringe is used, transfer the blood gently to a centrifuge tube that contains an appropriate amount of dry EDTA and mix thoroughly but gently by inverting 8 or 10 times. Label the tube and immediately plunge it into ice.

Although blood can be drawn from infants by essentially the same technique, it must be noted that the *maximum* amount that can be taken is 15 ml up to 2 months of age, 20 ml from 2–6 months and 30 ml from 6–12 months.

Post-natal cord blood can be collected by allowing blood to drip from the severed cord into EDTA-tubes, with continuous agitation to ensure that the anti-coagulant is thoroughly mixed. Note that although lipoproteins do not normally cross the placenta, the possibility of a maternal contribution to the cord blood sample may be increased if the clamping of the cord is delayed (Boyd, 1935; Dancis, 1962). It is therefore advisable to clamp the cord quickly and to collect the blood as soon as possible after the birth. Cord blood bilirubin is usually from 1–2 mg/dl (Davidson, 1941) and, at this level, may interfere with the estimation of plasma cholesterol. However, this difficulty can be overcome by adsorption of the bilirubin with zeolite (Moline, 1969).

Note that the foregoing description refers to the collection of small samples of blood for the analysis of lipoproteins. Large samples for preparative work should be obtained by the technique, and with the precautions used by the blood transfusion service.

There are many different procedures for taking blood from animals and the beginner will be well advised to consult a specialised manual, or someone who is experienced in handling the animal in question. In the case of animals like fish or reptiles, professional advice may well be essential. In general, the larger mammals can be bled by venepuncture using a syringe and needle, or even a blood transfusion set. The vein can be chosen to suit the convenience of the investigator. In sheep, goats etc., the jugular is usually suitable but in dogs and cats, the femoral is probably more often used. The ear veins of the rabbit have always been a popular route to this animal's blood because they are easily accessible and visible. However, other animals with large ears can also be bled in this way provided that the volume required is not too large. These techniques require only that the animal be adequately restrained, anaesthesia usually being unnecessary.

In the smaller mammals, with small veins, venepuncture is a more trying technique, particularly since it is necessary to bleed a number of animals to obtain enough plasma for an investigation. Heart puncture may then be the more attractive procedure, especially in the case of the guinea pig, which has an anatomy that is not well adapted to venepuncture. Many workers prefer to use heart puncture even for rabbits.

Note that blood must never be taken post-mortem.

Separation, storage and shipment of plasma

Separate the blood cells from the plasma by centrifugation at 1 500 *g* for 30 min at 0–4 °C. This should be done within 1 hour of collection. If a delay is unavoidable, it must not exceed 3 hours. Remove the plasma from the cells immediately after centrifugation, with a Pasteur pipette, and transfer it to a labelled vial. Add preservative solutions A, B and C at the rate of 0.1 ml to each 10 ml of plasma and mix thoroughly. (Note that we have assumed that EDTA was added in sufficient amount as anticoagulant to fulfil the requirement for an antioxidant. If serum is used, solution A should be modified by the addition of 3.7 g of EDTA.) Close the vial tightly and refrigerate in the dark.

If samples are to be shipped to another laboratory, purge the air-space with nitrogen before sealing the tube (cf. above). If a screw cap is used, it is prudent to secure it with tape. Lindgren (1975) recommends 'Scotch Magic Transparent Tape' which is stretched onto the tube and cap in a clockwise direction. Lindgren also recommends that the tube should be weighed so that leakage can be detected. Wrap each vial in tissue and pack them closely into a cardboard box so that they do not rattle. Alternatively, wrap the vials in a sheet of perforated styrofoam. Enclose the box or styrofoam package in a plastic bag which must then be tightly knotted. Pack crushed ice into a strong plastic bag, knot it and place it inside a second similar knotted bag. Sandwich the plasma samples between two such ice-packs, in a styrofoam box and fill any spare space with styrofoam packing, or with crumpled newspaper. Lindgren (1975) recommends packing the vials upright and marking the parcel to ensure that it is shipped in an upright position. This precaution is not necessary if the samples are sealed in ampoules. Finally, wrap and fasten the parcel securely. Be sure to address the shipment adequately and mark it to be refrigerated. Also, consult the schedules to ensure that the parcel is not held in store over a week-end. Above all, do not despatch any shipment unless the recipient has been warned in advance when to expect its arrival. There should then be no delay in its collection.

If it is decided to ship the samples frozen, pack them in the same way but use dry-ice as the refrigerant, which should not be sealed into bags.

Methods of concentrating lipoprotein solutions

When lipoproteins are prepared by ultracentrifugation and are collected from the top of the tube with a Pasteur pipette, the resulting solution can contain up to 100 mg of lipoprotein/ml. Preparations that are obtained by tube slicing, or by non-centrifugal procedures may, on the other hand, be relatively dilute and need to be concentrated. Because of the unstable nature of the lipoproteins, this is an operation that must be carried out by a mild, low-temperature procedure.

Concentration can often be usefully combined with a wash by a preparative ultracentrifugation in a solvent of appropriately increased density, after which it is best to collect the sample from the top of the tube with a pipette (Section 2.1.1.7). In some cases however, e.g. if the sample to be concentrated is small, it may be easier to concentrate LDL and HDL by centrifuging in a solvent of density 1.006 g/ml (overnight at 100 000 *g*). Under these conditions, the lipoproteins from a syrupy sediment on the bottom of the tube. The supernatant can be carefully aspirated off and the lipoprotein recovered. If small centrifuge tubes are used (2–3 ml), this technique allows microgram quantities of lipoprotein to be recovered in volumes of a few microlitres. Of course, this procedure is ineffective as a wash for the removal of proteins.

Unfortunately, ultracentrifugation is an expensive and time-consuming method of concentration, especially if only one or two small samples are involved, or if the samples are extremely dilute. Lyophilisation cannot be used because it would denature the lipoproteins. (Note however that the potential of lyophilisation in the presence of

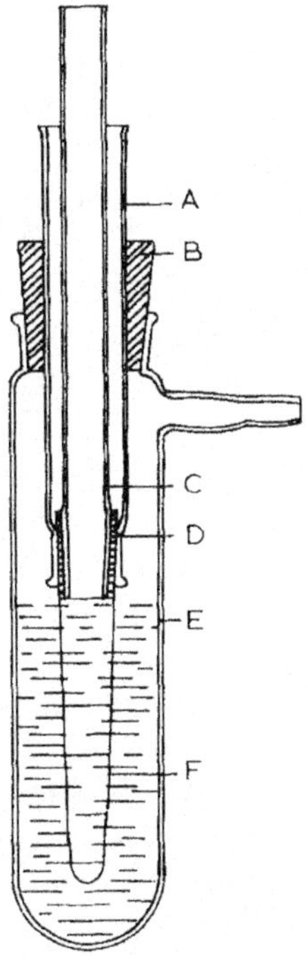

Fig. A2.1. Apparatus for concentration by vacuum dialysis (courtesy Membranfilter Gesellschaft, Göttingen, Germany). The outer glass mantle A is equipped with a ground joint articulating through the teflon cone D with the internal glass tube C. Between these two, the dialysis membrane F is clamped around the teflon cone D. The solution to be concentrated is placed in F and the outer chamber E is filled with buffer.

sucrose has not been properly investigated.) The techniques that are most widely used under these circumstances are based on the principle of ultrafiltration. Fortunately, the large size of the lipoprotein particles allows the use of highly permeable filtration membranes through which water will flow fairly rapidly. The technique in which the substance to be concentrated is knotted into a dialysis bag which is immersed in powdered, high-molecular weight polyvinylpyrrolidone or polyethyleneglycol, is simple and effective but is not recommended because contaminants penetrate the bag. Adsorbents that are commercially available, under various trade names, for the extraction of water in this way, also seem to suffer from the same defect although perhaps to a somewhat lesser degree. Moreover, in our experience, the same criticism applies to the 'Minicon' concentrator (Amicon Corpn.).

For dilute samples of modest size, pressure dialysis through a permeable membrane in the form of a thimble (Selectron ultrathimble; Schleicher-Schuell GmbH) is effective and is easy to perform at 4 °C (cf. also Clausen, 1981). The thimble is immersed in a suitable salt solution (Fig. A2.1) to which a vacuum is applied with a water-jet pump. Water and salts are drawn through the membrane, which can pass molecules of up to 100 000 in molecular weight. Because the membrane is vertical, the concentrated lipoproteins drain to the bottom of the thimble and can be collected in a volume of as little as 0.1 ml by means of a Pasteur pipette attached to a short length of fine-bore plastic tubing. If the thimbles are thoroughly washed in 0.15 M NaCl, they can be re-used several times.

Unfortunately, the Schleicher-Schuell 'thimbles' are not available in large sizes and, for the concentration of large volumes, the Micro-ProDiCon 'negative pressure micro protein dialysis concentrator' may be more attractive (Bio-Molecular Dynamics). This equipment works in essentially the same way as the 'thimble' but uses a long cylindrical membrane of greater surface area, which is attached to a collector that allows concentration to proceed to a small volume.

A widely used alternative to pressure dialysis is the stirred pressure ultrafiltration cell. These are made by the Amicon Corp. in sizes which have filters that range in diameter from 2.5 cm to 15 cm. If a large-pore membrane filter is used (e.g. XM100) a rapid concentration of the lipoproteins can be achieved. However, it is important to ensure that the magnetic stirrer in the chamber rotates at only about 1 rev/sec, or the lipoproteins may be denatured. It is also worth noting that it is more difficult to recover all of the concentrate from this type of cell than from the 'thimble', and that its final volume will probably be greater.

The extraction of lipid from lipoproteins

The techniques for the extraction of lipids from tissues, and their subsequent storage, are described in detail by Kates (1975) and this, together with the comprehensive review by Nelson (1975), may profitably be read in conjunction with this Appendix.

So far as is at present known, the components of the plasma lipoproteins are bound only by non-covalent forces i.e. by hydrophobic, electrostatic, hydrogen and van der Waals bonds. The separation of the lipid and the protein moieties can therefore be achieved by extraction with a mixture of polar and non-polar organic solvents. However, this 'delipidation' has much in common with gardening. Both are complex, poorly understood arts, in which different practitioners obtain much the same results by different, or even contradictory methods. This is well illustrated by the fact that some workers like to extract their lipoprotein preparations with mixtures of ethanol and ether, whereas others prefer chloroform–methanol. Likewise, some start with a dialysis against distilled water, while others favour 0.5 M NaCl, and still others prefer intermediate salt concentrations. In general, we suspect that few of these variations play a significant part in maintaining the integrity of the lipid-free protein, but they may have a bearing on the efficiency of the extraction process. In this context it must be explained that it is difficult to remove the last traces of lipid, particularly phospholipid, from the apo-lipoprotein. The most persistent extraction is needed if the residual phospholipid level is to be reduced below 1.0% of the protein. Unfortunately, the most efficient extractants for phospholipids are polar solvents like alcohol, that introduce the danger that sufficient water may be dissolved in the organic phase to promote the solution of appreciable quantities of the smaller apo-lipoproteins (Scanu and Edelstein, 1971). The analyst therefore has three courses open to him. He may dry the lipoprotein before extracting it, with the risk that it may be difficult to remove the last traces of lipid (cf. Scanu and Edelstein, 1971). He may extract a solution of lipoprotein with mixtures of ether and alcohol, in which case he can easily recover the precipitated protein, but must take special steps to recover any that dissolves in the organic phase. Lastly, he may consider extracting a solution of the lipoprotein with chloroform–methanol. This extract can easily be washed with water, which removes salts and peptides from the chloroform phase, but the high density of the solvent can make it difficult to collect the precipitate by centrifugation.

In general, there are two main reasons for wishing to separate the lipids from the apo-lipoproteins. Either it is a preliminary to the chemical analysis of the lipoprotein, or it is the first step in the preparation of the native apo-lipoproteins. In the first case, it is important that the integrity of the lipids is maintained, whereas denaturation of the proteins can be tolerated. For the preparation of the proteins however, it is clear that their native state must be preserved. Because of these differences in emphasis, two slightly different methods of extraction are used. Nonetheless, there are degradative processes that are common to all methods, namely, the enzymatic and the oxidative. Most of the enzymes that may be adsorbed to the lipoprotein particle are likely to be destroyed by high concentrations of alcohol and probably do not contribute significantly to degradation during the extraction procedure. By contrast, constant care must be exercised to prevent oxidation. All solvents should be peroxide free and should be stored under nitrogen. It is also advantageous to work at a low temperature, for example, Nelson (1975) recommends that the extractions should be made at 0 °C and that the solvents should be evaporated at 15 °C. This refinement becomes particularly important if long-chain polyunsaturated fatty acids are present, as in fish, or in human beings after some forms of dietary treatment. Antioxidants such as 2,6-di-*tert*-butyl-*p*-cresol

(BHT) can also be added to the solvents in a concentration of 0.005–0.05%. This prevents the oxidation of lipids and, according to Lee et al. (1981), may also help to control the degradation of the apo-lipoproteins. It is not clear to what extent the addition of BHT may lead to contamination of the isolated lipoproteins and interfere with their chemical analysis. However, if need be, it can be detected and removed from the extracted lipids by thin-layer chromatography on silica gel, using benzene as solvent.

Extraction for chemical analysis

The solvent most commonly used is a mixture of ethanol and ether. Both these components must be carefully re-distilled before use, since they normally contain sufficient lipid-like material to be of consequence in some lipid determinations. Unfortunately, wet ethanol–ether has the disadvantage that it will dissolve substantial amounts of salt as well as a little protein. The former can interfere with the running of thin-layer chromatograms and the lipoprotein preparation should be subjected to a preliminary dialysis against 0.15 N NaCl. This step is essential in the case of HDL preparations. The amount of protein that dissolves in the solvent mixture depends on the proportion of water that is present and it is essential to keep this below 3–4%. It is therefore important to start with a dry solvent i.e. to use absolute alcohol. In addition, the volume of the aqueous lipoprotein solution must not exceed 2–3% of that of the solvent. It may be possible to encourage the denaturation of the protein by heating the mixture on a water-bath, but the presence of ether will make it difficult to reach a really effective temperature. In practice, the amount of protein that is lost by solution in the extractant can be kept below 1% and, having regard to the generally low precision of the available methods for analysing lipoproteins, this may often be an insignificant source of error. Moreover, the presence of protein in the extract can always be checked by adding ether to reduce the proportion of ethanol to less than 40%. Dissolved protein will then be precipitated.

Reagents. Absolute ethanol, analytical grade ether from which peroxides are removed by shaking 1 litre with 100 ml of 20% $FeSO_4$ in 0.1 M H_2SO_4. Both solvents should be redistilled before use. A 0.15 M NaCl solution is made up as follows: 8.7 g of NaCl, 0.13 g of sodium azide and 0.372 g of EDTA in 1 litre of water, adjusted to pH 7.5 with $NaHCO_3$.

Method. Dialyse the lipoprotein solution against 0.15 M NaCl for 24 hours at 4 °C. Pipette 0.5–1.0 ml of this solution, which should contain about 10 mg of lipoprotein, into a stoppered tube and add 25 ml of ethanol–ether (3:1 v/v). Shake vigorously for 1 min (or use a vortex mixer) and then stand at room temperature of 24 hours. Centrifuge at 2 500 *g* for 30 min and recover the supernatant by aspiration. Wash the protein residue with the same solvent mixture and add the washings to the original extract. To recover any protein that is dissolved in the extract, add 8.75 ml of ether to each 10 ml of extract and stand for 24 hours at 4 °C. Any precipitate that is formed will contain some salt and may be difficult to centrifuge down. Finally, evaporate the extract to dryness under a stream of nitrogen at the lowest practicable temperature, which should not exceed 45 °C. It is best to complete the removal of water by the repeated addition and evaporation of small quantities of chloroform. Moreover, since dried lipids may be particularly susceptible to degradation, this extract should be analysed immediately. If this is not possible, store the lipids in chloroform–methanol (2:1 v/v), sealed under nitrogen, in the deep-freeze (cf. also Kates, 1975; Nelson, 1975). It may also be desirable to add an anti-oxidant such as BHT at this stage. Note that a sealed ampoule is the best storage vessel, although tubes with well-fitting ground-glass stoppers may be acceptable. Screw-cap vials that are fitted with Teflon or Polythene liners for the caps can also be used, but it is prudent to satisfy oneself that each vial is actually leak-proof. It is also worth noting that the stability of the lipids may be improved if the chloroform in which they are dissolved is purified by running over activated alumina immediately before use.

Some workers believe that the proportion of ethanol to ether in the extractant should be lowered to 3:2 for the delipidation of HDL. Although this will diminish the amount of protein that is dissolved in the first extract, it is probably better to change to the chloroform–methanol system.

It must also be noted that the protein residue will contain salt. Consequently, if a weighed amount of protein is needed, e.g. for carbohydrate estimation, the residue must first be extracted with water and re-dried.

The chloroform–methanol process. This is essentially the method of Folch et al. (1957), which has been the subject of valuable discussions by Radin (1969), Kates (1975) and Nelson (1975). Virtually no protein dissolves in this solvent but, because the mixture has a high density, it may be difficult to compact the protein adequately by centrifugation.

Method. Put 1 ml of lipoprotein solution into a stoppered tube and add 25 ml of chloroform–methanol (2:1 v/v). Shake vigorously and stand for 2 hours at room temperature. Recover the precipitated protein by centrifugation, or by filtering through a sintered glass filter. Wash the precipitate twice with 5 ml portions of chloroform–methanol (2:1). Combine the washings with the original filtrate and add 2.25 ml of water for each 10 ml of chloroform–methanol. Shake gently and set the tube in the refrigerator overnight. Then centrifuge lightly to ensure that separation of the phases is complete and remove the aqueous layer as completely as possible by aspiration. This fraction will contain salt and small polar molecules which may include some phospholipids e.g. lysolecithin. The chloroform phase can be washed with the aqueous phase made by adding water to chloroform–methanol in the proportions given above, and evaporated under nitrogen at a temperature not greater than 45 °C.

Extraction for preparative purposes

Although Helenius and Simons (1971) proposed to use detergents to remove the lipids from the lipoproteins, it is more usual to extract them with organic solvents. Both ethanol–ether and chloroform–methanol have their advocates, but the former is the most widely used, usually as some variant of the procedure described by Scanu and Edelstein (1971).

Since the usual object of preparative extraction is to recover the low molecular weight apo-lipoproteins in as nearly a native state as possible, the process must be carried out under the mildest conditions. Moreover, since some of these proteins are present in low concentrations, losses by solution in the extractant must be minimised. In some laboratories this has been done by starting the extraction with pure ether and continuing it with solvents that contain increasing proportions of ethanol (Shore and Shore, 1973; Lee and Alaupovic, 1974). These systems are biphasic, and unprecipitated protein remains in the aqueous phase. Alternatively, the extraction can start with pure alcohol, to which ether or chloroform is then added in excess. This gives a solvent mixture in which the apo-lipoproteins are almost insoluble. Both these procedures can yield a good quality product but the latter is the less tedious. It is not clear whether there is any real advantage to be gained from lyophilising the lipoprotein preparation before extracting it. If this course is adopted, the drying should be carried out on a very dilute solution, to ensure that the product is as finely divided as possible.

Herbert et al. (1977), who prefer to use mixtures of ether with methanol instead of ethanol, make the following recommendations for successful 'delipidation':

(1) the organic solvents must be kept below 0 °C
(2) excess bromide must be removed by dialysis
(3) the lipoprotein solution should be dialysed to 0.15 M NaCl, not against distilled water
(4) the precipitated protein should be centrifuged at the lowest feasible speed, to avoid the formation of a compacted pellet
(5) the protein should be dried as a thin film, not as a pellet

Although we agree with these precepts in general, it is our opinion that the isolated apo-lipoproteins should not be dried at all. If possible, they should be dissolved and processed immediately. An irksome property of apo-lipoprotein B, which makes preparations that contain it difficult to handle, is its tendency to form a gelatinous mass in the presence of moisture. This can be prevented from happening during the extraction by keeping the system strictly anhydrous during the later stages. In fact, Olofsson et al. (1980) extend this precept by recommending that the protein should never come into contact with pure hydrophilic solvents. However, gel formation is inevitable when the attempt is made to dissolve the protein, and the way to frustrate the formation of a virtually insoluble mass is to keep the finely divided particles separate during the hydration process. Drying in the form of a thin film may achieve this by mechanically dispersing the particles on the wall of the tube, albeit at the risk of producing irreversible changes during the drying process. It is probably better to add the solvent to a fine suspension of the protein in ether, which appears to surround the protein particles and keeps them from sticking together. If the ether is then allowed to evaporate slowly, conditions that favour the slow hydration and solution of the protein seem to result.

Ethanol–ether process

Reagents. Absolute ethanol; analytical grade diethyl ether, freed from peroxide as above; 0.03 M NaCl made by dissolving 1.74 g NaCl, 0.372 g of EDTA in water and adjusting to pH 7.5 with $NaHCO_3$.

The concentration of NaCl may not be critical. If it is very low, the flocculation of the protein after the first extraction may be prolonged. Some workers prefer to use 0.15 M NaCl.

Method. The lipoprotein solution will be prepared by ultracentrifugation (Chap. 2) and must be washed until it is free from all plasma proteins. Dialyse this concentrate against several changes of 0.03 M NaCl solution at 4 °C. The NaCl solution should be de-aerated and the dialysing vessel should be completely filled.

Adjust the concentration of the dialysed lipoprotein solution to contain not more than 10 mg of lipoprotein/ml and pipette not more than 2 ml of it into a 50 ml (125 mm × 30 mm) stoppered tube. Cool the tube and its contents in ice and then perform the following sequence of extractions:

(1) Add 20 ml of ethanol that has been pre-cooled to between −10

and −20 °C, and shake vigorously. Then add 30 ml of similarly pre-cooled ether, mix and stand for 24 hours at −15 °C. Collect the precipitate by centrifuging for 20 min at 850 *g*, at the lowest attainable temperature (−15 °C if possible). Be careful not to compact the protein into a pellet, but make sure that none adheres to the side of the tube. If this should happen, centrifuge a second time. Aspirate off the solvent as completely as possible.

(2) Disperse the protein by vigorous shaking in 50 ml of cold ether–ethanol mixture (1:3 v/v) and stand for 24 hours at −15 °C. Recover the protein as before.

(3) Repeat (2) with 50 ml of ether–ethanol (1:1 v/v) for an extraction time of 30 min.

(4) Substitute 50 ml of cold (−15 °C) ether–ethanol (3:1 v/v) in (3).

(5) Wash the protein precipitate twice with 50 ml of peroxide-free ether at −15 °C.

After the last wash, leave the protein suspended in 1–2 ml of ether. The protein is then best dissolved immediately in whatever aqueous solvent is chosen, but may be stored briefly as the ethereal suspension at −15 °C, preferably under nitrogen.

The chloroform–methanol process

Consecutive extractions with chloroform–methanol (2:1 v/v) can be made at low temperature, as described in the previous section. However, Olofsson et al. (1978, 1980) use a modified procedure that starts with lyophilised lipoprotein, on which the following is based. The lipoprotein solution must first be dialysed in five changes of de-aerated, double-distilled water at 4 °C. It is then freeze-dried and the solid transferred to stoppered centrifuge tubes at the rate of 0.5–1.0 g of apo-lipoprotein per tube, after which it is extracted according to the following sequence.

(1) Add 15 ml of dry, analytical grade chloroform that has been pre-cooled to between −10 and −20 °C. Shake well to disperse the solid as thoroughly as possible: this is important. Then add 30 ml of pre-cooled methanol, mix well and stand for 30 min at −10 to −20 °C. Shake the tube several times during this period. Collect the

protein residue by centrifugation at low temperature and aspirate off the solvent.

(2) Repeat (1).

(3) Extract the protein twice with 45 ml of cold chloroform–methanol (1:1 v/v) under the same conditions.

(4) Repeat (3) twice with 45 ml of cold chloroform–methanol (2:1 v/v).

(5) Wash the protein twice with 45 ml of cold, peroxide-free diethyl ether. After the final wash, leave the protein suspended in a small volume of ether, as in the previous method.

To dissolve the protein, add the chosen solvent e.g. 2 M acetic acid or 0.1 M $(NH_4)_2CO_3$, plug the tube with cotton wool and stand it in the refrigerator overnight for the ether to evaporate. Alternatively, bubble nitrogen through the suspension.

Extraction for quantitative immunochemical analysis

Some apo-lipoproteins, notably C-III and E, seem to be incompletely accessible to anti-sera if the plasma to be analysed is grossly hyper-triglyceridaemic. The following technique, which is a modification of the method of Cham and Knowles (1976), is a useful way of removing the excess lipid without precipitating the proteins. Since the samples to be analysed are usually small, a centrifuge that will accept small tubes is desirable. A good example is the Eppendorf Model 3200, which uses 1.5 ml polypropylene tubes with attached push-on caps.

Pipette 2 volumes of a mixture of *n*-butanol/di-isopropyl ether (15:85 v/v) into each tube and then add 1 volume of the serum. Mix thoroughly by rotating the tubes end over end for 30 min. Centrifuge for 6 min, discard the organic supernatant and remove the last traces of solvent from the aqueous phase by blowing a stream of N_2 over the surface. The volume of the sample is essentially unchanged by this extraction, which is applicable to isolated lipoprotein fractions as well as to plasma or serum.

APPENDIX 4

The preparation of stains

This appendix describes methods for preparing stains for locating lipoproteins or apo-lipoproteins after electrophoresis or immuno-diffusion in agar or agarose. There are several suitable dyes and the choice is to some extent a matter of personal preference. There are also many slightly different formulations for each stain but the following have been shown to give reliable results. In all cases, use only the best quality dye, preferably a grade that has been prepared specifically 'for electrophoresis'.

Stains for proteins

The staining of apo-lipoproteins after electrophoresis in polyacrylamide gel is dealt with in Section 6.4.2.1. The procedures described here are used mainly for the staining of immuno-precipitin lines, although they can also be used to locate the lipoproteins after a simple electrophoresis.

(1) Naphthalene Black 12B (Amidoschwartz; colour index No. 20470). This stain was widely used before the introduction of Coomassie Blue, which is markedly more sensitive. However, very satisfactory results can be obtained with Amidoschwartz and some workers prefer it. The following formulation is designed for staining precipitin lines and the solutions contain glycerol which prevents the dried agarose gel film from becoming brittle, and also gives an elegant sheen to the surface.

Dilute 60 ml of glacial acetic to 1000 ml with distilled water and dissolve 13.5 g of sodium acetate in the solution. Dissolve 1 g of

Naphthalene Black 12B in 850 ml of this buffer and add 150 ml of glycerol. Stir thoroughly to ensure that the viscous glycerol is completely dissolved. Filter the solution and keep it at room temperature in a dark bottle. It can be used until it fails to stain to an adequate intensity and becomes greyish.

The de-staining solution consists of 150 ml of glycerol dissolved in 850 ml of 2% acetic acid.

The prepared slide is immersed in the stain for at least 10 min, or longer if the precipitate is weak. Transfer the slide into the de-staining solution and agitate for 30 sec before transferring again into fresh de-staining solution for 30 min (or longer if the staining period was prolonged). Finally, wash the slides under the tap and dry them at room temperature.

(2) Coomassie Brilliant Blue R250 (colour index No. 42660). It has been rumoured that the manufacture of this dye will eventually cease, and some efforts have been made to find a replacement. For example, it is said that PAGE Blue 83 (BDH Ltd.) can be substituted directly in the formulations for Coomassie Blue that were originally proposed by De St. Groth et al. (1963) and by Chrambach et al. (1967). At the time of writing however, Coomassie Blue is still readily available.

Prepare a stock solution of the dye by stirring 1 g of Coomassie Blue R250 in 100 ml of water until it is completely dissolved. The working stain is freshly prepared by slowly adding one part of the stock solution to 20 parts of 12.5% trichloracetic acid, with continuous stirring. Stain the slides for 30–60 min and then transfer to a 10% solution of trichloracetic acid to de-stain the background. Finally, wash thoroughly and dry at room temperature.

Stains for lipids

The electrophoresis of plasma as described in Chapter 5 will only partially resolve the lipoproteins from the other plasma proteins and their unequivocal detection must depend on the presence of their lipid moiety. This is often done by staining the lipids with a fat-soluble dye, a technique that also makes it possible to distinguish precipitin lines that contain lipoprotein from those that do not. These fat stains are more difficult to prepare than the protein stains because they must be dissolved in a mixture of water with an organic solvent. Since the stain is usually saturated with dye, any change in the proportions of the solvent mixture, or in the temperature at which it is stored, will lead to the precipitation of particles of dye. These will adhere to the material that is being stained and, because they are often difficult to remove, will give results of poor appearance. It is therefore important to filter the stain solution before each use.

(1) Oil Red O (Colour index No. 26125. N.B. Hatch and Lees (1968) say that there is a dye of colour index No. 26120 that is also sometimes called Oil Red O, but which is useless as a lipoprotein stain.) A solution of this dye in 60% ethanol–water has been recommended by several laboratories and is the stain preferred for use in the Lipid Research Clinics Program (N.I.H. Manual, 1975).

It is made as follows: Dissolve 1.0 g of Oil Red O in 1.5 litre of absolute ethanol by heating in a flask fitted with a reflux condenser. When the solution is boiling, add 1.0 litre of water in five separate portions, allowing the mixture to regain the boiling point after each addition. Finally, reflux the mixture for at least one hour. Cool the solution to between 37 and 40 °C and never allow it to fall below 35 °C (if this should happen, reflux again for one hour and cool to 37 °C). Allow the freshly prepared stain to stand for at least 2 days for particles of dye to settle out.

According to Hatch and Lees (1968), the newly made stain, diluted four-fold with 60% ethanol, has an absorbance at 520 mμ of about 1.25 in a 1 cm cell. After 3 days at 37–40 °C, a four-fold dilution reads about 0.98 and will remain between 0.98 and 0.90 for a considerable time, during which it can be used. When the absorbance falls below 0.90, the stain should be discarded. Hatch and Lees (1968) also point out that the ethanol concentration can be monitored by measuring the specific gravity of the solution. The value for fresh stain is 0.891 at 37 °C.

(2) Sudan Black B (Colour index 26150). Noble (1968) makes a stock solution as follows. Shake 200 mg of dye in 100 ml of 60%

ethanol for 60 min at room temperature. Allow the mixture to stand overnight for the undissolved dye to settle out. A 1:100 dilution of this solution with 60% ethanol should have an optical density of 0.28–0.3 at 590 mμ in a 1 cm cell. Prepare the working solution by diluting the stock with an equal volume of 60% ethanol. This may be used until the optical density of a 1:100 dilution falls below 0.13. Keep these stains in stoppered bottles in the dark, at room temperature.

Good results can also be obtained with a stain made in the following way. Add 1 g of Sudan Black to 1000 ml of 60% ethanol, shake vigorously and then stand the mixture for 12 hours at 37 °C, in a closed vessel. Filter off the undissolved dye. This solution can be stored at 4 °C in a dark bottle and used for up to two months.

It is advantageous to add 0.1 ml of 30% NaOH to each 160 ml of stain just before it is used. This prevents the stained lipoproteins from fading on storage. Although the life of the stain is reduced by the addition of NaOH, in practice, the alkaline solution can be returned to the refrigerator and used for up to one month.

(3) The ozone–Schiff method. This procedure acts on the principle that unsaturated fatty acids are oxidised with ozone, to form aldehydes that are then detected with Schiff's reagent, The method has not been widely used, partly no doubt because it is not so simple as staining with a dye, and partly because it may take some trial and error to find the optimum conditions. Even at best, the colour tends to be relatively weak. The method may be of most value in staining cellulose acetate membranes.

The treatment with ozone should be done in a fume cupboard. Ozone can be produced by commercially available ozonisers, but is more often generated by the action of concentrated sulphuric acid on barium peroxide. Arrange supports in a beaker or tank (which should not be over-large) so that the acetate membranes can be suspended above a dish that contains 10–15 g of BaO_2. Add 20–25 ml of concentrated H_2SO_4 to the dish and seal the container with an air-tight cover. The time for which the strips should remain in the ozone will depend on such factors as the geometry of the equipment and the freshness of the BaO_2. From 10 to 30 min may be sufficient, though some have reported treatments of up to two hours.

After the appropriate time, remove the strips and rinse them for 1 min in 0.001 N NCl before transferring them to a bath of Schiff's reagent from which the light is excluded. (Schiff's reagent can be purchased ready-made, alternatively, use solution A from the following section on carbohydrate staining.) When the maximum colour has developed (up to 60 min at room temperature or up to 6 hours at 4 °C), wash the strips twice for 15 min in 0.5% HNO_3 and then for a further 15 min in 0.001 N HCl. Dry the strips at room temperature between sheets of filter paper.

Lipoproteins can also be detected on polyacrylamide gels in essentially the same way (Huismans and Felgenhauer, 1971).

Stain for carbohydrate

This technique can sometimes be of use in differentiating between those apo-lipoproteins that contain carbohydrate residues and those that do not. It has also been applied in studies of lipoprotein structure.

The following reagents are required (Crowle 1961):

Solution A. Dissolve 1.5 g of basic fuchsin in 500 ml of boiling distilled water and filter at 55 °C (this temperature is important). Cool the solution rapidly to 40 °C and add 3.75 g of sodium metabisulphite and 25 ml of 2 N HCl. Stir well and stand for 6 hours at 4 °C in a stoppered bottle. Then add 1.2 g of activated charcoal and shake hard for at least one minute. Filter off the charcoal as quickly as possible and store the solution at 4 °C in a dark bottle with a tight-fitting stopper.

Solution B.	1.0 g of periodic acid
	0.82 g of anhydrous sodium acetate
	Water to 100 ml
Solution C.	0.54 ml of glacial acetic acid
	0.89 g of anhydrous sodium acetate
	10.0 g of hydroxylamine hydrochloride
	Water to 100 ml
Solution D.	5 ml of 10% sodium metabisulphite in water
	5 ml of 2 N HCl

	90 ml of distilled water
	This solution must be freshly prepared
Solution E.	Solution D + 20% glycerol by volume

To stain an immuno-precipitin reaction, for example, proceed as follows: note that it is important to follow the instructions *exactly*.

(1) Immerse the slides in solution C for 15 min.

(2) Wash under running water for 15 min.

(3) Soak the slides in solution B for 10 min.

(4) Wash under running water for 10 min.

(5) Transfer the slides to solution A that has been diluted with an equal volume of water immediately before use. Continue this reaction for 30 min, or for longer if need be. It is important that, from this point onwards, the slides do not come into contact with water.

(6) Wash the slides three times, each of 2 min duration, in solution D.

(7) Follow with three washes, each of 60 min, in solution E. Finally, dry at room temperature.

The pre-staining of lipoproteins in plasma

Some of the difficulties associated with this process are discussed in Section 5.5.2. The stain most widely used for this purpose is that described by McDonald and Ribeiro (1959), which is prepared as follows:

Heat 100 ml of ethylene glycol to 100–110 °C and add 1 g of Sudan Black B. Stir well for 5 min and filter the hot solution through a Whatman No. 2 paper. When the solution is cold, filter it again through the same grade of paper. The solution, which has a final concentration of about 0.5%, can be kept in the dark for a month or more, in a closed container. *N.B.* Do not heat the glycol to more than 110 °C or a useless gel will be formed.

To stain the lipoproteins, add one part of Sudan Black solution slowly and with constant stirring, to five parts of plasma or lipoprotein solution. Stand the mixture at room temperature for at least 45 min and centrifuge briskly for 15 min before use. Alternatively, add one part of dye to two parts of plasma and keep the mixture in the refrigerator for at least 24 hours. Centrifuge before use.

Although this procedure can be used if the sample is to be applied in a gel, it may be better under these conditions, to use the technique of Naito et al. (1973) in which a more dilute dye solution is used. Prepare a stock solution of Sudan Black by dissolving 10 mg in 10 ml of ethylene glycol. Naito et al. recommend that this should be aged for two weeks in the dark before use. To make the working stain, dilute 1.2 ml of stock with 8.8 ml of ethylene glycol. This solution is mixed with the sample and the gel reagents as described in Section 5.5.2.1.

The photography of electrophoresis and immuno-diffusion gels

The method described here is probably the simplest, least expensive and most reliable way of recording the results of analyses in gels. The equipment and procedures are essentially the same for recording stained gels, whether they be rods or laminae, polyacrylamide or agarose. The photography of unstained immuno-diffusion gels requires a different kind of illumination, which we shall describe below.

Equipment. An automatic 'single lens reflex' camera which uses 35 mm film, e.g. the Olympus OM 10 with a 50 mm, f 1.8 objective. A copying stand which allows the camera height to be adjusted vertically, with the lens looking directly downwards. A +3 dioptre supplementary lens to fit on the front of the camera objective. An FL-W filter (or CC 30M filter) to fit in front of the supplementary lens. A lens shade to fit in front of the filter. Kodachrome 64 or Ektachrome 64 daylight colour transparency film.

For stained gels an X-radiograph illuminator can be used as a light source, e.g. two fluorescent lighting tubes (Phillips TL 20W/55 de luxe) mounted 15 cm apart beneath a translucent, white acrylic sheet.

For unstained immuno-precipitin lines, the illumination should be obliquely from below (Fig. 6.11) so that the lines are observed by scattered light. For these subjects a fast black-and-white film will be appropriate.

Procedure. The following details refer to the recording of stained polyacrylamide electrophoresis gels.

Cut a rectangular aperture 18 cm × 14 cm in a sheet of black card. This will form the frame that limits the field of the photograph. Place it over the diffuser of the illuminator and cover it with a sheet of glass.

If the field is not uniformly illuminated, put several sheets of thin tracing paper between the acrylic sheet and the black card, to act as an additional diffuser.

Note that the black mask must cover the whole of the illuminator except for the desired aperture or stray light will enter the camera lens and cause flare.

Mount the illuminator assembly on the copying stand below the camera, so that the film will be 41 cm above the diffuser. Under these conditions, using the apparatus described, the 18 cm × 14 cm field will nearly fill the exposed area of film and it is important to maintain this alignment to obtain a uniform magnification in all photographs. To ensure that the results of one experiment can be easily compared with those of another, it is equally important that all electrophoresis gels are run the same distance.

Put the polyacrylamide gel rods in clean, screw-topped test-tubes that are completely filled with water to eliminate air bubbles. Arrange these tubes on the glass plate over the aperture of the illuminator, with a transparent millimetre scale alongside to provide a permanent record of the distance the proteins have migrated. Extinguish the room lights to avoid reflections from the glass tubes, and focus the stained bands sharply in the viewfinder, using the normal camera mechanism (the setting will be roughly 10 m with the +3 dioptre lens). Set the iris diaphragm of the objective to f8. With the camera on its 'automatic exposure' setting, take three exposures of each set of gels with the film speed set at 64, 40 and 25 ASA respectively. In this way three transparencies with images differing in density will be obtained, which will ensure that both dense and faint bands are recorded as clearly as possible.

The 64 ASA Kodachrome or Ektachrome daylight film, when used with the FL-W filter and 'de luxe' or 'cool white' fluorescent lighting tubes, will give a background to the gels that is white or a neutral grey, with a correct recording of the colour of the stained bands. These films can be developed by Kodak, or the Ektachrome film can be rapidly developed by commercial processing laboratories. When enlarged paper prints are needed, they can be made in colour, directly from the transparencies, by Kodak. If the standard 9 cm × 13 cm or 13 cm × 18 cm prints are used, these enlargements will always be made at the same magnification. The former size is adequate for laboratory records but the larger format, in which the gels are virtually life-size, is better for publication. The cost of these colour prints is comparable with those made in monochrome.

If only a few photographs are to be recorded at infrequent intervals, the use of 35 mm film may lead to an inconveniently long delay between making the exposure and actually handling the print. It may then be worth considering the use of Polaroid 'instant' film to make prints directly. For this purpose, a suitable camera would be the Polaroid MP4. It is important, with this equipment, to maintain a constant film-to-lens distance for all gel photographs, to ensure that the magnification is the same for each. In this case, the image is focussed by moving the entire camera in the vertical plane. A green or yellow filter over the lens will be useful when recording bands stained with Coomassie Blue on black-and-white film.

APPENDIX 6

Radio-labelling

The technique used for the labelling of a lipoprotein or an apo-lipoprotein with radioactive iodine is partly determined by the use to which the product is to be put. For radioimmuno assays (RIA) and membrane binding studies, the tracer must be labelled to a high specific activity but must retain essentially all its immunological or binding properties. For metabolic studies, on the other hand, it is the biological integrity that must be retained and this more exacting requirement demands that not more than one atom of iodine is incorporated per molecule.

Labelling by the chloramine-T procedure (Greenwood et al., 1963) has been widely used for the preparation of tracers for RIA because it is simple and yields highly active products. Although it can lead to oxidative degradation, particularly in proteins that contain significant proportions of methionine or cystine, it does not generally result in an unacceptable loss of immunological properties. Iodine monochloride (McFarlane, 1958) has been said to cause less oxidative damage but, because of the presence of carrier iodine, it is more difficult to label the protein to a high specific activity. This alleged mildness has led some to prefer iodine monochloride to chloramine-T as the oxidising reagent when labelling lipoproteins. However, the procedure described by Hunter (1970) gives a considerable degree of control over the chloramine-T procedure and, if it is adhered to, the labelled product is not inferior to that obtained by the McFarlane method, for metabolic studies (Weech et al., 1978). The lactoperoxidase technique devised by Marchalonis (1969) has also been used with both lipoproteins and apo-lipoproteins but, as no marked advantages have been brought to light, it has not been widely adopted (cf. Fidge and Poulis, 1974). It should also be remarked that a number of solid state iodinating reagents are now available, one of which is an immobilised form of chloramine-B (Iodo-beads: Pierce Chemical Co.). At the time of writing however, there have been no detailed reports of their use with lipoproteins.

Apo-lipoprotein C-I cannot be satisfactorily labelled with radioactive iodine because it is deficient in tyrosine, but the other soluble apo-lipoproteins present no unusual difficulties and can be labelled by any of the conventional procedures. By contrast, intact lipoproteins are more difficult to label because their fragile nature, and the presence of many unsaturated lipids can lead to disproportionation and to the incorporation of significant amounts of iodine into the lipid moiety. The latter reaction can be particularly marked when the lipoprotein is not of human origin, since the proportion of unsaturated fatty acids in animals is often greater than that in man. Unfortunately, the choice of the method to be used for labelling lipoproteins is made more difficult by the inconsistent findings of many of the comparative studies that have been made (see Fidge, 1974; Weech et al., 1978). The reasons for this unpredictable behaviour are not understood, but there seems to be little doubt that the iodination is affected by subtle changes in the lipoprotein that occur during its preparation and subsequent manipulation. At present we can do no more than emphasise the importance of using lipoproteins that are freshly prepared from freshly drawn blood. Even under the most favourable conditions, some of the lipoprotein particles will undergo disaggregation during the iodination, and this process will continue if the labelled material is stored (Eaton and Kipnis, 1969; Schonfeld et al., 1974). It is therefore mandatory to purify the labelled preparation by gel filtration chromatography immediately before it is used (see below). Physical and immunochemical modification of apo-lipoprotein A-I has also been reported to occur during iodination (Steinberg et al., 1983).

Radiation damage during the storage of either lipoprotein or apo-lipoprotein is largely due to reaction with the radiolytic fission products of water, and can be diminished by the addition of a scavenger protein such as bovine serum albumin. A typical example of such a solvent is 3% albumin in 0.05 M sodium barbital buffer, pH 8.5.

(*a*) *Labelling of apo-lipoproteins.* When used as tracers in radio-immuno assay, apo-lipoproteins are labelled to a specific activity between 5 and 25 mCi/mg (i.e. about 4×10^8 d.p.s./mg). The following procedure is typical and may be adjusted either to suit the quantity of material to be treated, or to yield a product of different activity.

Note. Fainaru et al. (1977a) recommend that apo-lipoprotein E is labelled and stored in the presence of 0.1 M sodium decyl sulphate. This was said to improve the stability of the product and to improve the reproducibility of immuno-assays. It is not certain whether this is equally applicable to other apo-lipoproteins.

Reagents

0.05 M phosphate buffer, pH 8.0
Bovine serum albumin in phosphate buffer (BSA-phosphate): 3% albumin in phosphate buffer containing 0.02% sodium azide
Chloramine-T solution: 0.5 mg/ml in phosphate buffer
Sodium metabisulphate solution: 0.5 mg/ml in phosphate buffer
Carrier-free $Na^{125}I$: 1 mCi dissolved in 25 µl of phosphate buffer

Equipment. A 1 cm × 15 cm column of Sephadex G-50 equilibrated with BSA-phosphate.

Procedure. Dissolve the apo-lipoprotein in phosphate buffer to a concentration of 0.5 mg/ml and transfer 25 µl of the solution to a tube of about 2 ml capacity. Add 1 mCi of ^{125}I, followed by 25 µl of chloramine-T solution, and mix. After 60 sec, add 50 µl of *meta*-bisulphite solution, dilute with 1 ml of BSA-phosphate and remove the free iodide by passing down the G-50 column in BSA-phosphate. Store the resulting preparation at 4 °C and use within 4 weeks.

According to Schonfeld and Pfleger (1974), labelled apo-lipoprotein A-I undergoes a significant degradation during storage and should be purified by chromatography on a column of Sephadex G-75 not more than 24 hours before use. The first and the last of the 3 peaks that are produced should be discarded.

(*b*) *Labelling of lipoproteins.* Although the procedure to be described below has been found satisfactory for the iodination of both human and guinea pig LDL, it may be necessary to confirm that the conditions are equally applicable to other fractions, or to lipoproteins from other sources. This can be done by the method of Hunter (1970), as follows:

Reagents. Prepare the following solutions not more than 1 hour before use.

Phosphate-buffered saline (PBS): 0.15 M NaCl containing 50 mM Na_2HPO_4 and 1 mM EDTA, adjusted to pH 7.5 with HCl
Potassium iodide solution: several solutions of concentration between 0.1 mM (0.017 mg/ml) and 0.6 mM (0.1 mg/ml) KI in PBS
Sodium *N*-chloro-*p*-toluenesulphonamide (chloramine-T): 5 mg/ml in water
Solution A: 20 mg of sodium metabisulphite in 10 ml of water
Solution B: 50 mg of sodium metabisulphite in 10 ml of PBS
Bovine serum albumin solution (BSA): 5 g of albumin in 100 ml PBS
Trichloracetic acid (TCA): 20 g/100 ml water
Carrier-free $Na^{125}I$: 1 mCi in 0.20 ml of PBS

Equipment. The measurement of redox potential (Hunter 1970) requires a pH meter, set to measure the potential (in millivolts) between a platinum electrode immersed in the reaction mixture and a calomel reference electrode.

The chloramine-T reagent is added from a micrometer syringe or micro-burette into a reaction tube or vial, of about 2 ml capacity, which is mounted over a magneti stirrer.

Procedure. Into the reaction vessel, put 1 ml of PBS which contains a known quantity of lipoprotein, KI and $Na^{125}I$ e.g. 1 mg of apo-lipoprotein, 2.5 μg of KI and 1 mCi of ^{125}I. Set the voltmeter to read zero. Add a measured amount of chloramine-T solution (e.g. 5 μl) from the micrometer syringe, while stirring vigorously. Two minutes later, read the potential of the cell and withdraw a 10 μl sample of the reaction mixture. Add this sample to 1 ml of solution A in a 3 cm × 0.75 cm glass tube and follow it with 0.1 ml of BSA and 0.7 ml of TCA. Continue in this way, making further additions of chloramine-T until the redox potential reaches about 300 mV. Run all the samples that have been precipitated with TCA on a vortex mixer for 30 sec, stand them for 2 hours and then centrifuge for 3 min at 500 *g*. Remove the supernatant to a separate tube and wash the precipitate by centrifugation in 6% TCA. Count the radioactivity of the precipitate and of the combined supernatants.

To determine the activity bound to the apo-lipoprotein, extract the lipid from the precipitate by washing twice with 1.5 ml of ethanol/ether (3:1, v/v). Continue the first extraction overnight and the second for at least 4 hours, both at room temperature.

From the measurements of incorporated and un-incorporated activity, calculate the efficiency of iodination after each addition of chloramine-T. A plot of these values against redox potential will show that the binding of radioactivity ceases once the potential has reached a value that is characteristic of the system in use.

By repeating this experiment with different proportions of lipoprotein, KI and ^{125}I, it is possible to determine the conditions that give the most favourable incorporation of activity into the apo-lipoprotein, and the critical amount of chloramine-T that is needed to complete the reaction. Thus, in the case of human LDL, the conditions that are described below give a labelling efficiency of 65–80%, with some 95% of the label being bound to the apo-lipoprotein, the end-point of the titration occurring at about 200 mV.

Reagents. As above.

Equipment. In addition to the above, a 1 cm × 15 cm column of Sephadex G-50 equilibrated with PBS and a 1 cm × 100 cm column of Sephadex G-200 packed in one of the solvents described below.

Procedure. Prepare a solution of the lipoprotein in PBS, at a concentration of 1 mg of protein/ml. Transfer 1 ml to a small vial equipped with a magnetic stirrer and add 50 µl of 0.6 mM KI solution, followed by 1.0 mCi of ^{125}I. Stir briskly and add the chloramine -T reagent, through a fine plastic tube which dips below the surface of the reaction mixture, at the rate of 10 µl/min until the total has reached 40 µl. Then stop the stirrer and allow the mixture to stand for 5 min. Add 50 µl of solution B and remove the unbound iodide by passing the mixture down the Sephadex G-50 column in PBS. The ensuing treatment of the lipoprotein will then depend on the use to which it will be put:

(a) If it is intended for metabolic studies, the final purification must be carried out without delay and the experiment begun immediately after. Firstly, remove the products of degradation by chromatography on a column of Sephadex G-200 in PBS. Collect the peak that emerges at the void volume and concentrate it by pressure dialysis (Appendix 2) against 0.15 M NaCl. Continue the dialysis against this solvent for 18 hours at 4 °C and then sterilise the product by passing it through a 0.22 µm membrane filter (e.g. Millipore).

(b) If the labelled lipoprotein is to be used as a tracer in bench studies, it may be stored at 4 °C for up to 4 weeks in conventional solvents such as 0.15 M NaCl containing 0.3 M EDTA and buffered to pH 7.4 with 20 mM Tris/HCl (Goldstein and Brown, 1974). However, some workers prefer to add albumin to the solvent, since this not only has a preservative effect on native lipoproteins but also scavenges the products of radiolysis. Examples of such solvents are 0.05 M barbital buffer, pH 8.6, containing 3% bovine serum albumin (Schonfeld and Pfleger, 1974), and 0.13 M borate buffer, pH 8.0, containing 0.5% albumin, 0.05% EDTA and 0.05% sodium azide (Karlin et al., 1978). Nonetheless, some degradation of the labelled lipoprotein will occur during storage and it should be purified immediately before use, by passing it down a column of Sephadex G-200 that is equilibrated with the chosen storage buffer.

When labelled by this technique, LDL should contain approximately 0.5 g-atoms of iodine/mol of lipoprotein, assuming a particle weight of 3×10^6.

List of suppliers mentioned in the text

Amicon Corp.
21 Hartwell Avenue,
Lexington, MA 02173
U.S.A.

Applied Science Laboratories Inc.
P.O. Box 440,
State College, PA 16801
U.S.A.

Baker Chemical Co.
Phillipsburg, NJ 08865
U.S.A.

Bausch and Lomb Inc.
Rochester, NY 14602
U.S.A.

BDH Chemicals Ltd.
Broom Road,
Parkstone,
Poole, BH12 4NN
England

Beckman Instruments Inc.
Spinco Division,
1117 California Ave.
Palo Alto, CA 94304
U.S.A.

Becton-Dickinson and Co.
Rutherford, NJ 07070
U.S.A.

Bender and Hobein A.G.
Zürich
Switzerland

Bio-Molecular Dynamics,
P.O. Box 668,
Beaverton, OR 97005
U.S.A.

Bio-Rad Laboratories Inc.
2200 Wright Ave.
Richmond, CA 94804
U.S.A.

Boehringer Mannheim GmbH.
Sandhofer Strasse 116,
6800-Mannheim-31
Germany

Burdick and Jackson Laboratories Inc.
Muskegon, MI
U.S.A.

Cadbury-Schweppes Ltd.,
Birmingham
England

Chemetron,
Milano
Italy

Clay Adams Divn.
Becton Dickinson and Co.,
Persippany, NJ
U.S.A.

Corning Medical,
Palo Alto, CA 94304
U.S.A.

Difco Laboratory,
Detroit, MI 48232
U.S.A.

Drummond Scientific Co.
500 Parkway,
Broomall, PA 19008
U.S.A.

Dupont de Nemours and Co.
Wilmington, DE 19898
U.S.A.

Durrum Inc.
1228 Titan Way,
Sunnyvale, CA 94087
U.S.A.

Eastman Kodak Co.
Eastman Organic Chemicals,
Rochester, NY 14650
U.S.A.

Gelman Sciences Inc.
600 Sth. Wagner Road,
Ann Arbor, MI 48106
U.S.A.

Hoechst A.G.
D-62830 Frankfurt/Main-80
Germany

Hyland Therapeutics Divn.
Travenol Laboratories,
3300 Hyland Avenue,
Costa Mesa, CA 92626
U.S.A.

Iatron Laboratories Inc.
World-wide agents:
Newman-Howells Associates Ltd.,
Wolvesey Palace,
Winchester, S023 9NB
England

Immuno A.G.
Industriestrasse 72,
A-1220 Wien
Austria

Industrie Biologique Française S.A.
Gennevilliers
France

Instrumentation Specialities Co.
4700 Superior,
Lincoln, NE
U.S.A.

Kendal
Hospital Products Divn.
1 Federal Street,
Boston, MA02110
U.S.A.

LKB-Produkter AB,
S-16125 Bromma 1
Sweden

Mercer Chemical Corpn.
336 Bayview Avenue,
Amityville, NY 11701
U.S.A.

E. Merck AG,
Frankfurter Strasse, 250,
6100-Darmstadt-1,
Germany

Millipore Corp.
Bedford, MA 01730
U.S.A.

Nordic Immunological Laboratories,
Tilburg
The Netherlands

Anton Paar AG
A-8054 Graz
Austria

Packard Instrument Co.
2200 Warrenville Road,
Daniel Grove, IL
U.S.A.

Pharmacia Fine Chemicals AB,
Box 175,
S-75104 Uppsala 1
Sweden

Pierce Chemical Co.
Box 117,
Rockford, IL 61105
U.S.A.

Sartorius GmbH,
Postfach 19,
34-Göttingen
Germany

Schleicher-Schuell GmbH,
Postfach 4,
D-3354 Dassel
Germany

Scientific Furnishings Ltd.,
Poynton, Cheshire
England

Serva Feinbiochimica GmbH,
P.O. Box 105250
D-6900 Heidelberg 1
Germany

Shandon Scientific
Astmoor Industrial Estate,
Runcorn, WA7 1PR
England

Sigma Chemical Co.
P.O. Box 14508,
St. Louis, MO 63178
U.S.A.

Spectrum Medical Industries Inc.
60916 Terminal Annexe,
Los Angeles CA 90054
U.S.A.

Supelco Inc.
146 Water Street,
Bellefonte, PA 16823
U.S.A.

Transidyne General Corp.
Ann Arbor, MI 48106
U.S.A.

Vilene,
Halifax
England

Wellcome Reagents Ltd.,
299, Hither Green Lane,
London, SE13 6TL
England

Whatman Laboratory Products Ltd.
Springfield Mill,
Maidstone, Kent
England

Worthington Biochemical Corp.
P.O. Box 650,
Freehold NJ 07728
U.S.A.

References

ACKERS, G.K. and STEERE, R.L. (1967) in Methods in Virology (Maramorosch, K. and Koprowski, H., eds), Vol. 2, pp. 325–365, Academic, New York.

ACKMAN, R.G. (1981) in Methods in Enzymology (Colowick, S.P., Kaplan, N.O. and Lowenstein, J.M., eds), Vol. 72, pp. 205–252, Academic, New York.

ADAMS, G.H. and SCHUMAKER, V.N. (1969a) Ann. N.Y. Acad. Sci. *164,* 130–146.

ADAMS, G.H. and SCHUMAKER, V.N. (1969b) Anal. Biochem. *29,* 117–129.

ADAMS, G.H. and SCHUMAKER, V.N. (1970) Biochim. Biophys. Acta *202,* 315–324.

AGORASTOS, J., FOX, C., HARRY, D.S. and MCINTYRE, N. (1978) Clin. Sci. Molec. Med. *54,* 369–379.

ALAUPOVIC, P. (1968) in Progress in Biochemical Pharmacology (Miras, C.J., Howard, A.N. and Paoletti, R., eds), Vol. 4, pp. 91–109, Karger, Basel.

ALAUPOVIC, P. (1972) in Proceedings of the 19th Colloquium on the Protides of the Biological Fluids (Peeters, H., ed) pp. 9–19, Pergamon, Oxford.

ALAUPOVIC, P. (1980) in CRC Handbook of Electrophoresis (Lewis, L.A. and Opplt, J.J., eds), Vol. 1, pp. 27–46, CRC Press, Boca Raton, U.S.A.

ALAUPOVIC, P., CURRY, M.D., MCCONATHY, W. and FESMIRE, J. (1979) in Report of the High Density Lipoprotein Workshop, NIH publication No. 79–1661, pp. 227–240, U.S. Dept. of Health, Education and Welfare, Bethesda, U.S.A.

ALAUPOVIC, P., MCCONATHY, W.J. and FESMIRE J.D. (1983) in Report of apo-lipoprotein methods workshop (Lippel, K., ed), NIH publication, in press.

ALBERS, J.J. and HAZZARD, W.R. (1974) Lipids *9,* 15–26.

ALBERS, J.J., CABANA, V.G. and HAZZARD, W.R. (1975) Metabolism *24,* 1339–1351.

ALBERS, J.J., CABANA, V.G., WARNICK, G.R. and HAZZARD, W.R. (1975a) Metabolism *24,* 1047–1054.

ALBERS, J.J., WAHL, P.W., CABANA, V.G., HAZZARD, W.R. and HOOVER, J.J. (1976) Metabolism *25,* 633–644.

ALBERS, J.J., ADOLPHSON, J.L. and HAZZARD, W.R. (1977) J. Lipid Res. *18,* 331–338.

ALBERS, J.J., WARNICK, G.R. and CHEUNG, M.C. (1978) Lipids *13,* 926–932.

ALBERS, J.J., CHEUNG, M.C. and FLOREN, C-H. (1980) in Atherosclerosis V: Proc. 5th Int. Symp. on Atherosclerosis (Gotto, A.M., Smith, L.C. and Allen, B., eds), pp. 811–815, Springer, Berlin.

ALBERS, J.J., CHEUNG, M.C., EWENS, S.L. and TOLLEFSON, J.H. (1981) Atherosclerosis *39,* 395–409.

ALBRINK, M.J. (1959) J. Lipid Res. *1*, 53–59.

ALLEN, G. (1981) Sequencing of Proteins and Peptides, Elsevier Biomedical, Amsterdam.

ALLEN, R.J.L. (1940) Biochem. J. *34*, 858–865.

ALLINGTON, W.B., CORDRY, A.L., MCCULLOUGH, G.A., MITCHELL, D.E. and NELSON, J.W. (1978) Anal. Biochem. *85*, 188–196.

AMONS, R. and SCHRIER, P.I. (1981) Anal. Biochem. *116*, 439–443.

ANDERSON, J.T. and KEYS, A. (1956) Clin. Chem. *2*, 145–159.

ASSMAN, G., SMOOTZ, E., ADLER, K., CAPURSO, A. and OETTE, K. (1977) J. Clin. Invest. *59*, 565–575.

BACHORIK, P.S., WOOD, P.D., ALBERS, J.J., STEINER, P., DEMPSEY, M., KUBA, K., WARNICK, R. and KARLSSON, L. (1976) Clin. Chem. *22*, 1828–1834.

BACHORIK, P.S., WALKER, R., BROWNELL, K.D., STUNKARD, A.J. and KWITEROVICH, P.O. (1980) J. Lipid. Res. *21*, 608–616.

BAKER, H.N., DELAHUNTY, T., GOTTO, A.M. and JACKSON, R.L. (1974) Proc. Natl. Acad. Sci. U.S.A. *71*, 3631–3634.

BARTHOLOME, M., WIELAND, H. and SEIDEL, D. (1980) Clin. Chim. Acta *104*, 101–105.

BARR, S.I., KOTTKE, B. and MAO, S.J.T. (1981) Biochim. Biophys. Acta *663*, 491–505.

BAUER, N. and LEWIN, S.Z. (1960) in Physical Methods of Organic Chemistry (Weissberger, A., ed), Vol. 1, pp. 131–190, Interscience, New York.

BAUSSERMAN, L., HERBERT, P.N. and MCADAM, K. (1980) J. Exp. Med. *152*, 641–656.

BAUTOVICH, G.R., SIMONS, L.A., WILLIAMS, P.F. and TURTLE, J.R. (1975) Atherosclerosis *21*, 217–234.

BAXTER, G.P. and WALLACE, C.C. (1916) J. Amer. Chem. Soc. *38*, 70–105.

BEDFORD, D.K., SHEPHERD, J. and MORGAN, H.G. (1976) Clin. Chim. Acta *70*, 267–276.

BEISIEGEL, U. (1979) Isolierung und Characterisierung von Apolipoproteinen, Thesis (Dr. rer. physiol.), Philipps-Universität, Marburg, Germany.

BEISIEGEL, U. and UTERMANN, G. (1979) Eur. J. Biochem. *93*, 601–608.

BERENDS, G.T., DE JONG, J. and ZONDAG, H.A. (1972) Clin. Chim. Acta *41*, 187–198.

BERG, K. (1963) Acta Pathol. Microbiol. Scand. *59*, 369–382.

BERNFELD, P. (1955) Fed. Proc., Fed. Amer. Soc. Exp. Biol. *14*, 182.

BERNFELD, P., DONAHUE, V.M. and BERKOWITZ, M.E. (1957) J. Biol. Chem. *226*, 51–64.

BERNFELD, P., NISSELBAUM, J.S., BERKELEY, B.J. and HANSON, R.W. (1960) J. Biol. Chem. *235*, 2852–2859.

BIERMAN, E.L., HAYES, T.L., HAWKINS, J.N., EWING, A.M. and LINDGREN, F.T. (1965) J. Lipid Res. *7*, 65–72.

VAN DER BIJL, P. and VAN GENT, C.M. (1975) Clin. Chim. Acta *61*, 407–409.

BLOOR, W.R. (1916) J. Biol. Chem. *24*, 227–231.

BLUM, C.B., ARON, L. and SCIACCA, R. (1980) J. Clin. Invest. *66*, 1240–1250.

BONI, L. and HENDRIKX, A. (1980) Clin. Chim. Acta *102*, 41–47.

BORGSTROM, B. (1952) Acta Physiol. Scand. *25*, 101–107.

BOS, E.S., VAN DER DOELEN, A.A., DENISSEN, A.E.H.C., GOVERDE, B.C., MAGNA-

NI, H.N. and KOSTNER, G.M. (1983) Clin. Chim. Acta *127*, 383–390.
BOSANQUET, A.G., BICKERSTAFFE, R., VINEY, I.T. and FRASER, R. (1980) Med. Biol. Eng. Comput. *18*, 772–778.
BOWCOTT, J.H. and SCHULMAN, J.H. (1955) Z. Elektrochem. *59*, 283–289.
BOYD, E.J. (1935) J. Clin. Invest. *14*, 7–13.
BOYD, G.S. (1954) Biochem. J. *58*, 680–685.
BRADWELL, A.R. and BURNETT, D. (1975) Clin. Chim. Acta *58*, 283–290.
BRAGDON, J.H. and HAVEL, R.J. (1954) Science *120*, 113–114.
BREWER, H.B., LUX, J.E., RONAN, R. and JOHN, K.M. (1972) Proc. Natl. Acad. Sci. U.S.A. *69*, 1304–1308.
BREWER, H.B., SCHULMAN, R., HERBERT, P.N., RONAN, R. and WEHRLY, K. (1974) J. Biol. Chem. 249, 4975–4984.
BREWER, H.B., FAIRWELL, T., LARUE, A., RONAN, R., HOUSER, A. and BRONZERT, T. (1978) Biochem. Biophys. Res. Commun. *80*, 623–630.
BRIGGS, C.J., ANDERSON, D., JOHNSON, P. and DEEGAN, T. (1981) Ann. Clin. Biochem. *18*, 177–181.
BROWN, W.V., LEVY, R.I. and FREDRICKSON, D.S. (1969) J. Biol. Chem. *244*, 5687–5694.
BURSTEIN, M. (1962) C.R. Acad. Sci. *255*, 605–607.
BURSTEIN, M. (1964) Vox. Sang. *9*, 22–24.
BURSTEIN, M. and SAMAILLE, J. (1955) Compt. Rend. *241*, 664–665.
BURSTEIN, M. and SAMAILLE, J. (1958) Rev. franç. étud. chim. et biol. *3*, 780–781.
BURSTEIN, M. and SCHOLNICK, H.R. (1972) Life Sci. *11*, 177–184.
BURSTEIN, M., SCHOLNICK, H.R. and MORFIN, R. (1970) J. Lipid Res. *11*, 583–595.
BURSTEIN, M. and SCHOLNICK, H.R. (1973) in Advances in Lipid Research (Paoletti, R. and Kritchevsky, D., eds), pp. 67–108, Academic, New York.
CALVERT, G.D., YEATES, R.A. and ROEGER, D.C. (1979) Clin. Chim. Acta *97*, 135–142.
CARLSON, L.A. (1960) Clin. Chim. Acta *5*, 528–538.
CARLSON, L.A. (1963) J. Atherosclerosis Res. *3*, 334–336.
CASTLE, J.D., JAMIESON, J.D. and PALADE, G.E. (1975) J. Cell Biol. *64*, 182–210.
CATAPANO, A.L., JACKSON, R.L., GILLIAM, E.B., GOTTO, A.M. and SMITH, L.C. (1978) J. Lipid Res. *19*, 1047–1052.
CHAM, B.E. and KNOWLES, B.R. (1976) J. Lipid Res. *17*, 176–181.
CHAPMAN, M.J., GOLDSTEIN, S. and MILLS, G.L. (1978) Eur. J. Biochem. *87*, 475–488.
CHAPMAN, M.J., GOLDSTEIN, S., LAGRANGE, D. and LAPLAUD, P.M. (1981) J. Lipid Res. *22*, 339–358.
CHARD, T. (1981) Introduction to radio-immunoassay and related techniques, Elsevier Biomedical, Amsterdam.
CHARGAFF, E., ZIFF, M. and MOORE, D. (1941) J. Biol. Chem. *138*, 383–405.
CHARLWOOD, P.A. (1957) J. Amer. Chem. Soc. *79*, 776–781.
CHERVENKA, C.H. (1969) Manual of Methods for the Analytical Ultracentrifuge, Beckman Instruments Inc., Palo Alto, Calif. U.S.A.
CHEUNG, M.C. (1981) Am. J. Med. Technol. *47*, 545–549.
CHEUNG, M.C. and ALBERS, J.J. (1977) J. Clin. Invest. *60*, 43–50.

CHIN, H.P. and BLANKENHORN, D.H. (1968) Clin. Chim. Acta *20,* 305–314.

CHRAMBACH, A., REISFELD, R.A., WYCKOFF, M. and ZACCARI, J. (1967) Anal. Biochem. *20,* 150–154.

CHUNG, B.Y., WILKINSON, T., GEER, J.C. and SEGREST, J.P. (1980) J. Lipid Res. *21,* 284–291.

CHUNG, J. and NISHIDA, T. (1967) J. Lipid Res. *8,* 631–635.

CLAMP, J.R., BHATTI, T. and CHAMBERS, R.E. (1971) in Methods of Biochemical Analysis (Glick, D. ed), Vol. 19, pp. 229–344, Wiley, New York.

CLAUSEN, J. (1981) Immunochemical techniques for the identification and estimation of macromolecules, Elsevier Biomedical, Amsterdam.

COHN, E.J., GURD, F.R.N., SURGENOR, P.M., BARNES, B.A., BROWN, R.K., DEROUAUX, G., GILLESPIE, J.M., KAHNT, F.W., LEVER, W.F., LIU, C.H., MITTLEMAN, D., MOUTON, R.F., SCHMID, K. and UROMA, E. (1950) J. Amer. Chem. Soc. *72,* 465–474.

COLFS, B. and VERHEYDEN, J. (1967) Clin. Chim. Acta *18,* 325–334.

CONNERTY, H.V., BRIGGS, A.R. and EATON, E.H. (1961) Clin. Chem. *7,* 37–53.

CORNWELL, D.G. and KRUGER, F.A. (1961) J. Lipid Res. *2,* 110–134.

COX, A.C. and TANFORD, C. (1968) J. Biol. Chem. *243,* 3083–3087.

CRAWFORD, N. (1959) Clin. Chim. Acta *4,* 494–502.

CROWLE, A.J. (1961) Immunodiffusion, Academic, New York.

CURB, J.D., OVERTURF, M.L. and HARRIST, R.B. (1980) Atherosclerosis, *37,* 641–645.

CURRY, M.D., ALAUPOVIC, P. and SUENRAM, A. (1976a) Clin. Chem. *22,* 315–322.

CURRY, M.D., MCCONATHY, W.J., ALAUPOVIC, P., LEDFORD, J.H. and POPOVICH, M. (1976b)Biochim. Biophys. Acta *439,* 413–425.

CURRY, M.D., MCCONATHY, W.J. and ALAUPOVIC, P. (1977) Biochim. Biophys. Acta *491,* 232–241.

CURRY, M.D., GUSTAFSON, A., ALAUPOVIC, P. and MCCONATHY, W.J. (1978) Clin. Chem. *24,* 280–286.

CURRY, M.D., MCCONATHY, W.J., FESMIRE, J.D. and ALAUPOVIC, P. (1980) Biochim. Biophys. Acta *617,* 503–513.

CURRY, M.D., MCCONATHY, W.J., FESMIRE, J.D. and ALAUPOVIC, P. (1981) Clin. Chem. *27,* 543–548.

DANCIS, J. (1962) Am. J. Obstet. Gyn. *84,* 1749–1854.

DANIELSSON, B., JOHANSSON, B.G. and PETERSSON, B.G. (1973) Clin. Chim. Acta *47,* 365–369.

DANIELSSON, B., EKMAN, R. and JOHANSSON, B.G. (1978) Prep. Biochem. *8,* 295–319.

DAVIDSON, L.J. (1941) Am. J. Dis. Childh. *61,* 958–962.

DAVIS, B.J. (1964) Ann. N.Y. Acad. Sci. *121,* 404–427.

DELALLA, O. and GOFMAN, J.W. (1954) in Methods of Biochemical Analysis (Glick, D., ed), Vol. 1, pp. 459–478, Interscience, New York.

DELALLA, O.F., TANDY, R.K. and LOEB, H.G. (1967) Clin. Chim. *13,* 85–100.

DEL GATTO, L., LINDGREN, F.T. and NICHOLS, A.V. (1959) Anal. Chem. *31,* 1397–1399.

DELL, R.B., SCIACCA, R., LIEBERMAN, K., CASE, D.B. and CANNON, P.J. (1973) Circ. Res. *32,* 71–84.

DE ST. GROTH, S.F., WEBSTER, R.G. and DATYNER, A. (1963) Biochim. Biophys. Acta *71*, 377–391.

DITTMER, J.C. and WELLS, M.A. (1969) in Methods in Enzymology (Colowick, S.P., Kaplan, N.O. and Lowenstein, J.M., eds), Vol. 14, pp. 482–530, Academic, New York.

DYERBERG, J. and HJØRNE, N. (1970a) Clin. Chim. Acta *28*, 203–208.

DYERBERG, J. and HJØRNE, N. (1970b) Clin. Chim. Acta *30*, 407–413.

EATON, R.P. and KIPNIS, D.M. (1969) J. Clin. Invest. *48*, 1387–1396.

EDELSTEIN, C. and SCANU, A.M. (1980) in CRC Handbook of Electrophoresis (Lewis, L.A. and Opplt, J.J., eds), Vol. 1, pp. 89–101, CRC Press, Boca Raton, U.S.A.

EHNHOLM, C., GAROFF, H., RENKONEN, O. and SIMONS, K. (1972) Biochemistry *11*, 3229–3232.

EMES, A.V., LATNER, A.L., RAHTANI-NOBAR, M. and TAN, B.H.A. (1976) Clin. Chim. Acta *71*, 293–301.

ERIKSEN, N. and BENDITT, E.P. (1980) Proc. Natl. Acad. Sci. U.S.A. *77*, 6860–6864.

ESSEX, C.G., GRANT, E.H., SHEPPARD, R.J., SOUTH, G.P., SYMONDS, M.S., MILLS, G.L. and SLACK, J. (1977) Ann. N.Y. Acad. Sci. U.S.A. *303*, 142–155.

EWING, A.M., FREEMAN, N.K. and LINDGREN, F.T. (1965) in Advances in Lipid Research (Paoletti, R. and Kritchevsky, D., eds), Vol. 3, pp. 25–61, Academic, New York.

FAHEY, J.L. and MCKELVEY, E. (1965) J. Immunol. *94*, 84–90.

FAINARU, M., GLANGEAUD, M.C. and EISENBERG, S. (1975) Biochim. Biophys. Acta *386*, 432–443.

FAINARU, M., HAVEL, R.J. and FELKER, T.E. (1976) Biochim. Biophys. Acta *446*, 56–68.

FAINARU, M., FELKER, T.E., HAMILTON, R.L. and HAVEL, R.J. (1977) Metabolism *26*, 999–1004.

FAINARU, M., HAVEL, R.J. and IMAIZUMI, K. (1977a) Biochim. Biophys. Acta *490*, 144–155.

FAINARU, M. and DECKELBAUM, R.J. (1979) FEBS Letts. *97*, 171–174.

FALKO, J.M., SCHONFELD, G., WITZTUM, J.L., KOLAR, J.B., WEIDMAN, S.W. and STEELMAN, R. (1980) J. Clin. Endoc. Metab. *50*, 521–528.

FELISTE, R., DOUSSET, N. and DOUSTE-BLAZY, L. (1973) Clin. Chim. Acta *47*, 329–333.

FIDGE, N. (1974) Clin. Chim. Acta *52*, 5–13.

FIDGE, N.H. and POULIS, P. (1974) Clin. Chim. Acta *52*, 15–26.

FIESER, L.F. (1963) in Organic Syntheses (Rabjohn, N., ed), Vol. 4, pp. 195–201, Wiley, New York.

FISCHER, L. (1980) An Introduction to Gel Chromatography, Elsevier Biomedical, Amsterdam.

FISHER, W. (1970) J. Biol. Chem. *245*, 877–884.

FISHER, W.R., GRANADE, M.E. and MAULDIN, J.L. (1971) Biochemistry *10*, 1622–1629.

FISHER, W.R., HAMMOND, M.G. and WARMKE, G.L. (1972) Biochemistry *11*, 519–525.

FLESS, G.M. and SCANU, A.M. (1975) Biochemistry *14*, 1783–1790.

FLETCHER, M.J. and STYLIOU, M. (1970) Clin. Chem. *16,* 362–365.

FOLCH, J., LEES, M. and SLOANE-STANLEY, G.H. (1957) J. Biol. Chem. *226,* 497–509.

FOREMAN, J.R., KARLIN, J.B., EDELSTEIN, C., JUHN, D.J., RUBENSTEIN, A.H. and SCANU, A.M. (1977) J. Lipid Res. *18,* 759–767.

FORGEZ, P. and CHAPMAN, M.J. (1982) J. Biochem. Biophys. Methods. *6,* 283–296.

FORTE, G.M., NICHOLS, A.V. and GLAESER, R.M. (1968) Chem. Phys. Lipids *2,* 396–408.

FORTE, G.M. and NICHOLS, A.V. (1972) in Advances in Lipid Research (Paoletti, R. and Kritchevsky, D., eds), Vol. 10, pp. 1–41, Academic, New York.

FREEMAN, C.P. and WEST, D. (1966) J. Lipid Res. *7,* 324–327.

FRIEDEWALD, W.T., LEVY, R.I. and FREDRICKSON, D.S. (1972) Clin. Chem. *18,* 499–502.

FRINGS, C.F., FOSTER, L.B. and COHEN, P.S. (1971) Clin. Chem. *17,* 111–114.

FRUCHART, J.C., KORA, I., CACHERA, C., CLAVEY, V., DUTHILLEUL, P. and MOSCHETTO, Y. (1982) Clin. Chem. *28,* 59–62.

GAMBERT, P., LALLEMANT, C. and PADIEU, P. (1980) Clin. Chim. Acta *100,* 99–105.

GANROT, P.O. (1972) Scand. J. Clin. Lab. Invest. 29, Suppl. *24,* 39–47.

GERLACH, E. and DEUTICKE, B. (1963) Biochem. Z. *337,* 447–453.

GIANAZZA, E., PAGANI, M., LUZZANA, M. and RIGHETTI, P.G. (1975) J. Chromatog. *109,* 357–364.

GIBBONS, R.A. (1963) Nature (Lond.) *200,* 665–666.

GIBBONS, R.A. (1972) in Glycoproteins: their composition, structure and function. Pt. A (Gottschalk, A., ed), Vol. 5, pp. 34–140, BBA Library, Elsevier, Amsterdam.

GIBSON, J.C., RUBENSTEIN, A., BUKBERG, P.R. and BROWN, W.V. (1982) Arteriosclerosis *2,* 432a.

GIDEZ, L.I., MILLER, G.J., BURSTEIN, M., SLAGLE, S. and EDER, H.A. (1982) J. Lipid Res. *23,* 1206–1223.

GLOSSMANN, H. and NEVILLE, D.M. (1971) J. Biol. Chem. *246,* 6339–6346.

GODOLPHIN, W.J. and STINSON, R.A. (1974) Clin. Chim. Acta *56,* 97–103.

GOLDBERG, R.B., KARLIN, J.B., JUHN, D.J., SCANU, A.M., EDELSTEIN, C. and RUBENSTEIN, A.H. (1980) J. Lipid Res. *21,* 902–912.

GOLDBERG, R.J. (1953) J. Phys. Chem. *57,* 194–203.

GOLDSTEIN, J.L. and BROWN, M.S. (1974) J. Biol. Chem. *249,* 5153–5162.

GORDON, A.H. (1975) Electrophoresis of Proteins in Polyacrylamide and Starch Gels, Elsevier Biomedical, Amsterdam.

GOTTO, A.M., LEVY, R.I. and FREDRICKSON, D.S. (1968) Lipids *3,* 463–470.

GOTTO, A.M., LEVY, R.I., LUX, S.E., BIRNBAUMER, M.E. and FREDRICKSON, D.S. (1973) Biochem. J. *133,* 369–382.

GRABAR, P. and WILLIAMS, C.A. (1955) Biochim. Biophys. Acta *17,* 67–74.

GREENWOOD, F.C., HUNTER, W.M. and GLOVER, J.S. (1963) Biochem. J. *89,* 114–123.

GROPPER, L. (1964) Anal. Biochem. *7,* 401–428.

GULIK-KRZYWICKI, T., YATES, M. and AGGERBECK, L.P. (1979) J. Mol. Biol. *131,* 475–484.

GUSTAFSON, A., ALAUPOVIC, P. and FURMAN, R.H. (1965) Biochemistry *4,* 596–605.

HAMILTON, R.L., HAVEL, R.J., KANE, J.P., BLAUROCK, A.E. and SATA, T. (1971) Science *172,* 475–478.

HAMMOND, M.G. and FISHER, W.R. (1971) J. Biol. Chem. *246,* 5454–5465.

HANAHAN, D.J., EKHOLM, J. and JACKSON, C.M. (1963) Biochemistry *2,* 630–641.

HANCOCK, W.S., BISHOP, C.A., GOTTO, A.M., HARDING, D.R.K., LAMPLUGH, S.M. and SPARROW, J.T. (1981) Lipids *16,* 250–259.

HANIG, M. and SHAINOFF, J.R. (1956) J. Biol. Chem. *219,* 479–487.

HARDMAN, D.A. and KANE, J.P. (1980) Anal. Biochem. *105,* 174–180.

HARVIE, N.R. (1973) Anal. Biochem. *53,* 252–263.

HATCH, F.T., FREEMAN, N.K., JENSEN, L.C., STEVENS, G.R. and LINDGREN, F.T. (1967) Lipids *2,* 183–191.

HATCH, F.T. and LEES, R.S. (1968) in Advances in Lipid Research (Paoletti, R. and Kritchevsky, D., eds), Vol. 6, pp. 1–68, Academic, New York.

HATCH, F.T., LINDGREN, F.T., ADAMSON, G.L., JENSEN, L.C., WONG, A.W. and LEVY, R.I. (1973) J. Lab. Clin, Med. *81,* 946–960.

HAVEKES, L., HEMMINK, J. and DE WIT, E. (1981) Clin. Chem. *27,* 1829–1833.

HAVEL, R.J., EDER, H.A. a,d BRAGDON, J.H. (1955) J. Clin. Invest. *34,* 1345–1356.

HAY, C., ROOKE, J.A. and SKINNER, E.R. (1978) FEBS Lett. *91,* 30–33.

HEISKELL, C.L., FISKE, R.T., FLORSHEIM, W.H., TACHI, A., GOODMAN, J.R. and CARPENTER, C.M. (1961) Am. J. Clin. Path. *35,* 222–226.

HELENIUS, A. and SIMONS, S.W. (1971) Biochemistry *10,* 2542–2547.

HELLER, W. and PUGH, T.L. (1957) J. Colloid Sci. *12,* 294–307.

HENDERSON, L.O., LA GARDE, E. and HERBERT, P.N. (1980) Am. J. Clin. Path. *73,* 664–668.

HERBERT, P.N., SCHULMAN, R.S., LEVY, R.I. and FREDRICKSON, D.S. (1973) J. Biol. Chem. *248,* 4941–4946.

HERBERT, P.N., WINDMUELLER, H.G., BERSOT, T.P. and SHULMAN, R.S. (1974) J. Biol. Chem. *249,* 5718–5724.

HERBERT, P.N., BAUSSERMAN, L.L., HENDERSON, L.O., HEINEN, R.J., LAPIANA, M.J., CHURCH, E.C. and SCHULMAN, R.S. (1977) in The Lipoprotein Molecule, (Peeters, H., ed), pp. 35–56, Plenum Press, New York.

HEUCK, G.C., ERBE, I. and FLINT-HANSEN, P. (1983) Clin. Chem. *29,* 120–125.

HIRSCH, J. and AHRENS, E.H. (1958) J. Biol. Chem. *223,* 311–320.

HJERTEN, S. (1959) Biochim. Biophys. Acta *31,* 216–235.

HOLMQUIST, L. and CARLSON, K. (1977) Biochim. Biophys. Acta *493,* 400–409.

HOLMQUIST, L. and CARLSON, K. (1978) Anal. Biochem. *84,* 598–603.

HOLMQUIST, L. and BROSTRÖM, H. (1979) J. Biochem. Biophys. Methods *1,* 117–127.

HOLMQUIST, L., CARLSON, K. and CARLSON, L.A. (1978) Anal. Biochem. *88,* 457–460.

HOSLER, C.F. (1980) Clin. Chem. *26,* 176–177.

HUISMANS, B.D. and FELGENHAUER, K. (1971) Biochim. Biophys. Acta *248,* 330–332.

HULLEY, S.B., COOK, S.G., WILSON, W.S., NICHAMAN, M.Z., HATCH, F.T. and LINDGREN, F.T. (1971) J. Lipid Res. *12,* 420–433.

HUNTER, R. (1970) Proc. Soc. Exptl. Biol. Med. *133,* 989–992.

ISHIKAWA, T.T., BRAZIER, J.B., STEINER, P.M., STEWART, L.E., GARTSIDE, P.S. and GLUECK, C.J. (1977) Lipids *11,* 628–633.

IVERIUS, P.-H. (1972) J. Biol. Chem. *247,* 2607–2613.

JACKSON, R.L. and Gotto, A.M. (1972) Biochim. Biophys. Acta *285,* 36–47.

Jackson, R.L., Sparrow, J.T., Baker, H.N., Morrisett, J.D., Taunton, O.D. and Gotto, A.M. (1974) J. Biol. Chem. *249*, 5308–5313.

Jackson, R.L., Baker, H.N., Gilliam, E.B. and Gotto, A.M. (1977) Proc. Natl. Acad. Sci. U.S.A. *74*, 1942–1945.

Jagendorf, A.T., Patchornik, A. and Sela, M. (1963) Biochim. Biophys. Acta *78*, 516–528.

Jain, R.S. and Quarfordt, S.H. (1979) Life Sciences *25*, 1315–1324.

Jauhiainen, M., Laitinen, M., Penttila, I., Puhakainen, E. and Hietanen, E. (1983) Int. J. Biochem. *15*, 501–506.

Jeffrey, P.D., Nichol, L.W. and Smith, G.D. (1975) J. Biol. Chem. *250*, 533–541.

Jencks, W.P. and Durrum, E.L. (1955) J. Clin. Invest. *34*, 1437–1448.

Jensen, G.L. and Smith, S.C. (1976) Lipids *11*, 752–754.

Johansson, B.G. (1972) Scand. J. clin. Lab. Invest. *29*, suppl. 124, 7–19.

Johnson, P., Anderson, D., Muirhead, R.A. and Deegan, T. (1981) Biochem. Soc. Transactions *9*, 56.

Johnston, J.P. and Ogston, A.G. (1946) Trans. Faraday Soc. *42*, 789–801.

Kabara, J.J. (1962) in Methods of Biochemical Analysis (Glick, D., ed), Vol. 10, pp. 263–318, Interscience, New York.

Kahan, J. and Sundblad, L. (1969) Scand. J. clin, Lab. Invest. *24*, 61–68.

Kane, J.P. (1973) Anal. Biochem. *53*, 350–364.

Kane, J.P., Richards, E.G. and Havel, R.J. (1970) Proc. Natl. Acad. Sci. U.S.A. *66*, 1075–1082.

Kane, J.P., Sata, T., Hamilton, R.L. and Havel, R.J. (1975) J. Clin. Invest. *56*, 1622–1634.

Kane, J.P., Hardman, D.A. and Paulus, H.E. (1980) Proc. Natl. Acad. Sci. U.S.A. *77*, 2465–2469.

Kapitany, R.A. and Zebrowski, E.J. (1973) Anal. Biochem. *56*, 361–369.

Kapp, O.H. and Vinogradov, S.N. (1978) Anal. Biochem. *91*, 230–235.

Karlin, J.B., and Rubenstein, A.H. (1979) in The Biochemistry of Atherosclerosis (Scanu, A.M., Wissler, R.W. and Getz, G.S., eds), pp. 189–227, Marcel Dekker, New York.

Karlin, J.B., Juhn, D.J., Starr, J.I., Scanu, A.M. and Rubenstein, A.H. (1976) J. Lipid Res. *17*, 30–37.

Karlin, J.B., Juhn, D.J., Scanu, A.M. and Rubenstein, A.H. (1978) Eur. J. Clin. Invest. *8*, 19–26.

Karlin, J.B., Juhn, D.J., Goldberg, R.B. and Rubenstein, A.H. (1979) in Report on High-density lipoprotein methodology workshop, NIH publication No. 79-1661, pp. 219–222, US Dept. of Health, Education and Welfare, Bethesda, MD, U.S.A.

Kashyap, M.L., Srivastava, L.S., Chen, C.Y., Perisutti, G., Campbell, M., Lutmer, R.F. and Glueck, C.J. (1977) J. Clin. Invest. *60*, 171–180.

Kashyap, M.L., Srivastava, L.S., Hynd, R.A., Gartside, P.S. and Perisutti, G. (1981) J. Lipid Res. *22*, 800–810.

Kates, M. (1975) Isolation, analysis and identification of lipids, Elsevier Biomedical, Amsterdam.

Kenny, A.P. (1952) Biochem. J. *52*, 611–619.

Koeselman, H.B., Lewis, B. and Pilkington, T.R.E. (1961) J. Atherosclerosis Res.

1, 85–88.

KOHN, J. (1961) Nature (Lond.) *189*, 312–313.

KOREN, E., MCCONATHY, W.J. and ALAUPOVIC, P. (1982) Biochemistry *21*, 5347–5351.

KOSTNER, G. (1974) Biochim. Biophys. Acta *336*, 383–395.

KOSTNER, G.M. (1976a) in Low Density Lipoproteins (Day, C.E. and Levy, R.S., eds), pp. 229–269, Plenum Press, New York.

KOSTNER, G.M. (1976b) Clin. Chem. *22*, 695.

KOSTNER, G.M. (1977) in Proceedings of 25th. Colloquium on Protides of the Biological Fluids (Peeters, H., ed), pp. 83–90, Pergamon, Oxford.

KOSTNER, G.M. and HOLASEK, A. (1969) Biochim. Biophys. Acta *188*, 157–159.

KOSTNER, G.M. and HOLASEK, A. (1970) Lipids *5*, 501–504.

KOSTNER, G.M. and ALAUPOVIC, P. (1972) Biochemistry *11*, 3419–3428.

KOSTNER, G.M. and HOLASEK, A. (1972) Anal. Biochem. *46*, 680–683.

KOSTNER, G.M., ALBERT, W. and HOLASEK, A. (1969) Hoppe-Seyler Z. physiol. Chem. *350*, 1347–1352.

KOSTNER, G.M., DEFISCH, A., ALBERT, W. and HOLASEK, A. (1972) Monat. Chem. *103*, 1695–1704.

KOSTNER, G.M., PETEK, W. and HOLASEK, A. (1974) Clin. Chem. *20*, 676–681.

KRATKY, O., LEOPOLD, H. and STÄTINGER, H. (1969) Z. angew. Phys. *4*, 273–277.

KRATKY, O., LEOPOLD, H. and STÄTINGER, H. (1973) in Methods of Enzymology (Hirs, C.H.W. and Timasheff, S.N., eds), Vol. 27, pp. 98–110, Academic, New York.

KUKSIS, A., MYHER, J.J., MARAI, L. and GEHER, K. (1975) J. Chromatog. Sci. *13*, 423–430.

KUNITAKE, S.T., LOH, E., SCHUMAKER, V.N., SHUN KIN MA, KNOBLER, C.M., KANE, J.P. and HAMILTON, R.L. (1978) Biochemistry *17*, 1936–1942.

KUNITAKE, S.T., MCVICAR, J.P., HAMILTON, R.L. and KANE, J.P. (1982) Arteriosclerosis *2*, 442a.

KUSHWAHA, R.S., HAZZARD, W.R., WAHL, P.W. and HOOVER, J.J. (1977) Ann. Intern. Med. *87*, 509–516.

LANGAN, T.A., DURRUM, E.L. and JENCKS, W.P. (1955) J. Clin. Invest. *34*, 1427–1436.

LAPLAUD, P.M., BEAUBATIE, L. and MAUREL, D. (1980) J. Lipid Res. *21*, 724–738.

LAURELL, C-B. (1966) Anal. Biochem. *15*, 45–52.

LAURELL, C-B. (1972) Scand. J. Clin, Lab. Invest. *29*, suppl. 124, 21–37.

LEE, D.M. (1976) in Low Density Lipoproteins (Day, C.E. and Lecy, R.S., eds), pp. 1–47, Plenum Press, New York.

LEE, D.M. (1980) Biochem. Biophys. Res. Commun. *95*, 1663–1672.

LEE, D.M. and ALAUPOVIC, P. (1970) Biochemistry *9*, 2244–2252.

LEE, D.M. and ALAUPOVIC, P. (1974a) Biochem. J. *137*, 155–167.

LEE, D.M. and ALAUPOVIC, P. (1974b) Atherosclerosis *19*, 501–520.

LEE, D.M., VALENTE, A.J., KUO, W.H. and MAEDA, H. (1981) Biochim. Biophys. Acta *666*, 133–146.

LEE, J.C. and TIMASHEFF, S.N. (1981) J. Biol. Chem. *256*, 7193–7201.

LEES, R.S. (1970) Science *169*, 493–495.

LENARD, J. (1971) Biochem. Biophys. Res. Commun. *45*, 662–668.

LEVY, R.T. and FREDRICKSON, D.S. (1965) J. Clin, Invest. *44*, 426–441.

LEWIS, L.A., GREEN, A.A. and PAGE, I.H. (1952) Am. J. Physiol. *171*, 391–400.

LIM, C.T., CHUNG, J., KAYDEN, H.J. and SCANU, A.M. (1976) Biochim. Biophys. Acta *420*, 332–341.

LINDAHL, U., CIFONELLI, H.A., LINDAHL, B. and RODEN, L. (1965) J. Biol. Chem. *240*, 2817–2820.

LINDGREN, F.T. (1974) in Fundamentals of Lipid Chemistry (Burton, R. and Guerra, F.C., eds), pp. 470–510, Bio-Science Publishing Division, Webster Groves, Missouri, U.S.A.

LINDGREN, F.T. (1975) in Analysis of Lipids and Lipoproteins (Perkins, E.G., ed), pp. 204–224, American Oil Chemists Soc., Champaign, IL, U.S.A.

LINDGREN, F.T., NICHOLS, A.V., FREEMAN, N.K., WILLS, R.D., WING, L. and GULLBERG, J.E. (1964) J. Lipid Res. *5*, 68–74.

LINDGREN, F.T., FREEMAN, N.K., EWING, A.M. and JENSEN, L.C. (1966) J. Amer. Oil Chemists Soc. *43*, 281–285.

LINDGREN, F.T., JENSEN, L.C. and HATCH, F.T. (1972a) in Blood lipids and lipoproteins: quantitation, composition and metabolism (Nelson, G.J., ed), pp. 181–274, Wiley, New York.

LINDGREN, F.T., JENSEN, L.C., WILLS, R.D. and STEVENS, G.R. (1972b) Lipids *7*, 194–201.

LINDGREN, F.T., ADAMSON, G.L., JENSEN, L.C. and WOOD, P.D. (1975) Lipids *10*, 750–756.

LINDGREN, F.T., SILVERS, A., JUTAGIR, R., LAYSHOT, L. and BRADLEY, D.D. (1977) Lipids *12*, 278–282.

LINDGREN, F.T., ADAMSON, G.L. and KRAUSS, R.M. (1980) in CRC Handbook of Electrophoresis (Lewis, L.A. and Opplt, J.J., eds), Vol. 1, pp. 229–247, CRC Press, Boca Raton, U.S.A.

LOPES-VIRELLA, M.F., STONE, P., ELLIS, S. and COLWELL, J.A. (1977) Clin. Chem. *23*, 882–884.

LOVELOCK, J.E. (1957) Proc. Roy. Soc. (Lond.), Ser. B, *147*, 427–433.

LOWE, C.R. (1979) An Introduction to Affinity Chromatography, Elsevier Biomedical, Amsterdam.

LOWRY, O.H., ROSEBROUGH, N.J., FARR, A.L. and RANDALL, R.J. (1951) J. Biol. Chem. *193*, 265–275.

LUX, S.E., JOHN, K.M. and BREWER, H.B. (1972) J. Biol. Chem. *247*, 7510–7518.

MACHEBOEUF, M. (1929) Bul. Soc. Chim. Biol. *11*, 268–281.

McConathy, W.J. and ALAUPOVIC, P. (1973) FEBS Lett. *37*, 178–182.

McConathy, W.J. and ALAUPOVIC, P. (1974) FEBS Lett. *41*, 174–178.

McConathy, W.J. and ALAUPOVIC, P. (1976) Biochemistry *15*, 515–520.

MCDONALD, H.J. and RIBEIRO, L.P. (1959) Clin. Chim. Acta *4*, 458–459.

MCFARLANE, A.S. (1958) Nature (Lond.) *182*, 53–54.

MCTAGGART, F., SENNITT, C.M., DIXON, R.M., WILLIAMS, J.M. and STONE, M.C. (1978) Symp. on Precipitation Techniques as an Aid to the Study of Biochemical Abnormalities Related to Atherosclerosis, pp. 2–13, ICI Pharmaceuticals Divn., Alderley Park, SK10 4TG, England.

MAGNANI, H.N. (1976) Biochim. Biophys. Acta *450,* 390–401.

MAGNANI, H.N. and ALAUPOVIC, P. (1972) Clin. Chim. Acta *38,* 405–411.

MAHLEY, R.W. and WEISGRABER, K.H. (1974) Biochemistry, *13,* 1964–1969.

MAHLEY, R.W., WEISGRABER, K.H. and FRY, D.L. (1973) Fed. Proc. *32,* 294.

MAHLEY, R.W., WEISGRABER, K.H., INNERARITY, T., BREWER, H.B. and ASSMAN, G. (1975) Biochemistry, *14,* 2817–2823.

MALINS, D.C., WEKELL, J.C. and HOULE, C.R. (1965) J. Lipid Res. *6,* 100–105.

MALMENDIER, C.L., CHRISTOPHE, J. and AMERYCKX, J.P. (1979) Clin. Chim. Acta *99,* 167–176.

MALMENDIER, C.L., PAROUTAUD, P. and AMERYCKX, J. (1980) FEBS Lett. *109,* 43–44.

MANCINI, G., CARBONARA, A.O. and HEREMANS, J.F. (1965) Immunochem. *2,* 235–254.

MANGOLD, H.K. and MALINS, D.C. (1960) J. Amer. Oil Chemists Soc. *37,* 383–390.

MAO, S.J.T. and KOTTKE, B.A. (1980) Clin. Chem. *26,* 1369.

MAO, S.J.T. and KOTTKE, B.A. (1980a) Biochim. Biophys. Acta *620,* 447–453.

MAO, S.J.T., GOTTO, A.M. and JACKSON, R.L. (1975) Biochemistry *14,* 4127–4131.

MAO, S.J.T., BHATNAGER, P.K., GOTTO, A.M. and SPARROW, J.T. (1980) Biochemistry *19,* 315–320.

MARCEL, Y.L., BERGSETH, M. and NESTRUCK, A.C. (1979) Biochim. Biophys. Acta *573,* 175–183.

MARCEL, Y.L., VEZINA, C., EMOND, D. and SUZUE, G. (1980) Proc. Natl. Acad. Sci. U.S.A. *77,* 2969–2973.

MARCH, S.C., PARIKH, I. and CUATRECASAS, P. (1974) Anal. Biochem. *60,* 149–152.

MARCHALONIS, J.J. (1969) Biochem. J. *113,* 299–305.

MARGOLIS, S. (1967) J. Lipid Res. *8,* 501–507.

MARGOLIS, S. and LANGDON, R.G. (1966) J. Biol. Chem. *241,* 469–476.

MARIER, J.A. and ROSE, D. (1964) Anal. Biochem. *7,* 304–314.

MARKWELL, M.A.K., HAAS, S.M., TOLBERT, N.E. and BIEBER, L.L. (1981) in Methods in Enzymology (Colowick, S.P. and Kaplan, N.O., eds), Vol. 72, pp. 296–303, Academic, New York.

MAULDIN, J. and FISHER, W.R. (1970) Biochemistry. *9,* 2015–2020.

MAURER, H.R. (1971) Disc Electrophoresis and related techniques of polyacrylamide gel electrophoresis, Walter de Gruyter, Berlin.

MELISH, J.S. and WATERHOUSE, C. (1972) J. Lipid Res. *13,* 193–200.

MESELSON, M., STAHL, F.W. and VINOGRAD, J. (1957) Proc. Natl. Acad. Sci. U.S.A. *43,* 581–589.

MICHAELIS, L. (1931) Biochem. Z. *234,* 139–151.

MILCH, L.J., WEINER, N. and ROBINSON, L.G. (1955) J. Phys. Chem. *59,* 948–951.

MILLER, N.E., THELLE, D.S., FØRDE, O.H. and MJØS, O.D. (1977) Lancet (*1*) 965–967.

MILLER, J.P., MAO, S.J.T., PATSCH, J.R. and GOTTO, A.M. (1980) J. Lipid Res. *21,* 775–780.

MILLS, G.L. (1977) FEBS Letts. *84,* 171–173.

MILLS, G.L., CHAPMAN, M.J. and MCTAGGART, F. (1972) Biochim. Biophys. Acta *260,* 401–412.

MILLS, G.L., TAYLAUR, C.E. and CHAPMAN, M.J. (1976) Clin. Sci. Molec. Med. *51*, 221–231.

MILLS, G.L., TAYLAUR, C.E., CHAPMAN, M.J. and FORSTER, G.R. (1977) Biochem. J. *163*, 455–465.

MILLS, G.L., TAYLAUR, C.E. and MILLER, A.L. (1979) Clin. Chim. Acta *93*, 173–180.

MILLS, G.L. and WILKINSON, P.A. (1962) Clin. Chim. Acta *7*, 685–693.

MOLINE, C. and BARRON, E.J. (1969) Clin. Chem. *15*, 521–526.

MORDASINI, R.C. and RIESEN, W.F. (1980) J. Clin. Chem. Clin. Biochem. *18*, 917–920.

MORING-CLAESSON, I. (1956) Ark. Kemi *10*, 47–56.

MUCKLE, T.J., EDWARDS, J.A. and AUCKLAND, P. (1977) Clin. Chem. *23*, 2302–2305.

MYERS, L.H., PHILLIPS, N.R. and HAVEL, R.J. (1976) J. Lab. Clin. Med. *88*, 491–505.

NAITO, H.K., WADA, M., ERHART, L.A. and LEWIS, L.A. (1973) Clin. Chem. *19*, 228–234.

NARAYAN, K.A. (1975) in Analysis of Lipids and Lipoproteins (Perkins, E.G., ed), pp. 225–249, American Oil Chemists Society, Champaign, Illinois, U.S.A.

NELSON, G.J. (1975) in Analysis of Lipids and Lipoproteins (Perkins, E.G., ed), pp. 1–22, American Oil Chemists Society, Champaign, Illinois, U.S.A.

NESTRUCK, A.C., SUZUE, G. and MARCEL, Y.L. (1980) Biochim. Biophys. Acta *617*, 110–121.

NEUBECK, W. and SEIDEL, D. (1975) Clin. Chem. *21*, 853–856.

NEUBECK, W., WIELAND, H., HABENICHT, A., MÜLLER, P., BAGGIO, G. and SEIDEL, D. (1977) Clin. Chem. *23*, 1296–1300.

NEVILLE, D.M. (1971) J. Biol. Chem. *246*, 6328–6334.

NIH: Manual of Laboratory Operations, Lipid Research Clinics Program. Vol. 1: Lipid and Lipoprotein Analysis. Publication No. 75–628, U.S. Dept. of Health, Education & Welfare, Bethesda, U.S.A.

NOBLE, R.P. (1968) J. Lipid Res. *9*, 693–700.

NOBLE, R.P., HATCH, F.T., MAZRINAS, J.A., LINDGREN, F.T., JENSEN, L.C. and ADAMSON, G.L. (1969) Lipids *4*, 55–59.

NICOLAIDES, N., FU, H.C. and ANSARI, M.N.A. (1970) Lipids *5*, 299–307.

NURY, F.S. and SMITH, R.B. (1957) Clin. Chem. *3*, 110–115.

OLOFSSON, S-O., MCCONATHY, W.J. and ALAUPOVIC, P. (1978) Biochemistry *17*, 1032–1036.

OLOFSSON, S-O., BOSTRÖM, K., SVANBERG, U. and BONDJERS, G. (1980) Biochemistry *19*, 1059–1064.

ONCLEY, J.L. (1941) Ann. N.Y. Acad. Sci. *41*, 121–137.

ONCLEY, J.L. (1969) Biopolymers *7*, 119–132.

ONCLEY, J.L., WALTON, K.W. and Cornwell, D.G. (1957) J. Amer. Chem. Soc. *79*, 4666–4671.

ONONOGBU, I.C. and LEWIS, B. (1976) Clin. Chim. Acta *71*, 397–402.

OPPLT, J.J. (1980) in CRC Handbook of Electrophoresis (Lewis, L.A. and Opplt, J.J., eds), Vol. 1, pp. 151–182, CRC Press, Boca Raton, U.S.A.

OPPLT, J.J. and OPPLT, M.A. (1980) in CRC Handbook of Electrophoresis (Lewis, L.A. and Opplt, J.J., eds), Vol. 1, pp. 47–86, CRC Press, Boca Raton, U.S.A.

ORNSTEIN, L. (1964) Ann. N.Y. Acad. Sci. *121*, 321–349.

OSBORNE, J.C. and BREWER, H.B. (1977) Advances in Protein Chemistry (Anfinsen, C.B., Edsall, J.T. and Richards, F.M., eds), Vol. 31, pp. 253–337, Academic, New York.

OSTLUND-LINDQUIST, A-M. and BOBERG, J. (1979) Clin. Sci. *56,* 99–100.

OUCHTERLONY, O. (1953) Act. Path. Microbiol. Scand. *32,* 231–240.

PAGE, I.H. and MOINUDDIN, M. (1962) Circulation *25,* 651–652.

PAGNAN, HAVEL, R.J., KANE, J.P, and KOTITE, L. (1977) J. Lipid Res. *18,* 613–622.

PARMELEE, D.C., TITANI, K., ERICSSON, L.H., ERIKSEN, N., BENDITT, E.P. and WALSH, K.A. (1982) Biochemistry, *21,* 3298–3303.

PARRA, M.G. (1976) Ph. D. Dissertation, Philipps-Universität, Marburg, Germany.

PARRA, M.G., DE GARCIA, M.M., WIEGANDT, H. and KOSTNER, G.M. (1977) Proc. 25th Colloquium Protides of the Biological Fluids (Peeters, H., ed), pp. 411–414, Pergamon Press, Oxford.

PATSCH, J.R., SAILER, S., KOSTNER, G., SANDHOFER, F., HOLASEK, A. and BRAUNSTEINER, H. (1974) J. Lipid Res. *15,* 356–366.

PATSCH, J.R., PATSCH, W., SAILER, S. and BRAUNSTEINER, H. (1976) Biochim. Biophys. Acta *434,* 419–427.

PATSCH, W., SCHONFELD, G., GOTTO, A.M. and PATSCH, J.R. (1980) J. Biol. Chem. *255,* 3178–3185.

PERKINS, E.G. (1975) Analysis of Lipids and Lipoproteins, American Oil Chemists Society, Champaign, Illinois, U.S.A.

PETERSON, E.A. (1981) Cellulosic Ion Exchangers, Elsevier Biomedical, Amsterdam.

PETERSON, G.L. (1979) Anal. Biochem. *100,* 201–220.

PFAFFINGER, D., EDELSTEIN, C. and SCANU, A.M. (1983) J. Lipid Res. *24,* 796–800.

PICARD, J. and VEISSIERE, D. (1970a) C.R. Acad. Sci. *270,* 1845–1848.

PICARD, J. and VEISSIERE, D. (1970b) Clin. Chim. Acta *30,* 149–155.

POLZ, E., WURM, H. and KOSTNER, G.M. (1981) Artery, *9,* 305–315.

POWNALL, H.J., PAO, Q., ROHDE, M. and GOTTO, A.M. (1978) Biochem. Biophys. Res. Commun. *85,* 408–414.

PRATT, J.J. and DANGERFIELD, W.G. (1969) Clin. Chim. Acta *23,* 189–201.

PRINCE, L.M. (1977) Microemulsious. Academic, New York.

RADDING, C.M. and STEINBERG, D. (1960) J. Clin. Invest. *39,* 1560–1569.

RADIN, N.S. (1969) in Methods in Enzymology (Colowick, S.P., Kaplan, N.O. and Lowenstein, J.M., eds), Vol. 14, pp. 245–254. Academic, New York.

RADOLA, B.J. (1973) Ann. N.Y. Acad. Sci. *209,* 127–143.

RAS, M.R., MASDEN, S., FRISON, J.C. and RUBIES-PRAT, J. (1978) Clin. Chem. *24,* 840–841.

RAY, B.R., DAVISSON, E.O. and CRESPI, H.L. (1954) J. Phys. Chem. *58,* 841–846.

REARDON, M.F., POAPST, M.E., UFFELMAN, K.D. and STEINER, G. (1981) Clin. Chem. *27,* 892–895.

REDGRAVE, T.G., ROBERTS, D.C.K. and WEST, C.E. (1975) Anal. Biochem. *65,* 42–49.

REES, A.W., LEWIS, E.A. and DE BUYSERE, M.S. (1974) Anal. Biochem. *62,* 19–65.

REICHL, D., MYANT, N.B., PFLUG, J.J. and RUDRA, D.N. (1977) Clin. Sci. Molec. Med. *53,* 221–226.

REISFELD, R.A., LEWIS, U.J. and WILLIAMS, D.E. (1962) Nature (Lond.) *195,* 281–283.

REISNER, A.H., NEMES, P. and BUCHOLTZ, C. (1975) Anal. Biochem. *64*, 509–516.
REMAN, F. and VERMOND, A. (1978) Clin. Chim. Acta *87*, 387–394.
RESSLER, N., SPRINGGATE, R. and KAUFMAN, J. (1961) J. Chromatog. *6*, 409–415.
RICHARDS, E.G., TELLER, D. and SCHACHMAN, H.K. (1971) Anal. Biochem. *41*, 189–214.
RICHARDS, E.G., TELLER, D.C., HOAGLAND, V.D., HASCHMEYER, R.H. and SCHACHMAN, H.K. (1971) Anal. Biochem. *41*, 215–217.
RIGHETTI, P.G. (1983) Isoelectric focussing: theory, methodology and applications, Elsevier Biomedical, Amsterdam.
RIGHETTI, P.G., PAGANI, M. and GIANNAZZA, E. (1975) J. Chromatog. *109*, 341–346.
RITTLAND, S. (1975) Scand. J. Gastroenterol. *10*, 5–15.
RODBARD, D. and LEWALD, J.E. (1970) Act. Endocrin., Suppl. 147, 79–103.
RODBARD, D., BRIDSON, W. and RAYFORD, P.L. (1969) J. Lab. Clin, Med. *74*, 770–781.
RONAN, R., KAY, L.L., MENG, M.S. and BREWER, H.B. (1982) Biochim. Biophys. Acta *713*, 657–662.
ROSSENEU, M., VERCAEMST, R., STEINBERG, K.K. and COOPER, G.R. (1983) Clin. Chem. 427–433.
RUDEL, L.L., LEE, J.A., MORRIS, M.D. and FELTS, J.M. (1974) Biochem. J. *139*, 89–95.
RUDMAN, D., GARCIA, L.A. and HOWARD, C.H. (1970) J. Clin. Invest. *49*, 365–372.
SATA, T., HAVEL, R.J., KOTITE, L. and KANE, J.P. (1976) Proc. Natl. Acad. Sci. U.S.A. *73*, 1063–1067.
SAUAR, J., Ritland, S., Home, R. and HORN, R. (1978) Clin. Chim. Acta *88*, 461–467.
SCANU, A.M. and HUGHES, W.L. (1960) J. Biol. Chem. *235*, 2876–2883.
SCANU, A.M., POLLARD, H. and READER, W. (1968) J. Lipid Res. *9*, 342–349.
SCANU, A.M. and EDELSTEIN, C. (1971) Anal. Biochem. *44*, 576–588.
SCANU, A., TOTH, J., EDELSTEIN, C. and STILLER, E. (1969) Biochemistry *8*, 3309–3316.
SCHACHMAN, H.K. (1957) in Methods in Enzymology (Colowick, S.P. and Kaplan, N.O., eds), Vol. 4, pp. 32–103, Academic, New York.
SCHACHMAN, H.K. (1959) Ultracentrifigation in Biochemistry, Academic, New York.
SCHONFELD, G. and PFLEGER, B. (1974) J. Clin. Invest. *54*, 236–246.
SCHONFELD, G., LEES, R.S., GEORGE, P.K. and PFLEGER, B. (1974) J. Clin. Invest. *53*, 1458–1467.
SCHONFELD, G., FRICK, M.S. and BAILEY, A.P. (1976) J. Lipid Res. *17*, 25–29.
SCHONFELD, G., CHEN, J-S., MCDONNELL, W.F. and JENG, I. (1977) J. Lipid Res. *18*, 645–655.
SCHONFELD, G., GEORGE, P.K., MILLER, J., REILLY, P. and WITZTUM, J. (1979) Metabolism *28*, 1001–1010.
SCHUH, J., FAIRCLOUGH, G.F. and HASCHEMEYER, R.H. (1978) Proc. Natl. Acad. Sci. U.S.A. *75*, 3173–3177.
SCHULMAN, R.S., HERBERT, P.N., WEHRLY, K. and FREDRICKSON, D.S. (1975) J. Biol. Chem. *250*, 182–190.
SEARCY, R.L. (1967) Clin. Chim. Acta *15*, 73–76.
SEIDEL, D., ALAUPOVIC, P. and FURMAN, R.H. (1969) J. Clin. Invest. *48*, 1211–1223.

SEIDEL, D., SCHMITT, E.A. and ALAUPOVIC, P. (1970) German Med. Monthly *15,* 671–675.

SEIGLER, L. and WU, W.T. (1981) Clin. Chem. *27,* 838–841.

SHELBURNE, F.A. and QUARFORDT, S.H. (1974) J. Biol. Chem. *249,* 1428–1433.

SHELBURNE, F.A. and QUARFORDT, S.H. (1977) J. Clin. Invest. *60,* 944–950.

SHORE, V.G. and SHORE, B. (1973) Biochemistry *12,* 502–507.

SHORE, V.G., SHORE, B. and LEWIS, S.B. (1978) Biochemistry *17,* 2174–2179.

SHORE, V.G., GARCIA, R.E., PENN, A.L. and SHORE, B. (1980) in CRC Handbook of Electrophoresis (Lewis, L.A. and Opplt, J.J., eds), Vol. 1, pp. 103–125, CRC Press, Boca Raton, U.S.A.

SIEBER, A. and BECKER, W. (1974) Clin. Chim. Acta *50,* 153–155.

SIMMONS, P. (1971) Clin. Chim. Acta *35,* 53–57.

SKIPSKY, V.P. and BARCLAY, M. (1969) in Methods in Enzymology, (Colowick, S.P., Kaplan, N.O. and Lowenstein, J.M., eds), Vol. 14, pp. 530–598, Academic, New York.

SKIPSKY, V.P., BARCLAY, M., BARCLAY, R.K., FETZER, V.A., GOOD, J.J. and ARCHIBALD, F.M. (1967) Biochem. J. *104,* 340–352.

SMITH, E.B. and SLATER, R. (1970) Atherosclerosis *11,* 417–438.

SMITH, E.B. and SLATER, R.S. (1972) Lancet (*1*) 463–469.

SMITH, I. (1976) Chromatographic and Electrophoretic Techniques, Vol. 2, Heinemann Medical Books, London.

SNIDERMAN, A., TENG, B. and JERRY, M. (1975) J. Lipid Res. *16,* 465–467.

SPERRY, W.M. and WEBB, M. (1950) J. Biol. Chem. *187,* 97–106.

SRINIVASAN, S.R., ELLEFSON, R.D., WHITAKER, C.F. and BERENSON, G.S. (1978) Clin. Chim. *24,* 157–161.

STEELE, B.W., KOEHLER, D.F., AGAR, M.M., BLASZKOWSKI, T.P., KUBA, K. and DEMPSEY, M.E. (1976) Clin. Chem. *22,* 98–101.

STEINBERG, K.K., COOPER, G.R., GRAISER, S.R. and ROSSENEU, M. (1983) Clin. Chem. *29,* 415–426.

STEPHENS, R.E. (1975) Anal. Biochem. *65,* 369–379.

STERNBERG, J.C. (1977) Clin. Chem. *23,* 1456–1464.

STONE, M.C., THORP, J.M., MILLS, G.L. and DICK, T.B.S. (1970) Clin. Chim. Acta *30,* 809–828.

STONE, M.C., THORP, J.M., MILLS, G.L. and DICK, T.B.S. (1971) Clin. Chim. Acta *31,* 333–354.

SWANEY, J.B., BRAITHWAITE, F. and EDER, H.A. (1977) Biochemistry *16,* 271–278.

TAN, M.H., ERROL, G.W., GLEASON, R.E. and SOELDNER, J.S. (1973) N. Engl. J. Med. *289,* 416–419.

THOMAS, K.K. and GILBERT, L.I. (1968) Arch. Biochem. Biophys. *127,* 512–521.

THOMPSON, G.A. and KAPOULAS, V.M. (1969) in Methods in Enzymology (Colowick, S.P., Kaplan, N.O. and Lowenstein, J.L., eds), Vol. 14, pp. 668–678, Academic, New York.

THOMPSON, G.R., BIRNBAUMER, M.E., LEVY, R.I. and GOTTO, A.M. (1976) Atherosclerosis *24,* 107–118.

THORP, J.M., HORSFALL, G.B. and STONE, M.C. (1967) Med. biol. Engng. *5,* 51–56.

TORO-GOYCO, E. (1958) Ph. D. Dissertation, Harvard University, Cambridge, Mass. U.S.A.

TRIPLETT, R.B. and FISHER, W.R. (1978) J. Lipid Res. *19*, 478–488.
TUSZYNSKI, G.P. and WARREN, L. (1975) Anal. Biochem. *67*, 55–65.
UTERMANN, G. (1975) Hoppe-Seyler's Z. Physiol. Chemie *356*, 1113–1121.
UTERMANN, G. and BEISIEGEL, U. (1979) Eur. J. Biochem. *99*, 333–343.
VAN HANDEL, E. and ZILVERSMIT, D.B. (1957) J. Lab. Clin. Med. *50*, 152–157.
VERBRUGGEN, R. (1975) Clin. Chem. *21*, 5–43.
VESTERBERG, O. and SVENSSON, H. (1966) Act. Chem. Scand. *20*, 820–834.
VISSER, L. and BLOUT, E.R. (1971) Biochemistry *10*, 743–752.
WADA, M., NAITO, H.K., EHRHART, L.A. and LEWIS, L.A. (1973) Clin. Chem. *19*, 235–239.
WALKER, W.H.C. (1977) Clin. Chem. *23*, 384–402.
WARNICK, G.R. and ALBERS, J.J. (1978) J. Lipid Res. *19*, 65–76.
WARNICK, G.R. and ALBERS, J.J. (1978b) Clin. Chem. *24*, 900–904,
WARNICK, G.R. CHEUNG, M.C. and ALBERS, J.J. (1979) Clin. Chem. *25*, 596–604.
WEBER, K. and OSBORN, M. (1969) J. Biol. Chem. *244*, 4406–4412.
WEBER, K. and KUTER, D.J. (1971) J. Biol. Chem. *246*, 4504–4509.
WEECH, P.K. (1981) Clin. Chem. *27*, 357–358.
WEECH, P.K. and MILLS, G.L. (1978) Biochem. J. *175*, 413–419.
WEECH, P.K., MCTAGGART, F. and MILLS, G.L. (1978) Biochem. J. *169*, 687–695.
WEECH, P., MCCONATHY, W.J., ALAUPOVIC, P. and FESMIRE, J. (1980) in Proc. Fifth Intnl. Symposium on Atherosclerosis (Gotto, A.M., Smith, L.C. and Allen, B., eds), pp. 804–806, Springer, New York.
WEEKE, B. (1973) in A Manual of Quantitative Immuno-electrophoresis (Axelsen, N.H., Krøll, J. and Weeke, B. eds), pp. 15–56, Universitetsforlaget, Oslo.
WEHR, C.T., CUNICO, R.L., OTT, G.S. and SHORE, V.G. (1982) Anal. Biochem. *125*, 386–394.
WEINSTEIN, Y., GIVOL, D. and STRAUSBACH, P-H. (1972) J. Immunol. *109*, 1402–1404.
WEISGRABER, K.H., MAHLEY, R.W. and ASSMANN, G. (1977) Atherosclerosis *28*, 121–140.
WEISGRABER, K.H., BERSOT, T.P. and MAHLEY, R.W. (1978) Biochem. Biophys. Res. Commun. *85*, 287–292.
WEISGRABER, K.H., TROXLER, R.F., RALL, S.C. and MAHLEY, R.W. (1980) Biochem. Biophys. Res. Commun. *95*, 374–380.
WEISGRABER, K.H., RALL, S.C. and MAHLEY, R.W. (1981) J. Biol. Chem. *256*, 9077–9083.
WHICHER, J.T. and BLOW, C. (1980) Ann. Clin. Biochem. *17*, 170–177.
WIEJA, J.G. and SMITH, C.J. (1976) Anal. Biochem. *74*, 636–637.
WIELAND, H. and SEIDEL, D. (1978) Inn. Med. *5*, 290–297.
WILLIAMS, C.A. and CHASE, M.W., (1968) in Methods in Immunology and Immunochemistry (Williams, C.A. and Chase, M.W., eds), Vol. 2, pp. 249–270, Academic, New York.
WILLIAMS, F.G., PICKELS, E.G. and DURRUM, E.L. (1955) Science *121*, 829–830.
WILSON, D.E. and SPIGER, M.J. (1973) J. Lab. Clin. Med. *82*, 473–482.
WINTER, A., PERLMUTTER, H. and DAVIES, H. (1975) LKB Application Note 198, LKB-Produkter AB, Bromma, Sweden.

WOLLENWEBER, J. and KAHLKE, W. (1970) Clin. Chim. Acta *29,* 411–420.

WONG, R.A., BANDERO, P.G., JENSEN, L.C., PAN, S.S., ADAMSON, G.L. and LINDGREN, F.T. (1977) J. Lab. Clin. Med. *89,* 1341–1348.

ZILVERSMIT, D.B. (1963) in Biochemical Problems of Lipids (Frazer, A.C., ed), BBA Library, Vol. 1, pp. 257–263, Elsevier, Amsterdam.

ZILVERSMIT, D.B. (1969) in Structural and Functional Aspects of Lipoproteins in Living Systems (Tria, E. and Scanu, A.M., ed), pp. 329–368, Academic, New York.

ZILVERSMIT, D.B., SISCO, P.H. and YOKOYAMA, A. (1966) Biochim. Biophys. Acta *125,* 129–135.

Subject index

Acrylamide, hazards of 187
Agar, for LP-X analysis 204
Agarose 182, 373, 441
Albumin
  adsorption of 454
  purification of 175
  as scavenger 484
Alpha-lipoprotein 3, 7, 78, 82, 95, 119, 171–173, 181, 194
Amidoschwarz 472
Ammonium sulphate 85
Ampholytes 220, 255, 402
Antioxidants 115, 409, 464
Antisera 292
Apo-lipoproteins 13–17
  analysis 124
  chemical composition 244, 245
  degradation of 385
  distribution 127, 214, 422
  electrophoresis 252–288, 389
  estimation by
    electroimmuno-assay 427, 440–445
    electrophoresis 258, 435–438
    immunodiffusion 426, 438–440
    immuno-nephelometry 432
    radio-immunoassay 429–432, 445–448
  estimation of total 247–249
  functions 16
  identification 252–295
  molecular weight estimation 270–272, 283
Apo-lipoproteins (*continued*)
  polymorphs viii, 17, 244, 245
  radiolabelling 484
  reference standards 433–435
  separation from lipids 228
  solvents for 387, 392, 394
  staining of 253–258
  terminal amino acids 244–245
Apo-lipoprotein A-I 16, 244
  distribution 127, 422
  estimation of 421–448
  identification of 276–286
  isolation 394–403
Apo-lipoprotein A-II 16, 244
  distribution 422
  estimation 421–448
  identification 276–286
  isolation 403
Apo-lipoprotein A-III 17
Apo-lipoprotein A-IV 17, 244
  identification 276–286
  isolation 404–405
Apo-lipoprotein B 16, 102, 244
  chemical estimation of 249–252
  degradation of 385
  distribution of 127
  estimation of 421–448
  identification of 286–288
  isolation of 405–410
  molecular weight 287
Apo-lipoproteins C-I, C-II and C-III 16, 244

Apo-lipoproteins C-I, C-II and C-III *(continued)*
distribution 127, 422
estimation 421–448
identification 276–286
isolation 410–416
Apo-lipoprotein D 16, 244
distribution 422
estimation 421–448
identification 276–286
isolation 416–418
Apo-lipoprotein E 16, 244
distribution 127, 422
estimation 421–448
identification 276–286
isolation 418–420
Apo-lipoprotein F 16, 245, 277, 283
identification 276–286
isolation 421
Apo-lipoprotein G 17
Apo-lipoprotein H 17, 245
Apo-lipoprotein SV-D 245, 277, 283, 284

Beta-lipoprotein 3, 7, 78, 84, 95, 119, 171–173, 181
Blood
collection of 454–457
post-natal cord 456
n-Butanol/di-isopropyl ether 471
Butylated hydroxytoluene 115, 409

Carbohydrates
in apo-lipoproteins 244
estimation of 295–299
staining for 256, 476
Cellulose acetate membranes 178–180, 387
Chloramine-T 482, 484–487
Chloramphenicol 408, 435, 454
Chloroform/methanol 467, 470
Chloroform, purification 466
Cholesterol, estimation of 228, 237
Cholesteryl esters 233, 235
Chromatography
gas phase 235, 242, 296–299
HPLC 243, 389
thin-layer 231–234, 300–307
silicic acid columns 230
Chromatography of
neutral lipids 231–233
phospholipids 233
total plasma lipids 230, 235
Concanavalin A 103
Conductivity 114, 413
Constitutive polypeptides 15
Coomassie Blue 473
Cronar film 182, 440

Dextran sulphate 91, 205
Diazomethane 241
Density
adjustment 26–37
effect of centrigufation 21
measurement 38–42
of plasma 20
Density gradients 64–78, 156–163
determination of profile 75–77
fractionation of 74
preparation of 70–73
Dialysis tubing 386
Di-isopropylfluorophosphonate 450
5,5′-Dithiobis(2-nitrobenzoic acid) 18, 450, 453, 455
Dry-weight determination 224

Electron microscopy 114, 128, 129, 166–170
Electrophoresis
agarose 180–185
cellulose acetate 178–180
for estimation of lipoproteins 367–377
paper 174–178

polyacrylamide gel
Electrophoresis *(continued)*
polyacrylamide gel
of apo-lipoproteins 246, 252–288, 389
of lipoproteins 185–202
Electrophoretic mobility 3, 7
Ellman reagent 18, 450, 454, 455
Ethanol/ether 465–467, 469–470
Ethylenediaminetetra-acetic acid 450

Fatty acid esters, estimation 222
Fatty acid estimation 222, 241
Flotation rate 3, 10, 120
concentration dependence 131, 146, 322, 340
estimation 145–149

Gel electrophoresis
of apo-lipoproteins 259–276
of lipoproteins 180–202
Gel filtration chromatography 95, 98, 163–166
Glass beads 105
Glycerides, estimation of 228, 238–240
Glycine-rich peptide 17
Glycoproteins 17, 256
β_2-Glycoprotein-I 17, 245, 277, 283, 284
Guanidine hydrochloride 386, 392, 394

HDL 8
centrifugal assay 324–345
determination of HDL cholesterol 358–363
electrophoretic assay 366–377
isolation 59, 72, 90–93
properties 3
Heparin 83, 101, 205
HPLC 243, 389
Hydrated density 8, 151–156, 340
temperature coefficient 22
Hydrocarbons 221, 242
Hydroxyapatite 105, 110, 417

Immunoadsorbents 98
Immuno-assay
of apo-lipoproteins 438–448
of lipoproteins 380–383
Immunodiffusion 288–295
Immuno-electrophoresis
of LP-X 203–205
one-dimensional 202
two-dimensional 210–219
Immuno-precipitation, enhancement of 289
Isoelectric focussing 135, 189
of apo-lipoproteins 273–276, 284, 390, 399–403, 414–416, 420
of lipoproteins 219
Isoelectric point
of apo-lipoproteins 244, 245, 284
of lipoproteins 171
Isopropanol 250

Johnston–Ogston effect 320–323, 340–344, 347–349

LDL
centrifugal assay 345–351
determination of LDL cholesterol 363–366
electrophoretic assay 366–377
immuno-assay 381
isolation 58, 72, 88–93
properties 3
Lectin 193
Liebermann–Burchard method 237
Light scattering 122, 351–357
Lipids
estimation of 236–242
extraction from lipoproteins 228, 463–471
qualitative analysis 228–236
Lipoproteins
chemical analysis 123–125, 129, 228–307, 460–462

Lipoproteins *(continued)*
  concentration of 460–462
  degradation of 450–454
  estimation
    centrifugal 315–350
    electrophoretic 310, 367–377
    gravimetric 224, 309
    immuno-assay 380–383
    nephelometric 351–357
    precipitation 358–367
    refractometric 226–228, 313–315
  families 12, 15
  flotation rate 148
  isoelectric point 171
  partial specific volume 153–156
  particle size analysis 129, 133, 149–151, 163–165, 170
  particle weight estimation 149–153
  primary 12
  purity 117
  radio-labelling 484–487
  refractive increment 227
  secondary 12
  stabilising of 114–116
  storage of 114–116, 450–454
  thermal expansivity 22
α-Lipoprotein, *see* Alpha-lipoprotein
β-Lipoprotein, *see* Beta-lipoprotein
Lp(a) 13, 110
LP-X 13, 111, 128, 203–210
  estimation of 378–379

Molecular weights
  of apo-lipoproteins 244–245, 270–272, 283
  of lipoproteins, *see* Particle weight
Mono-alkyl diacyl-glycerols 221, 233, 235, 242

Naphthalene Black 472
Nephelometry 351–357
Neuraminidase 285–286
Neville method 269

Oil Red O 178, 474
Oxidative degradation 18, 408–410, 450–451, 464

Partial glycerides 221, 233, 235
Partial specific volume 151, 153–156
Particle size 11, 120, 121, 129, 133, 149–151, 163–165, 170
Particle weight 120, 149–153, 163–165
Pasteur pipettes 50
Pevikon 107
Phenylmethyl sulphonyl fluoride 18, 450
Phospholipid estimation 222, 240
Photodensitometry 258, 368–376, 435–438
Photographic recording 479–481
Plasma
  freezing of 454
  storage of 18, 458–459
Polyacrylamide gel 185–202, 250–276
Polydispersity 4, 126, 130
  estimation by
    chromatography 133–135, 165
    electron microscopy 129, 170
    electrophoresis 135–138, 173, 196, 198, 210–216
    ultracentrifugation 130–133, 156–163
Polyethylene glycol 85, 361, 461
Polyvinylpyrrolidone 85, 461
Pre-β-lipoproteins 3, 7, 95, 171–173, 181
Proline-rich peptide 17
Propyl gallate 115
Pycnometry 38–41

Radio-labelling 482–487
Refractive increment 227
Refractometry 75, 226–228

Saponification 236
Schiff reagent 180, 257, 475
Sephadex fraction V-D 245, 277, 283, 284
Serum amyloid A protein 245, 277, 283, 284
Sieve coefficient 164
Sodium azide 18, 450
Sodium decyl sulphate 391
Sodium dodecyl sulphate 391
Sodium phosphotungstate 83, 90
Stabilising agents 115
Stokes equation 121, 150, 165
Sucrose 115, 159
Sudan Black 178, 474, 477

Terminal amino acids 244, 245
Tetramethyl urea 250
Thimerosal 18, 450
Threonine-poor peptide 17, 245, 277, 283
Triglycerides 233

Ultracentrifugation (analytical) 130
  calibration 327
  cells 325
  density gradients 132
Ultracentrifugation (analytical) *(continued)*
  estimation of lipoproteins 315–350
  flotation rate estimation 145–149
  hydrated density estimation 153–156
  particle diameter estimation 150–151
  particle weight estimation 149–153
  polydispersity by 132, 141–145
Ultracentrifugation (preparative)
  polydispersity by 156–163
  rotors and tubes 43–47, 69
  zonal 77
Ultrafiltration 461–462
Urea 252, 386, 392, 394

VLDL 8, 119
  centrifugal assay 345–351
  determination of VLDL cholesterol 363–366
  electrophoretic assay 366–377
  isolation 57, 72, 88–93
  nephelometry 351–357
  properties 3
  subfractionation 60–63

Wax esters 221, 235
Weber and Osborne method 268

Printed in Dunstable, United Kingdom

91900950R00298